# UFOs in Reality

## Programmed Aerospace Monitors of Our Species

*T.R. Dutton*

authorHOUSE®

*AuthorHouse™ UK Ltd.*
*500 Avebury Boulevard*
*Central Milton Keynes, MK9 2BE*
*www.authorhouse.co.uk*
*Phone: 08001974150*



*First published by AuthorHouse 03/9/2011*

*ISBN: 978-1-4567-7159-1*

Dedicated to Marion, whose patience, tolerance, support and active participation have made this adult-lifetime quest possible.

# Contents

## PART 1 ESTABLISHMENT OF A TESTABLE SCIENTIFIC EXPLANATION

## PART 2 TESTING THE ASTRONAUTICAL THEORY (A.T.)

## PART 3 EXTRA-EXTRAORDINARY EVENTS & CONCLUSION

# Author's Foreword

This book is an account of a broadly based piece of research, with almost all aspects of the UFO phenomenon taken into consideration. The work and discoveries described, herein, will probably pre-empt the eventual declassification of UFO investigations undoubtedly carried out, since the 1950s, by the world's intelligence agencies. Though the account is largely autobiographical, it describes not only my own work and experiences, but also my interactions with many other people who have, by their experiences, contributed to the research.

As a professional aerospace engineer with longstanding interest, knowledge and expertise in that field, my story begins with a technological study of local UFO reports from the Greater Manchester area during 1967. The reports selected for consideration were those describing strange aerial craft (SAC). After six years, the study became global in scope and grew into a piece of astronautical (space-activities) research. Some fifteen years later, this had produced a vital key to the mysteries; a testable scientific theory, derived from the global data, that became known as the Astronautical Theory (AT) for SAC events. From there, my travels took me to interview eye-witnesses about their experiences, many of which came to be incorporated into the study's database. At an early stage in the investigation I became aware that the objects being described by witnesses were real and artificially-contrived, but utilised technology unable to be humanly reproduced. It began to seem that there existed a wider physical reality, not yet probed by mainstream science and, therefore, unable to be incorporated into present-day human engineering.

As this book will reveal, the AT has been used successfully to explain the nature of many SAC encounters, among them some very famous ones, and to demonstrate the accurately programmed nature of the events.

Throughout the course of more than forty years' research much attention has been given, by the media and many authors, to the probability of official cover-ups. The drip-feed releases of Ministry of Defence UFO files in Britain and the public disclosures of ex-US Military personnel, now released from security restrictions over there, have refuelled the clamour for official pronouncements. As the reader will soon discover, this book provides good reasons for official cover-ups and denials. There is little room for doubt that defence and intelligence agencies throughout the world will have carried out work similar to mine and concluded in a similar way --- that human activity is being constantly monitored by very advanced technology, originating from somewhere out there in space.

One can only hope that the publication of this book will distract public attention away from officialdom and allow such a serious topic to be discussed soberly and without ridicule, in both public and academic arenas. As has been demonstrated on many occasions during the past twenty years, the AT facilitates direct observation of the visiting craft. Several amateur sky-watching groups and at least one astronomer have used its predictions successfully and have informed me of their successes. The existence of this work is known internationally.

**T.R. Dutton. © 2010**

# PART 1

## Establishment of a Testable Scientific Explanation

# PHASE 1:
# The Prelude

## CHAPTER 1

## Space-age Engineering

My career within the British aircraft industry began in September 1959, after being accepted as a permanent member of the Special Projects Office of the AVRO (A.V.Roe & Co., Ltd.) Weapons Research Division (WRD). This was based at the scenic Woodford airfield, situated in the green fields of Cheshire, some 10 miles (16 km) south of the city of Manchester. The Special Projects Office was part of a larger department headed by Mr. (later Professor) John E. Allen. He had decided to make good use of my expressed interest in future spacecraft and new propulsion possibilities. A demonstrated ability to visualise and to draw complicated imaginary concepts in three dimensions was another talent he had appreciated. (Computer graphics facilities were still in the realms of fantasy then). During my four years under his direction, John Allen gave me every opportunity to develop my knowledge and talents. He was an advocate for a British Space Programme and was directing his department to investigate future possibilities that would allow Britain to participate actively, within its means, in the exploration of space.

Those were the very early years of the Space Age, when the Americans and Soviets were committing vast resources to the Cold War 'Space Race' --- and having many failures, even on the launch pads. It was quite clear to us all that more effective and economical ways of launching things into space had to be found before Britain could produce its own launchers in quantity. An attempt was being made to adapt the Blue Streak IBM for satellite launching, but after a few trials, the Government of that time eventually withdrew its support for that project and, also, for the less well-known Saunders-Roe Black Knight and Black Arrow rocket launchers.

In contrast, space efforts within the Special Projects Office were being concentrated in very different directions.

Prior to my presence, John Allen had conceived a provisional cheap launcher for small satellites, which would be carried aloft by a Vulcan bomber before release. That was a practicable possibility. Having already designed, produced and tested the air launched, liquid-fuelled, Blue Steel supersonic cruise missile, the Company had already acquired the necessary skills and expertise to produce an air-launched rocket with a very different purpose.

Looking towards the future, we were encouraged to investigate rocket launchers using 'exotic' fuels (to reduce the physical size of the craft), recoverable stages, and very advanced 'aerospaceplanes'. The aerospaceplanes were winged vehicles. They were intended to take off from conventional runways and then to climb to high altitude using advanced turbojet engines, whilst accelerating to Concorde speeds (Mach 2). From there, ramjets would accelerate them to still higher altitudes and to speeds greater than Mach 5. Finally, rockets would power the rapidly climbing craft into orbit. On leaving its orbit, an aerospaceplane would glide back through the atmosphere (like the Space Shuttle) and land on an airfield. Only the fuel and the satellite would be left behind, instead of all those expensive throw-away rocket stages.

Through his efforts to promote Space-mindedness in Britain, John Allen became known as an expert on such matters by influential people at the Granada Television Studios in Manchester. He was often consulted about the latest developments in the Space Race and, to help them produce a programme about the distant future, he was asked for spacecraft designs that represented our thinking on future manned missions to the Moon and to Mars. Much of that task was allocated to me. With advice on aerodynamics, magnetohydrodynamics (MHD), astrophysics, celestial mechanics, advanced micro-'g' propulsion units (for use in space) and advanced structures, freely available from the various specialised experts within the department, I was able to produce schemes for those missions. The spacecraft conceived for the Mars mission used nuclear rocket propulsion and one of them looked very similar to the Jupiter Mission ship in the film "2001 - A Space Odyssey", which was produced some years later. All aspects of the missions were considered, especially, how the crew members were going to be accommodated, sustained and protected (from radiation and meteors, for example) during periods of many months in the hostile environment of the Solar System. It was inspiring stuff for a young forward-looking engineer to begin his career with — but, alas, all too far ahead in terms of the available technology and funding. (NASA abandoned its National Aerospaceplane project, and the planned Mars Mission is still some years away into the future.)

In 1963, after becoming part of the Hawker Siddeley Dynamics Group, the entire department was disbanded. Having just settled in that desirable area with a wife and small son, it suited my requirements to try to remain at that site for a few more years. Even though I knew I would be stepping down from the main promotional ladders, I accepted an offer of a post in the Hawker Siddeley Aviation (H.S.A.) Wind Tunnels Department as an engineer/analyst. The tunnels were situated on the other side of the airfield at Woodford. Therein, I made new friends, gained new knowledge and expertise, and lived with the hope that, one day, I might be able to progress further with my envisaged career. That was my situation when, in 1967, strange things began to happen in the skies above the entire Manchester area.

## The 1967 Happenings.

During the late summer period of 1967, people reported having seen strange aerial craft, often with multicoloured lights, flying low, or hovering, over residential areas. As Autumn approached, the number of reports appearing in local newspapers increased rapidly. I began to take an interest in those reports, because many of the witnesses were convinced that they had seen artificial objects that had not been produced on Earth. The newspapers often referred to them as 'Flying Saucers', even when the description given in the article described something nothing like a saucer. I decided to carry out a low-key investigation of the reports, using the newspaper cuttings. Aerial craft were my professional business and my four years' 'crash course' in astronautics (the science of travel in Space) was still very fresh in my mind. Little did I know that that spare-time study would go on --- and on --- and on --- for more than forty years, to the present day!

What follows, in this Phase 1 section, is my report of that early investigation, which expanded quickly to include UFO reports from all over Britain. The Phase 1 study lasted for six years.

# CHAPTER 2

# The things in the skies

*During the Cold War years, the British Ministry of Defence (MoD) always asserted that its interest in Unidentified Flying Objects (UFOs) was limited. Its sole purpose was to determine whether any reports of that kind represented potential military threats to the UK. In effect, if the objects described by witnesses were considered not to be Soviet surveillance craft, then the reports were simply filed away and virtually forgotten. The author believes that, even today, the full story has yet to be revealed. In his view, in order to support MoD's perfectly legitimate position, its technical departments would have been tasked with thoroughly investigating each report in detail. Only then would it have been possible for the MoD publicly to dismiss the defence significance of UFO reports.*

*In the following chapter, the author describes his own professional investigations of individual reports. He has good reasons to believe that MoD's technical staff would have carried out similar studies and would have probably arrived at the same conclusions.*

## Investigations Begin

The descriptions of the strange aerial craft (SAC), seen in the air above the Manchester area during 1967, varied considerably. Clearly, the first question to be addressed was whether or not the reports made any engineering sense. Were these objects just figments of imaginations fired by the initial rumours? Were they real but commonplace things, say, like newspapers carried along by a strong wind and, at night, reflecting the lights of the city below? Or, were the witnesses correct in thinking that they were being surveyed by craft from another planet, somewhere out there in space? There was only one way to find out --- the reports had to be analysed in detail.

One thing seemed to be very clear to me --- whatever they were, they would certainly **not** be experimental military aircraft flying at such low altitudes over residential areas. The risks of failure and disaster on the ground below would be too high. Nor did I take seriously the suggestion that perhaps the Soviets had managed to produce spy-craft of that nature, which had avoided detection by the NATO radar systems. My analysis of the reports soon confirmed that there were no obvious explanations for the things described. They had often frightened the witnesses by their complete silence and their controlled manoeuvres, at times, just over the housetops. They had also displayed coloured lights and, most eerily, some had seemed to have had an all-over glow in the darkness.

I drew coloured sketches of selected craft and listed their characteristics and manoeuvres ---- their arrivals and departures, changes of body colour and of the lighting configurations as the objects moved about.

The list of Manchester sightings was soon to be supplemented by similar events being reported in Staffordshire, in the Stoke-on-Trent area, some 35 miles (56 Km) to the south of Manchester. The same things were happening there too. Just occasionally a national newspaper would report a sighting in the South of England, but North West England seemed to be where the activity was being concentrated. Most commonly, eye-witnesses described objects that were discs with rounded domes on top, which I categorised as 'DISC/DOME'. Very few resembled saucers. In fact, they didn't seem to resemble any sensible kind of aircraft at all! They were more like a collection of hats and the lids of kitchen utensils, yet they apparently displayed such agility and rapidity in the air that no man-made aircraft (or missile) could have possibly hoped to match them.

When I obtained a copy of a report **[1]** in 1968, more detailed information on the Staffordshire events became available to me. My earlier observations were reinforced by that excellent piece of documentation produced by two amateur astronomers. The objects seemed to be real enough. This was confirmed in various ways, but one of the most tangible proofs of their reality came from this Staffordshire report of an event that occurred on the night of 2nd September, 1967 :-

"*No sound* [engine noise?] *was heard by any of the witnesses during the whole sighting, although* [one of the witnesses] *said "it was like a wind when it came over"*.

*"Four of the witnesses were asked to make a drawing of what they had seen. The four simple outline drawings* [produced by them] *all show the same object - a disc, surmounted by a dome. All witnesses were agreed that the disc was a dull orange/yellow while the dome was bright red..* [One witness] *said the red dome was stationary, but the disc seemed to be spinning."*

The report also stated that the object had travelled low over the houses before appearing to land in a field some 400 to 800 yards (metres) away from them. One boy witness said that, after the landing, the bright red dome went out "like a light" but that the orange disc had turned more yellow before fading away. The fading process had taken between one and three minutes. There seemed to be little doubt that the residents of that new housing estate had seen a flying craft of some kind. But, if so, it was obviously using propulsion techniques unknown to me or, I suspected, to anyone else in the aerospace business. The boy's account was very revealing. The entire structure of the vehicle had seemed to be electrically energised, at least during its deceleration and descent into the field. After a landing had been accomplished, the power source had seemed to have been switched off, as the boy had suggested.

The police had been called and three policemen arrived at the estate some thirty minutes after the event. By then, some of the venturesome boys had already begun to explore the fields in the direction of the object's location. The police sent the boys to their homes and then began to continue the search, without success. They were returning and joking about the affair when a light was seen ascending from the distant fields by one of the residents. Everyone turned to look and all saw a big white light, like a distant glowing orb, climb into the night sky above the low hills. After reaching an estimated altitude of about 300 feet (100 metres) above the top of the hills, it stopped, and then may have tilted before disappearing *"like the picture when the television set is switched off"*. (When an old black-and-white TV was switched off, the picture collapsed into a central bright spot, before the screen became blank).

All this substantiated my already well-established suspicion that we were dealing with sci-fi technology, such as we could only speculate about. If the things being observed were really solid objects, they seemed to be quantum-mechanically propelled ---- which meant that the Laws of Newton, Aerodynamics and Thermodynamics (which govern all human aerospace engineers' efforts) were apparently incidental to their operations.

So, were they solid? That was the burning issue to be resolved.

The vital first clue came from the witness who had said *"it was like a wind when it came over"*

Gliders and sailplanes have very smooth, streamlined, shapes. These are necessary to minimise the air resistance of unpowered aircraft created to ascend in rising air currents, like soaring birds. They are designed to slice through the air and cause the smallest possible amount of commotion (turbulence) in the air trailing behind them (wakes). A turbulent wake means lost energy. It translates into a resistance to motion, called 'Drag'. Even so, when one of those very efficient and graceful aircraft passes low, overhead, perhaps during a landing, a loud 'SWOOSH' is heard. This noise is created by unavoidable disturbances in the air caused by the rapid motion of the glider and vortices shed from the wingtips.

Solid objects that do not have smoothly and gently curved (streamlined) shapes are called 'bluff' by airflow specialists (aerodynamicists) --- and such shapes, when speeding through the air, leave a lot of disturbance trailing behind them. Consider what this means. If a low-flying glider produces an audible swooshing noise, what kind of noise would a gliding, bluff, disc/dome shape produce when flying overhead? My expectation was that it would create a blustering buffeting noise, rather like the noise we hear in our ears when we're walking head-on into a fairly-strong wind. The witness who had been standing directly beneath the flightpath of that glowing dome from the skies had given me the vital clue. That flying thing that had landed in Staffordshire had been almost certainly SOLID! At least, it had been so during its presence in that area. Will-o-the- wisps and hallucinations do not cause wind noises.

Very soon after the 1967 events period, the task of SAC categorisation was begun. I had, also, by then, referred to various books on the topic. The shapes being described ranged from wingless cigar-shaped cylinders --- through my range of flying hats and lids --- to the occasional triangle and rectangle. All the objects had seemed to display lights in different arrangements at night, and many of them had seemed to glow in the dark. When seen in daylight, the cigar and disc shaped things were described as being metallic silver or grey in appearance. But daylight sightings were rare occurrences. I decided that the most common disc-shaped objects would be given most of my attention. A sample of my coloured drawings of the craft described by witnesses is shown by Fig.1. Eventually the craft were categorised as shown in Fig.2

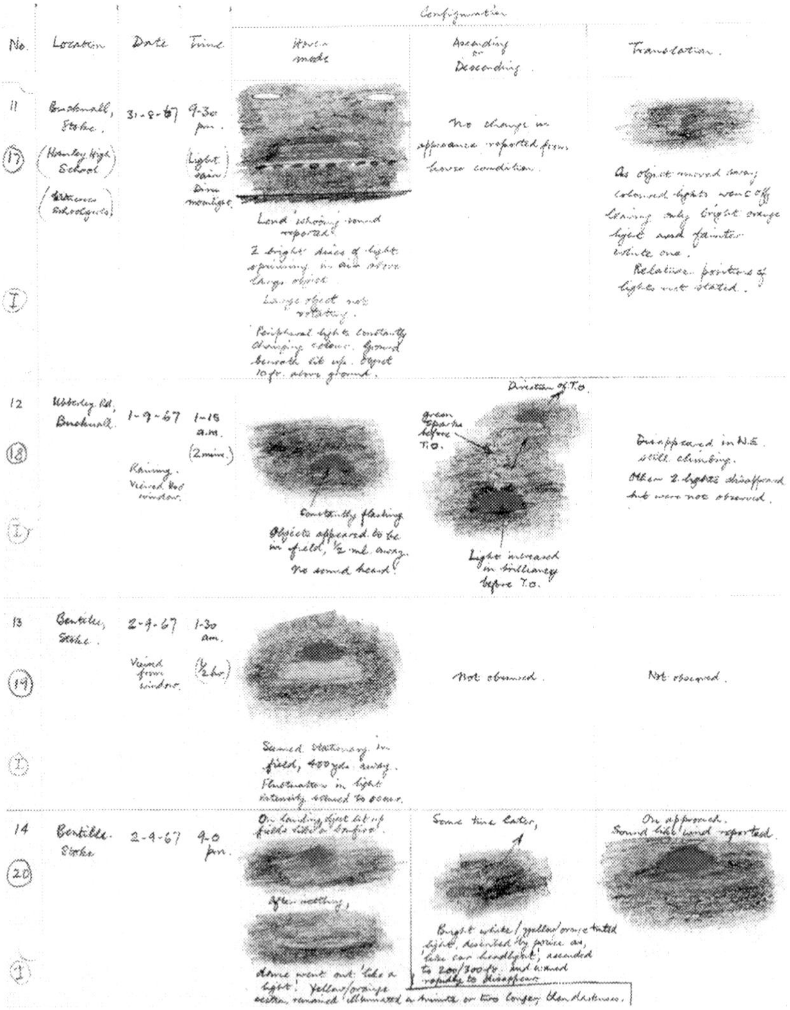
Configuration
No.
Location
Date
Time
Hover mode
Ascending or Descending
Translation.
11
Bucknall, Stoke.
(Hanley High School)
31-8-67
9-30 pm.
(Light rain. Dim moonlight.)
No change in appearance reported from hover condition.
Loud whooshing sound reported.
2 bright discs of light spinning in air above large object.
Large object not rotating.
Peripheral lights constantly changing colour. Ground beneath lit up. Object 10 ft. above ground.
As object moved away coloured lights went off leaving only bright orange light and fainter white one.
Relative positions of lights not stated.
12
Ubberley Rd, Bucknall.
1-9-67
1-15 a.m.
(2 mins.)
Raining.
Constantly flashing
Objects appeared to be in field, ½ ml. away.
No sound heard!
Direction of T.O.
green sparks before T.O.
Light increased in brilliancy before T.O.
Disappeared in N.E. still climbing.
13
Bentilee, Stoke.
2-9-67
1-30 am.
Viewed from window.
(½ hr.)
Not observed.
Not observed.
Seemed stationary in field, 400 yds. away.
14
Bentilee. Stoke
2-9-67
9-0 pm.
Some time later,
On approach. Sound like wind reported.

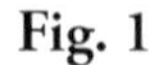

**Fig. 1**

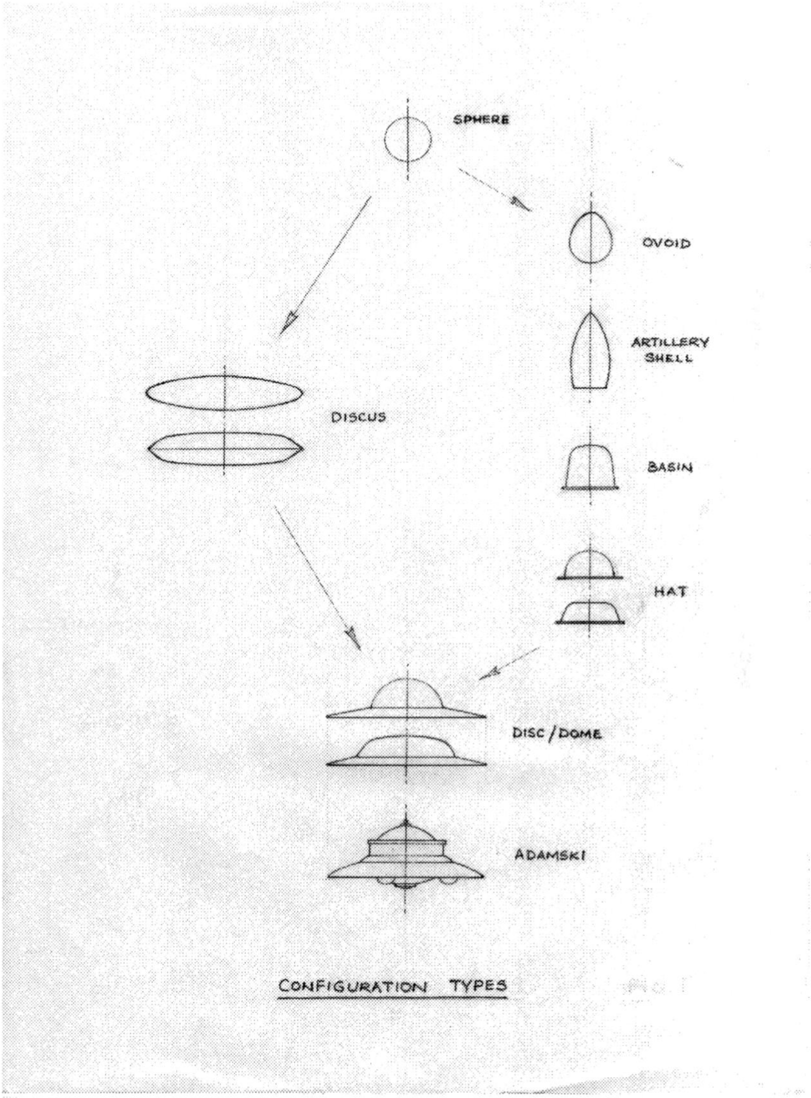
SPHERE
OVOID
ARTILLERY SHELL
DISCUS
BASIN
HAT
DISC / DOME
ADAMSKI
CONFIGURATION TYPES

**Fig. 2**

The simplest shape reported had been a small, spinning, sphere. It had been seen in daylight and had been thought to have been spinning rapidly about its vertical axis. This provided the simplest shape in a sequence of vertical-axis craft described during that period. As the right-hand side of Fig.2 shows, the little sphere had derivatives, which had ranged from ovoids to hat shapes with small brims. But smooth, discus-like shapes had also featured occasionally, as shown on the left-hand side of the diagram. These were effectively squashed spheres and were the most aerodynamic of all the vertical-axis craft. The disc/dome types were aerodynamically more advanced and larger than the hat shapes, and could be regarded as having specialised roles differing from the localised surveying roles seemingly played by all the items on the right of the diagram. The (allegedly) occupied craft portrayed by the controversial Adamski photographs (1952) has been added to demonstrate its similarity with the disc/domes reported during 1967, in Britain. It is also interesting that estimates of size for that type given by eye-witnesses had been generally similar to those given by George Adamski.

## Propulsion Implications.

The characteristics of those SAC described by the witnesses, had been extremely puzzling. They could be summarised like this:-

1. They had produced little or no noise, even at close quarters;
2. When hovering close to the ground they had created only localised disturbances in the trees or other vegetation directly beneath them;
3. Some form of self-glow had been sometimes evident;
4. During hovering, rotation of all or part of the structure about the vertical axis had been commonly observed;
5. Radio and TV reception had sometimes been interrupted by their presence.

Much of this evidence pointed towards there having been an electrodynamic propulsion system that had been common to all the vertical-axis craft. This implied that the different shapes had not been determined by the propulsion system, but by the specialised roles of the various configurations. It seemed logical that spinning of all or part of the structures could have provided spin-stabilisation during low-speed and hovering modes of operation.

The phenomenal accelerations and intense radiant emissions associated with the objects suggested that the propulsion system could produce a very high rate of energy release; and implied that the total energy 'charge' contained within it was enormous. The energy charge seemed to be capable of slow, controlled, release or of being suddenly discharged on demand. This conjured up the idea of a small nuclear power unit, providing the power, which had been coupled, in some way, with some kind of electrical condenser. But how such a system could have been employed to produce continuous motion in the sky was, and still is, beyond my understanding. Any field effects produced by known electrical systems are, by nature, short-range interactions with other surfaces in proximity. Quantum-mechanical reaction motors are limited to ion and photon rockets, conceived to supply micro-'g' accelerations for exploration craft operating in low gravity conditions, out in space.

All this seemed to support Leonard Cramp's suggestion **[2]** that the only kind of propulsion system that seemed to be compatible with the reported performance capabilities --- and general lack of forceful emissions --- is one in which all the atoms of the craft (and any occupants) are acted on simultaneously by artificially-produced and directional internal forces. Cramp went on to observe that the only example of the kind of propulsion, just discussed, is provided in Nature, by gravity. All the atoms/molecules of a solid body are persuaded to accelerate simultaneously, and in the same direction, when the body is allowed to fall freely in a gravitational field. But a gravity field is regarded as a distortion of space and time produced by the presence of a mass and, currently, we have no means of creating such a field artificially. But consideration of those things led me into other thoughts.

In Nature, we know that forceful interactions between matter and radiation occur randomly --- for example, when a body absorbs heat. In that circumstance, the atoms and molecules of the body absorb heat quanta (energy packets) and are pushed this way and that as they do so. The result is vibration of the structure at molecular level, which manifests as a rise in the body's temperature and, possibly, self-glow.

These thoughts led me to consider the possibility that someone 'up there' had discovered some way of organising such a process, so that all the atoms/molecules responded in the same direction to the absorbed quanta. Conceivably, the result would be a 'solid-state' propulsion unit, energised by some kind of 'exciter' unit, the emission of the energy being controlled by switching units on command. Given this capability, the direction of the applied forces producing acceleration, deceleration and control, could then be determined by the portions of the structure energised and the direction of the emissions triggered. As a consequence of the need for the entire craft and its contents to be maintained in a state of electrical excitation during all flight modes, structural fluorescence might be produced, the colour of the light emitted depending on the quantum energy levels of the bound electrons in the structure. Then, consideration of the way in which the molecules of hydrogen compounds can be brought into alignment by placing them within a 'magnetic bottle', to form the basis for a proton magnetometer, led me to the thought that perhaps powerful magnetic fields might provide the basis for my conceptual 'solid state' propulsion system. (Powerful magnetic disturbances have been registered during Close Encounters.) However, as I had no means available for testing the concept, I decided to leave innovative propulsion ideas for the future.

Before leaving this section, I will set on record my early thoughts about the functions of those disc appendages, which had seemed to be inessential for propulsion purposes. It had seemed to me that those annular appendages could have had useful functions, such as:-

a. Spin stabilisation of a non-rotating centre-body;
b. Storage of electrical energy;
c. Air or liquid cooling of the propulsion unit;
d. Cabin heat shielding during hypersonic travel through the atmosphere;
e. Aerodynamic surfaces (very unlikely);
f. Installation provision for sensors;
g. Directional microwave emitters and receivers, possibly to sense whether the ground below the craft was sufficiently level to facilitate a landing.

## Flight Characteristics.

The solid-state 'whole body' propulsion system we have just conjectured would, presumably, allow power to be switched in various ways to produce accelerations in any desired direction and to provide pulses for vehicle stabilisation and control. Given those capabilities, flight manoeuvres beyond any achievable by current aircraft would become possible. The Harrier aircraft and the new JSF (Joint Strike Fighter) probably provide the closest approximations to such capabilities than any other aircraft yet devised by the aerospace industry, but like any other mechanical devices their agility is limited by the laws of Newtonian mechanics. They are noisy and lumbering when hovering and their ability to manoeuvre at speed is limited by the inertial loads on both structure and pilot. In contrast, the pilot of a craft propelled by a solid-state propulsion system, in which all the molecules of the craft and its contents receive coherent impulses simultaneously, would be able to accelerate and decelerate the craft, at currently unthinkable rates, in any direction. However, there might be other physical restrictions on such a craft.

Any solid craft operating within the atmosphere must progress through a viscous medium --- air. Air may not be as sticky as treacle but it is nevertheless viscous. It sticks to the surface of objects in motion, creating a 'boundary layer', which produces aerodynamic skin-friction drag. Also, when the airflow cannot close in quickly enough behind a moving body, the boundary layer breaks away and pressure (or suction) drag is created. There are other reports on record, besides the Staffordshire event described previously, supporting the idea that the unidentifiable strange aerial craft (SAC) classified as UFOs are, to some extent, affected by aerodynamic phenomena. We will examine just two of the possibilities.

**Skipping Motion :** Speeding SAC have sometimes been described as having appeared to 'skip' rather like flat stones skimming over the surface of a pond. This could indicate that the unidentifiable craft are prone to a known aerodynamic instability called 'phugoid motion', which can affect conventional aircraft. After encountering a small atmospheric disturbance, an aircraft can begin to rise and fall above and below its mean flight path at a steady, usually slow, rate until the disturbance has been damped out. With disc/dome configurations such an oscillation might be caused by periodic vortex-shedding from the disc's edges or by eddies trailing behind the dome. Both these influences would produce oscillating changes of surface pressure that could cause the craft to steadily rise and fall whilst ostensibly in level flight.

**Oscillating or Spiralling Vertical Descents**: Several reports are on record of disc and disc/dome objects descending in the manner of falling leaves. They have seemed to 'pendulum' from side to side or spiral down, in a stable manner, with the axes of the bodies remaining nearly vertical throughout their descents.

A coin dropped, flat, into an aquarium can be persuaded to reproduce the motions just described, as can a simple cardboard disc dropped in air. If the disc is allowed to fall without rotation about its centre, it will oscillate from side to side. If it is given spin on release, it will spiral to the ground. However, I was not convinced that the disc/dome and bell-shaped configurations described by some witnesses would behave in the same way. The only way I could think of to test this doubt was to build a model having one of those shapes. But how was I to determine the actual proportions of the model and to distribute the mass of the object in a sensible manner? In the absence of anything better, I decided to use the proportions of an Adamski bell-shaped craft, as determined by Cramp in **[2]**. An approximate representative distribution of mass was then attempted by assuming that the bulk of the propulsion unit might have been situated under an empty cylindrical 'cabin'; that the dome above the cabin contained more propulsion equipment; and that the bell-shaped 'disc' contained waveguides or much ducting and low-density items. The resulting model, shown by Fig. 3, represented a full scale craft which had had an estimated diameter of some 35 feet (10.67m.). Of course, as previously explained, the full-size dimensions and masses were only speculative, but it was pleasing to discover during free-fall tests, in still air, that the model performed exactly like the objects in those UFO reports. The scaled speeds of descent were roughly consistent with the descent rates reported, a spiralling descent being marginally less rapid than an oscillating one. The estimated full-scale descent rates, based on the measurements taken with the model, were 106 ft/sec (32.3 m/sec) and 117 ft/sec (35.7 m/sec), respectively. Even very low rates of spin produced spiralling. Throughout the tests, the model remained stable and the motion showed no signs of becoming divergent.

**Fig. 3**

During the design and manufacture of the model, I tried to reproduce all the outstanding features revealed by the Adamski photographs. Falling into this category were the shaped cut-outs in the dark central body beneath the craft, which partially surrounded the three projecting hemispheres. These suggested to me that perhaps the lower centre-body might have been designed to move up and down relative to the bell structure to which the hemispheres were attached. Consideration, next, of the manner in which the cylindrical 'cabin' projected from the bell structure, and of the projecting rim surrounding the cabin's upper edge, further suggested that the cabin might have been able to be retracted into the bell structure in other modes of flight. It seemed to me to be very feasible that the craft had been designed with 'variable geometry' features to enable it to adapt to various flight conditions. These revealing details tended to validate the Adamski photographs for me. In my professional pursuits, I had learnt to recognise the possible significance of interesting features revealed by photographs of new aircraft and missiles, for which details had not been released. Unfortunately, many self-appointed sceptics and cynics, undeterred by their ignorance of such things, have always been keen to dismiss the photographs as being clever hoaxes.

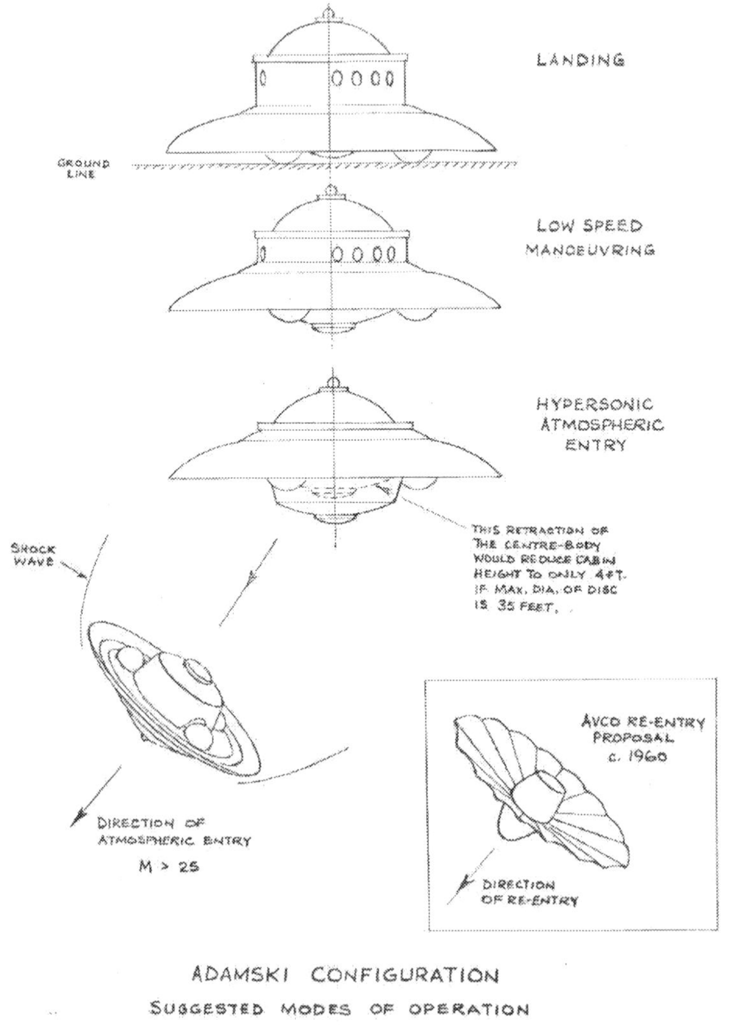


**Fig 4**

In Fig. 4 I have tried to depict the various configurations the craft could have adopted. The first diagram shows how it might have appeared had it actually settled onto the ground. The diagram below it depicts the configuration shown by the photographs, which might have been appropriate for low-speed manoeuvring with a view towards a possible landing. The third diagram demonstrates how the bell structure could have moved upwards to shield the central cabin, perhaps during high speed flight through the atmosphere when, also, the lower centre-body might have been retracted upwards, as shown, to reduce drag or for some other purpose. Since the object was alleged to be a personnel carrier, consideration of the cabin height remaining for any occupants in this fully retracted configuration indicated that it might have been only about 4 feet (1.2 metres). The diagrams at the bottom of Fig.4 show how closely the fully-retracted configuration resembled an early Avco re-entry capsule concept. It seemed just possible, if the supposed occupants of the craft lay prone on couches --- mounted on what would later become the cabin's ceiling --- that such a craft could be delivered into the atmosphere, directly, from space, in the manner shown, upsidedown. Presumably, after the craft had entered the lower layers of the atmosphere, the powerplants would have facilitated controlled rotation of the craft into the landing configuration. The occupants could have then un-strapped themselves, legs first, and proceeded to lower themselves, feet first, onto the floor --- even as the lower centre-body moved outwards to extend the height of the re-orientated cabin.

The above procedure is feasible, but I have to add that I do not believe it is one generally followed, given the advanced nature of the technology being constantly displayed in our skies. But I have tried to show that there could be much more to those Adamski photographs than anyone else seems to have guessed. Recently, I discussed the photographs with a retired radar expert and asked him --- supposing the craft depicted had been real --- what the purpose of that lower structure might have been. After careful consideration he made this assessment:-

*The central body could have contained microwave emitters, beaming high frequency, pulsed, E-M energy into that parabolic undersurface of the bell structure, which would then have beamed the pulses vertically downwards. From analysis of the differences in the reflection times of the pulses transmitted from different segments of the periphery of the craft, it would then be possible to know whether the ground below the craft was flat enough to land on.*

So, there's another good reason for that strange configuration --- one which might explain why, during the alleged encounter in rocky terrain, the Adamski craft did not land but continued to hover. The most likely application for this craft would have been one of descent to, and ascent from, the Earth's surface. This was no cruising craft. In my view, it would have been the equivalent of that Lunar Excursion Module (LEM), which was designed for the specialised task of taking the Apollo astronauts to and from the Moon's surface.

## References

[1] Pace, A.R. Stanway, R.H. 'UFOSs, Unidentified, Undeniable', February, 1968 Newchapel Observatory, Stoke-on-Trent, Staffordshire.

[2] Cramp, L.G. 'Space, Gravity and the Flying Saucer' (book) T. Werner Laurie, Ltd. 1954

# CHAPTER 3

# MAP LOCATIONS

Was there any method hidden in the distribution of the sightings of 1967? That was another of the questions addressed in the early days of my research. Fifty impressive sightings were selected for the exercise and the locations of these were plotted on a small-scale map of mainland Britain. The result is shown by Fig. 5.

**Fig. 5**

A band, only 35 miles (56 km) wide, running magnetic north-south, contained all the events in the North West and Midlands of England. Another band, twice as wide, enveloped the remaining events in the South and South West. This was set at right-angles to the north-south strip. Lines drawn through the centres of the two bands crossed at the site of that very famous prehistoric construction, Stonehenge! Could that be significant? Initially, I recoiled from the idea. It was much too close to Erick von Daneken's speculations for comfort --- so, initially, I dismissed it as being pure coincidence. However, that simple exercise showed me how apparently organised the activity had been.

So --- what was so special about those sites that might have been of interest to someone with an eagle's eye-view of Britain? That was the next poser to be answered.

Switching my attention to large-scale Ordinance Survey maps of the affected areas, the search began for large man-made features of the landscape which might be of interest to someone visiting this planet from afar. Round each sighting location, a circle of 1 mile (1.6 km) radius was drawn and the features within that circle were noted. Reservoirs, power stations, military areas, airfields, railway stations and sidings, major roads, large residential areas, industrial sites and so on, were initially considered. Then, as an afterthought --- after remembering the Stonehenge intersection --- I decided to include marked ancient sites. The outcome of the exercise is shown by Fig.6.

Much to my amazement, ancient landmarks came out with top marks, especially those roads built during the period of Roman occupation. 77 % of the selected 1967 event sites had been within 1 mile of known ancient sites. The next features to be favoured had been those associated with transportation, with the newly opened M6 motorway figuring very prominently. In fact, it was discovered that the north-south distribution band followed the path of the M6 from Preston in the North-West down to Stafford in the Midlands. That was as far as the construction had gone in 1967. All along that 70 miles (113 km) stretch of motorway, major new flyover junctions were being constructed to link it with the adjacent towns and cities. And both the Manchester and the Stoke-on-Trent areas were in a state of dramatic transformation as a result of that and other major activities.

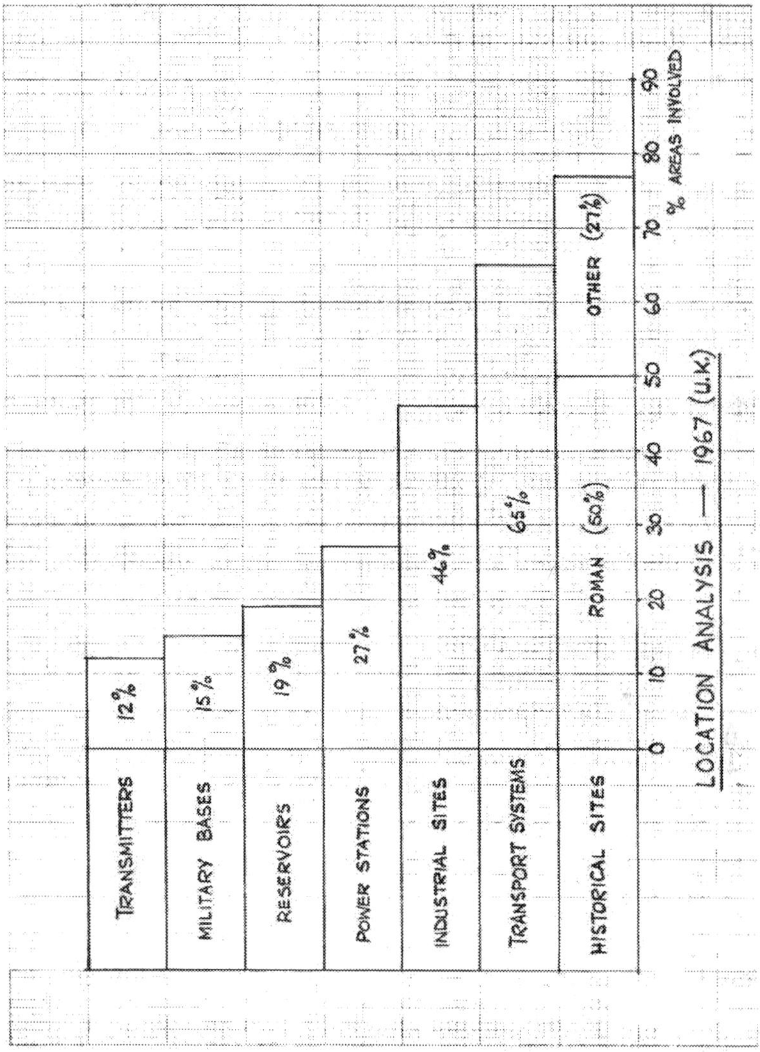


**Fig. 6**

In Manchester and adjacent Salford, vast areas of small terraced housing were being swept away and the residents were being re-housed in tower blocks. Those stark skyscrapers were sprouting like petrified trees out of the desolation of the rubble deserts where, once, people had lived. It was a shock to behold at ground-level, but it would have been a fascinating scene from space.

It was realised that the sites study had not been a truly statistical exercise because the relative abundance of each feature had not been determined at outset. Nevertheless, the results seemed to have some significance --- especially so, since the inclusion of ancient sites had been an afterthought. The outcome would have been quite different if they had not been included. After studying the details of the exercise again, it seemed to me to be possible that the probes had been programmed to explore and record all new developments in the vicinity of ancient landmarks, because new roads, motorway flyovers, housing estates and schools often featured in the 1967 sample. This thinking prompted the speculation that perhaps *human development had been monitored throughout the centuries since the original landmarks were created* --- a suspicion which seems to have been validated in various ways since then. (Other authors have made similar assertions since the late 1960s, so it seems important to set my speculation in correct historical context --- ie. it pre-empted those later writings.)

The 1968 findings on ancient sites were not widely known until I presented them at the 2nd BUFORA National Research Conference in 1976. At first I had regarded them as being too speculative, but during my continued monitoring of British UFO reports from 1967 to 1973, the association of such events with known ancient sites became commonplace. During an eighteen months period between mid 1971 and December 1972, responding to a request from BUFORA, I took on the role of Area Investigator of reported UFO events in N.W England. Resulting from this, the Autumn period of 1972 saw me travelling to and from the Lower Broughton area of Salford, Greater Manchester, from my location in North Cheshire. Salford has a long history and, very quickly, it became apparent to me that the events being reported from there were occurring in places of historical interest. In fact, my investigations revealed that the associated visible relics of the Industrial Revolution and the period of Victorian expansionism overlay sites of ancient settlements. Here is an example:-

On one of my investigative visits I noticed that a certain road in that industrial area bore the name 'Camp Street'. Enquiries made, later, by telephone, to Salford Museum, confirmed for me the area of interest had once been occupied by the Romans. It was further revealed to me that the nearby Bury New Road, one of the arterial routes running north from Manchester, followed the course of an ancient Roman road.

So, there again, were the truly ancient links with the new developments which were then in evidence --- a new refuse disposal depot and a large student's hostel which was still in the early stages of construction at that time. After carrying out a number of studies of that kind, *I concluded that the chance of seeing a UFO (probe) must be greatly increased for people living in any place where large scale construction developments were taking place --- and especially so, if the site also had ancient connections.* During that period of great changes in and around Manchester, it seemed to me that the operators of those probes probably knew more about the nature of the developments than the people living in the areas involved.

All this led me to another idea. *Perhaps the automated probes navigated by visual reference to landmarks.* They may have initially arrived here during prehistoric times. When the Romans arrived and began occupying much of Britain , the key artificial landmarks had been theirs and the prehistoric sites, which had been established long before their time of occupation. Throughout that ancient period of surveillance, the available landmarks had been programmed into the navigation systems of the terrain-following probes, and had been referenced regularly by them. It followed that whenever one of the vital landmarks came to be modified by later generations of humankind, the probes' map references had to be updated with some urgency.

This method of visual mapping for navigation purposes is used by certain wasps and bees, and it has been developed in recent years for military use by terrain-following attack aircraft and stealth bombers. It seems probable that certain extraterrestrials developed the technique a long time ago and are still using it.

To conclude this section, I want to demonstrate that the orderly geographical distribution of UFO events during 1967 were not unique. Another intense outbreak of activity occurred in Britain during 1971. Fig. 7 shows the distribution of those events. Many of them seemed to have been definitely linked to the course of the extended M6 motorway and to the developments taking place in the areas adjacent to Manchester.

OBJECTS SEEN OVER SEA
LIVERPOOL
LEEDS
MANCHESTER
BIRMINGHAM
LONDON
U.K.
SIGNIFICANT SIGHTINGS — 1971

**Fig. 7**

# CHAPTER 4

# Balls of coloured fire (1967 – 1973)

The study of British events continued until 1973. During those six years from 1967, I had recorded strange 'fireball' events that had been reported from time to time. Typically, they had resembled large balls of glowing plasma (electrified air or gas), red or green or blue and usually of great brilliance, which had been observed during fine weather periods. When seen travelling at high speeds across the countryside, they had never seemed to have any meteor-like tail or trail. At times they flew along very slowly and in a controlled sort of way. Sometimes they were very large, as big as the Moon. I remember investigating a unique report from a man in mid-Cheshire. He had been motoring home on a country road late at night, when a huge example had sped, low in the sky, across his path. He had been severely shocked by the event. The object had disappeared behind a small wood and he had expected to witness a gigantic explosion --- which had not occurred. When I interviewed him I asked for some idea of the size of the object compared with something in his room. He pointed to a large brass plaque hanging on a wall only 6 feet (2 metres) away. That plaque was about 18 inches (50 cm) in diameter! Many of us would have died of fright!

## Spring Equinox Events, 1973

A set of outstanding events, in daylight, which I investigated, could have been evidence of a near-miss by a sizeable asteroid. I could find no official report on it but it was certainly a very significant event --- or rather, series of events. Although I carried out a comprehensive investigation of the eyewitnesses' reports, I will just summarise my findings for this record, since the happenings were unlike the other cases to be dealt with in this section.

On the morning of 23rd March, 1973 --- a beautiful sunny morning with clear visibility, at some time during the period 08:20 and 08:25 hrs. BST, a series of spectacular 'fireball' events were reported to have occurred above an area of England extending from Stafford in the Midlands to somewhere north-west of Manchester. A colleague at Woodford airfield in north-east Cheshire, Mr. Cyril Kay, was one of the witnesses. From his first-floor office window he had seen a brilliant green object streaking northwards at high altitude, to the west of his position. Such was the speed of the object, it had given the impression of being a rapidly-moving green streak, progressing horizontally. Reports from two other witnesses located in the Manchester area appeared in The Manchester Evening News that evening. These prompted me to investigate the reports, because they opened up the possibility of establishing the flight path and speed of the object. I visited the Manchester witnesses, took bearings and elevation angles from those sites and combined them with those given to me by my colleague. The result was quite staggering. The estimated velocity of the green object had been in excess of 43,000 mph (69,000 km/hr)! Furthermore, from the elevation angles and triangulations from the three sites it seemed that its altitude had been only 16,000 ft.! So it looked as though the Earth might have narrowly escaped an almighty collision on that morning.

Soon afterwards I was informed that fireball events had occurred in the Midlands, virtually at the same time. This information came to me from Mr. Anthony Pace, co-author of Ref.1 who was, by that time, BUFORA's Research Director. He had interviewed the witnesses in that area and, subsequently, he supplied me with that information to enable me to carry out an overall study and to produce a report.

It became quite clear to me that several different objects had been seen during a five-minute time interval that morning. In the Stoke-on-Trent region a shower of apparently insubstantial fireballs had been reported. The paths followed by these would have taken them over Newcastle-under-Lyme to descend into fields close to the M6 motorway. There had been no reports of explosive collisions and no meteoric material was ever found. This led me to think that they had been balls of glowing plasma and that they may have been produced in the wake of that primary object seen from the Manchester area. That idea was given further credibility by a report of another meteoric object seen overflying Stafford. If a fragment of the primary object had broken away somewhere in the region of Colchester on the East Coast and that fragment had then progressed over Stafford, the division of the ionised wake of the primary object might have produced self-contained balls of spinning plasma, which then slowed and finally descended over Stoke-on-Trent to dissipate their stored energy on collision with the ground.

Overall, I concluded that those Spring Equinox events had been caused by a very rare natural occurrence. My guess was we had been very fortunate that morning, because, to have survived the rigours of a such a hypervelocity pass through the dense layers of our atmosphere, that primary object would have had to have been huge when it first entered our atmosphere! Something the size of Snowdon mountain perhaps?

## The Fireball Cycle

During my study of British UFO reports for the period 1967-73, plasma-like fireball events were analysed separately. This was because I suspected that there might be natural explanations for them. One such possibility that came to mind was that they might originate from space, the Sun's corona being one conceivable source. Would it be possible, I wondered, for a ball of spinning plasma to be hurled out from the Sun and then follow the Earth's magnetic field lines before being projected into our atmosphere? I did a little exercise of that kind, but as it was outside my normal range of applications, I decided not to persist with it. Instead, I turned my attention to investigating whether there were any date connections with UFO activity of the vehicular kind. In some cases this seemed to be a possibility, but the main pointer came from a very simple exercise, which is reproduced as Fig. 8

After superimposing the occurrence dates for all the years of the British study in the manner shown, it became apparent that the phenomenon was cyclic. Plasma-ball events had generally occurred within +/- 6 days of 10 equally-spaced dates throughout the year. However, in no single year was a complete set recorded. The dates are shown in Fig.8 and, below, they are listed and compared with the dates of well-known periodic meteor showers.

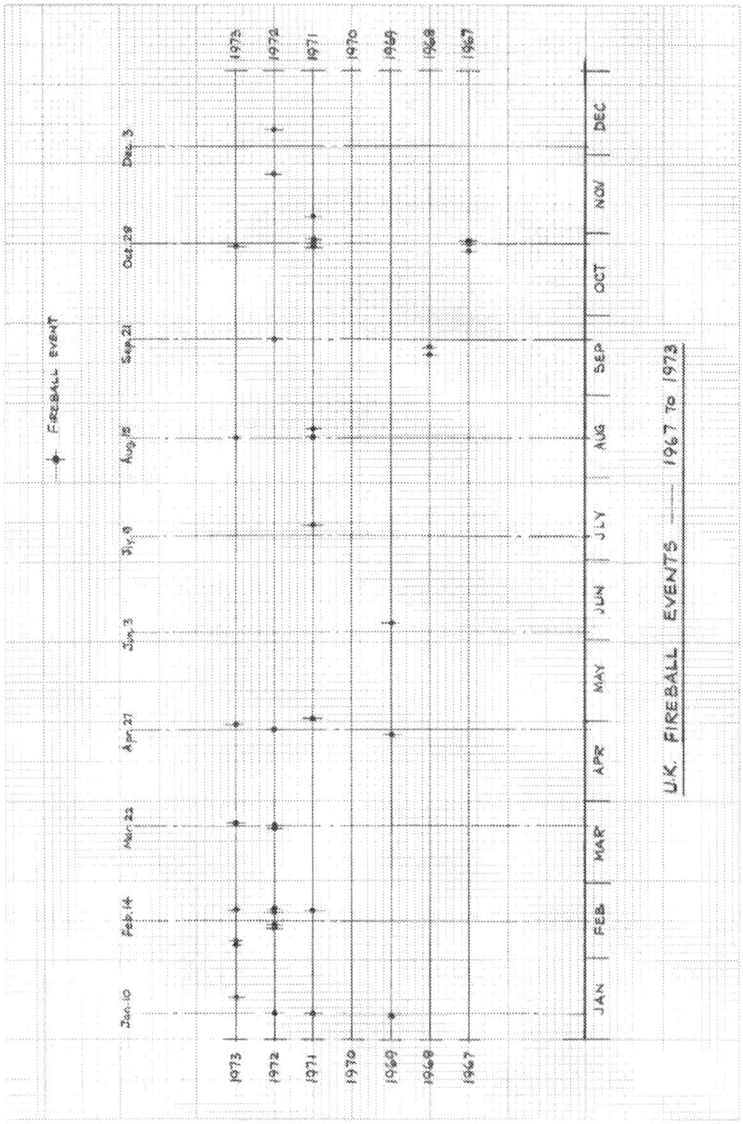
FIREBALL EVENT
Jan.10
Feb.14
Mar.22
Apr.27
Jun.3
Jly.8
Aug.18
Sep.21
Oct.28
Dec.3
1973
1972
1971
1970
1969
1968
1967
JAN
FEB
MAR
APR
MAY
JUN
JLY
AUG
SEP
OCT
NOV
DEC
U.K. FIREBALL EVENTS — 1967 TO 1973

**Fig. 8**

| **Mean 'Fireball'Dates** | **Meteor Dates** | **Meteor Shower** |
|---|---|---|
| 10th January 3rd | -4th January | Quarantids |
| 14th February | ---- ---- | |
| 22nd March | ---- ---- | |
| 27th April 19th | – 22nd April | Lyrids |
| ----- | 1st – 13th May | Aquarids |
| 3rd June | ---- ---- | |
| 9th July | ---- ---- | |
| 15th August | 22nd July – 17th August | Perseids |
| 21st September | ---- ---- | |
| ---- | 15th - 25th October | Orionids |
| 28th October | 26th Oct – 16th Nov | Taurids |
| ---- | 5th – 17th November | Leonids |
| 3rd December | 9th –13th December | Geminids |

As can be seen, most of those mean 'fireball' dates fitted neatly between the periods of intense annual meteoric bombardment. This observation and the cyclic nature of the events seemed to suggest that the phenomenon might be intelligently controlled and that *the links between it and the vehicular activity might be more than mere coincidence.*

The dates of this cycle were serendipitously used to good effect when the SAC study became global in scope. This is the work presented in Phase 2.

# PHASE 2:
# Links with the SUN and the STARS

*"And there shall be signs in the sun, ..... and in the stars; ..........."*
*Luke Ch 22, vs.25*

*This second phase of the story provides information essential to an understanding of how the Astronautical Theory came into being. It explains the logic used and the astronomical basis on which the theory stands. For some readers this could be the most difficult section of the book. It will require concentration and a real determination to understand. For that I must apologise, but* ***herein lies the evidence for programmed surveillance activity from space, a discovery which demonstrates the existence, in our environment, of extraterrestrial technology far in advance of our own.****. That is why Chapters 5, 6 and 7 could not be merely placed in an appendix --- so please read on.*

## CHAPTER 5

## A WORLD OF VISITATIONS

### Data Collection and Initial Processing.

The encouraging results from the U.K investigations study inspired me to look at the global situation. Only in that way would it be possible to recognise and define any kind of overall astronautical activity. I needed ready and abundant sources of detailed reports gathered from all over the world and over a long period of time.

Before going on to explain how I collected data for the global study, it is important to stress that many reports of Unidentified Flying Objects (UFOs) can be explained after careful examination by experienced observers of the skies. Therefore, the acronym 'UFO', in my view, should stand for Unidentifiable Flying Objects and such were the reports I sifted out from the sources I examined to provide me with data. I restricted my selection to reports of **Strange Aerial Craft (SAC), or strange lights in the sky behaving in a controlled way and, sometimes, as if they were attached to an unseen craft.**

Just as if fate took me in hand at that point in my investigations, a source of data was suddenly offered to me. After an informal presentation of my work had been presented to a Manchester-based research group, DIGAP (Direct Investigation Group for Aerial Phenomena), I was approached by one of those present in the meeting --- Mr. Peter Rogerson, a librarian. He was producing an international catalogue of Close Encounter events that covered a period from the mid-nineteenth century to the present day. It was called INTCAT and it was being issued in serialised form in the Merseyside (Liverpool area) magazine, MUFOB --- and he would send me free copies of the magazine as they were issued! That very generous offer was just what was needed to launch me into a new and exciting phase of the study.

Peter was faithful to his word and my data file grew steadily, month by month. Each catalogued report was read carefully and only the most seemingly-genuine ones were selected for the database. Here are two early examples :-

"2 November 1885, dawn, Scutari [Uskudar], Turkey. A luminous object circled the harbour, at altitude 5-6m. It was seen as a blue-green flame which illuminated the whole town. The object then plunged into the sea, after making several circles above the ferryboat pier. Duration 1.5min."

"12 November 1887, Cape Race (Atlantic Ocean). A huge sphere of light was observed rising out of the ocean by witnesses aboard the "Siberian". It rose to an altitude of 16m., flew against the wind and came close to the ship, then "dashed off" towards the SE. Duration 5 min. "

Unfortunately, a change in the location of the editorial office caused the issue of MUFOB to be suspended for a long period of time. Editorial policy had changed when the magazine eventually came back into circulation, causing the publication of INTCAT to cease. This meant that the supply of chronologically-ordered data had terminated with reports from September 1954. Another source of data had to be found to speed up the creation of my data file.

Again, fate took a hand, this time through the timely intervention of the then-Secretary of DIGAP, the late Mrs Joan Nelstrop. Remembering my need for data, she handed me a copy of a catalogue of several thousands of reports prepared by an organisation calling itself the National Centre for UFO Reports, based in Staffordshire. It was a wonderful piece of timing --- another case of a need being instantly met. The process of sifting through all those reports and selecting the most appropriate ones was a very spare-time consuming business. The period covered by the catalogue was March 1946 to the end of 1971.

From these two sources, only some 450 cases were chosen for processing. The Latitude and Longitude co-ordinates for these events were next plotted onto a graph representing the globe --- and various efforts were made to look for meaningful links between the points on the graph; that is, links which might have indicated that an astronautical activity from space might have produced them. These exercises were leading nowhere until I selected a set of reports of objects entering or leaving seas or lakes. At first, I represented the locations of these 'Water Events' as generous blobs of plasticine on my young son's small globe atlas and I was pleased to find I could link them by a series of approximate Great Circles. I then plotted those points on a graph representing the globe and defined the Great Circles (GCs) more accurately. Fig. 9 shows the picture obtained and the inclinations of the GCs to the Earth's equator. This seemed to be quite a breakthrough, but GCs are not in themselves indicators of activity from space. Even so, the discovery inspired the next stage in the investigation.

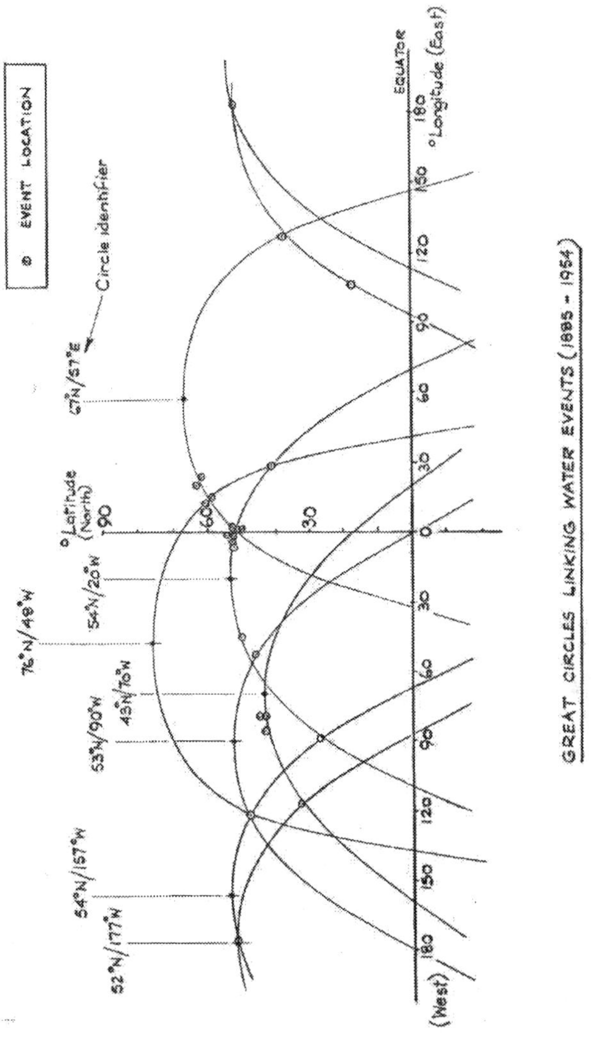


**Fig. 9**

## The Search for Approach Paths from Space

For readers with little or no knowledge of astronomy, Fig. 10 and the following information will aid understanding of all that follows.

For the purposes of observational astronomy, the Earth is envisaged to be rotating about its North-South (polar) axis at the centre of a huge ball, the Celestial Sphere, and the stars are regarded as being painted onto the inside of that ball. As the Earth rotates on its axis, from West to East, once every 23 hrs. and 56 mins. inside the ball, the stars directly overhead change progressively, rising in the East and setting in the West.

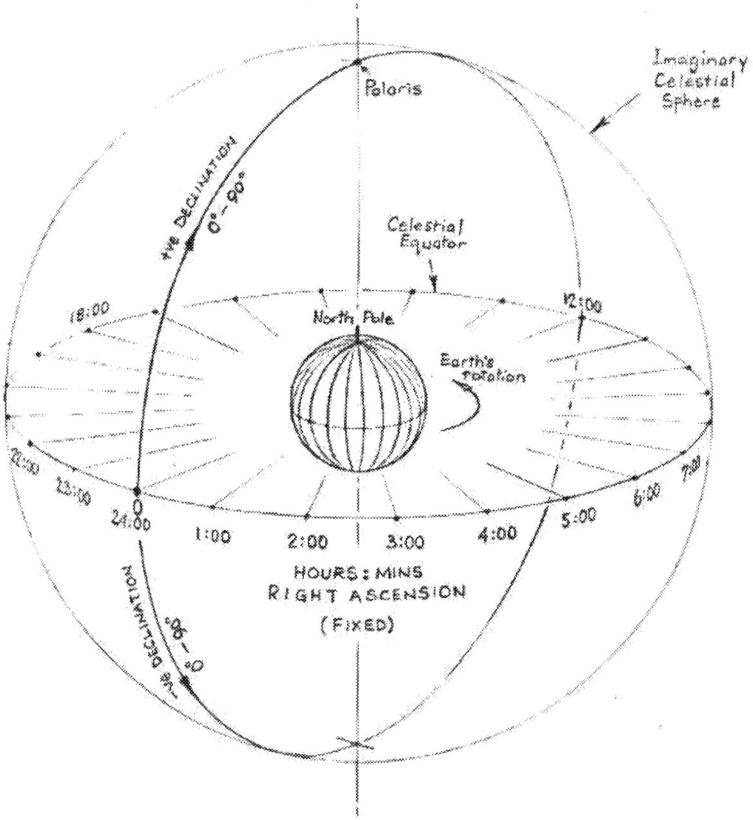


The Celestial Sphere

**Fig. 10**

On Earth we locate places by their latitude and longitude co-ordinates relative to the equator and the Greenwich (north-south) meridian. Similarly, astronomers envisage the ball (sphere) of the sky to be given a Celestial Equator (lined up with the Earth's equator) and **a reference (north-south) meridian located by the position of the Sun among the stars at noon during the Spring Equinox**. The star currently above the Earth's North Pole (Polaris, the Pole Star) marks, approximately, the Celestial North Pole for astronomers. Equivalent latitude, relative to the celestial equator, is called **Declination** and the equivalent longitude divisions are measured in **hours of Right Ascension, RA. (1 hour of RA represents 15° of rotation of the sky) and this is measured in hours Right Ascension from the reference (Spring Equinox) meridian, in an easterly direction.** In this way, the stars, constellations and distant galaxies have known (fixed) positions in the sky and can be readily identified. The planets and other bodies in our solar system, being much closer to us, are seen to move relative to this fixed background of stars during the course of any year. Our time on Earth is determined by the time taken by a position on the Earth's surface to move from Noon one day to Noon on the following day, and this takes 24 hours of **Mean Solar Time**. (Notice that this is longer by 4 minutes than the time taken for one revolution of the Earth relative to the stars **(Star** or **Sidereal Time)** and this difference results from the movement of the Earth in its orbit round the Sun.)

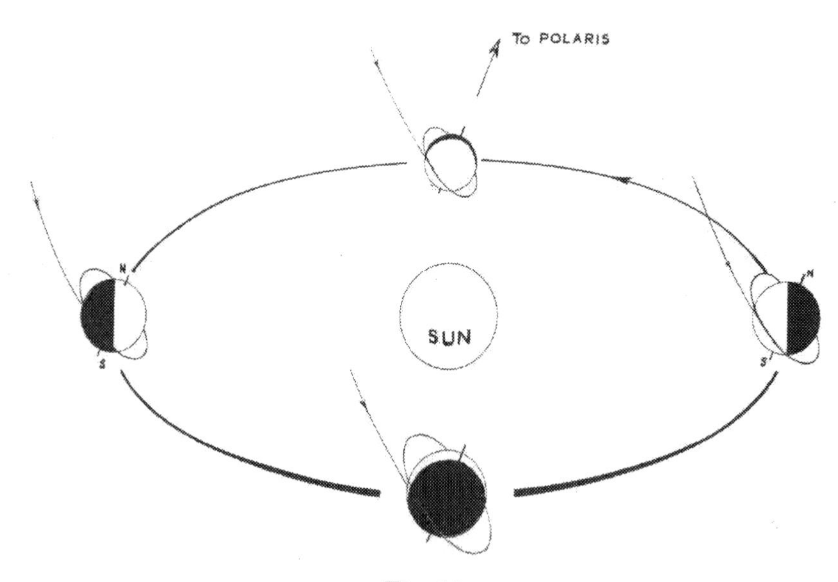

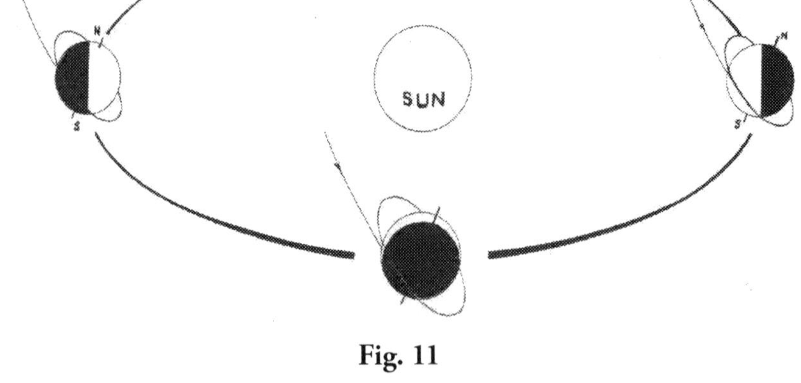


**Fig. 11**

Fig.11 is a diagrammatic representation of the Earth in its orbit round the Sun. As stated in the previous paragraph, in comparison with the movement of the Earth in orbit, the stars are so far away from us that they seem to be fixed in the surrounding sky, no matter where the Earth is placed in its orbit on any given day of the year. The simplest scenario I could envisage was one in which ET craft might approach the Earth from the same direction in space, whatever the time of year, and might then enter into a fixed orbit round the Earth to facilitate visitations to the planet's surface for a limited period of time. If the plane of the sought-for orbit is considered to be fixed relative to the fixed stars, then a spacecraft following such a path round the world, for **a few hours,** would be seen to pass through the same background of stars on each pass over a given latitude on the Earth directly beneath it. These principles were then used to search for evidence that that kind of surveillance from space had resulted in *sightings of SAC (probes) at locations directly beneath such an orbit.* From the diagram, notice how the orientation of the hypothetical ET craft's orbit, relative to **the Earth's terminator** (which marks the separation of day and night), would be affected by the time of year. Therefore, the time of day or night, such an orbiter would be observed from **a given** location directly below it, would depend on the date of the observation.

Clearly, then, the exercise about to be described needed the **timing** of each event, as well as its date and location. That requirement effectively reduced the useable data sample to 368 cases. The numerical breakdown of these events in Northern and Southern Hemispheres is shown by Table 1. They were predominantly in the Northern Hemisphere.

**TABLE 1**

| Nominal Date | Northern Hemisphere | Southern Hemisphere | Totals |
|---|---|---|---|
| Jan 10th | 27 | 7 | 34 |
| Feb 14th | 20 | 9 | 29 |
| Mar 22nd | 31 | 9 | 40 |
| Apr 27th | 33 | 7 | 40 |
| Jun 3rd | 24 | 7 | 31 |
| July 9th | 31 | 5 | 36 |
| Aug 15th | 42 | 1 | 43 |
| Sept 21st | 35 | 2 | 37 |
| Oct 28th | 50 | 7 | 57 |
| Dec 3rd | 19 | 2 | 21 |
| Totals | 312 | 56 | 368 |

In order to minimise the time required for the exercise, (which throughout, was NOT computerised) another expedient was adopted --- the timings of events would be adjusted so that they would correspond to **Mean Solar Time** at each location on ten specific dates of the year. ***The equally-spaced fireball dates , as listed previously in Chapter 4, were chosen for that purpose.*** The reason for this decision needs to be understood. It has to be explained that events on the Earth were about to be linked to the stars in a meaningful astronomical way. Rather than having to examine every case on whatever day of the year it had occurred, which would have involved a lot of manual calculations, I decided instead to adjust the Local Time **(Mean Solar Time)** of each event by four minutes for every day its date was different from the nearest nominal (fireball) date. This would ensure that the same stars would be overhead at the ADJUSTED time on the nominal date as were overhead at the REAL time on the actual date.

The Phase 1 study had indicated that the small craft witnessed in Close Encounters had been too small to be space cruisers. It followed that they had been delivered and would be retrieved by large cruisers, probably established in **short-term** orbits or **partial** orbits. The search was about to begin for evidence to substantiate that idea.

***My hope was that by linking the stars that were immediately overhead at the time of each Close Encounter event, the linked points would be found to form arcs of Great Circles fixed in space, perhaps indicating the existence of a number of fixed orbital paths.***

# CHAPTER 6

# Celestial connections

*".... how many ways are above the firmament,.....?"*
*ESDRAS II Ch. 4 v. 5*

## First Indicators

The search for the astronomical and astronautical connections had begun in earnest. Having corrected all the data in the way described in Chapter 5, and lacking any computer aids, the next thing to do was to plot out some graphs. Figs.12(a) and 12(b) show the results of that exercise. As explained previously, the search was on for bits of Great Circle hoops in the sky. The curved lines passing through the points on these graphs show that **I found them**!

The points on each graph refer to the **Latitude (Declination in the sky above the location)**, plotted on the vertical scale, and **Mean Solar Time** (measured on the horizontal scale) of each selected Close Encounter (CE) event. The curved lines represent arcs of Great Circles, distorted and split by being shown against a **flattened Celestial Sphere**, as drawn on flat graph paper. From these graphs some very important observations could be made. They indicated that orbits with ten distinct inclinations to the Equator had been used (including those revealed by the Water Events exercise), sometimes over long periods of years ---- and that some of those orbits, shown at different times of year, were linked to the same stars. This latter clue was particularly exciting, because there was a possibility that it might indicate where the visitors were coming from. So, that was the next idea to be explored.

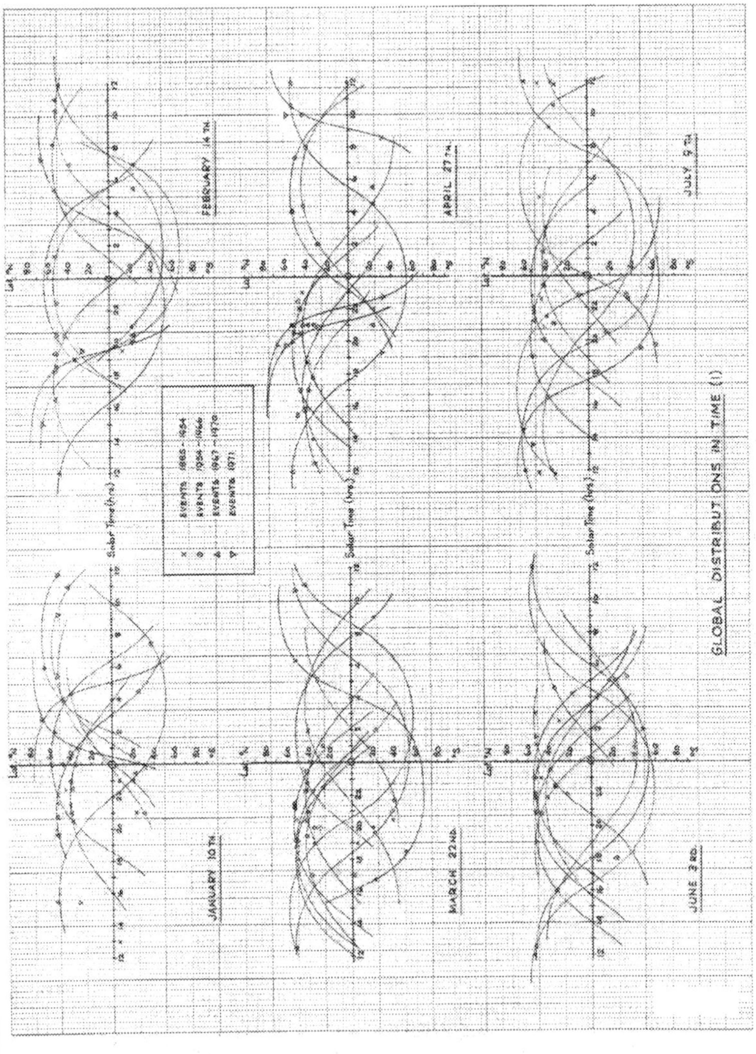
FEBRUARY 14TH
APRIL 27TH
JULY 9TH
JANUARY 30TH
MARCH 22ND
JUNE 3RD
GLOBAL DISTRIBUTIONS IN TIME (1)

**Fig.12a**

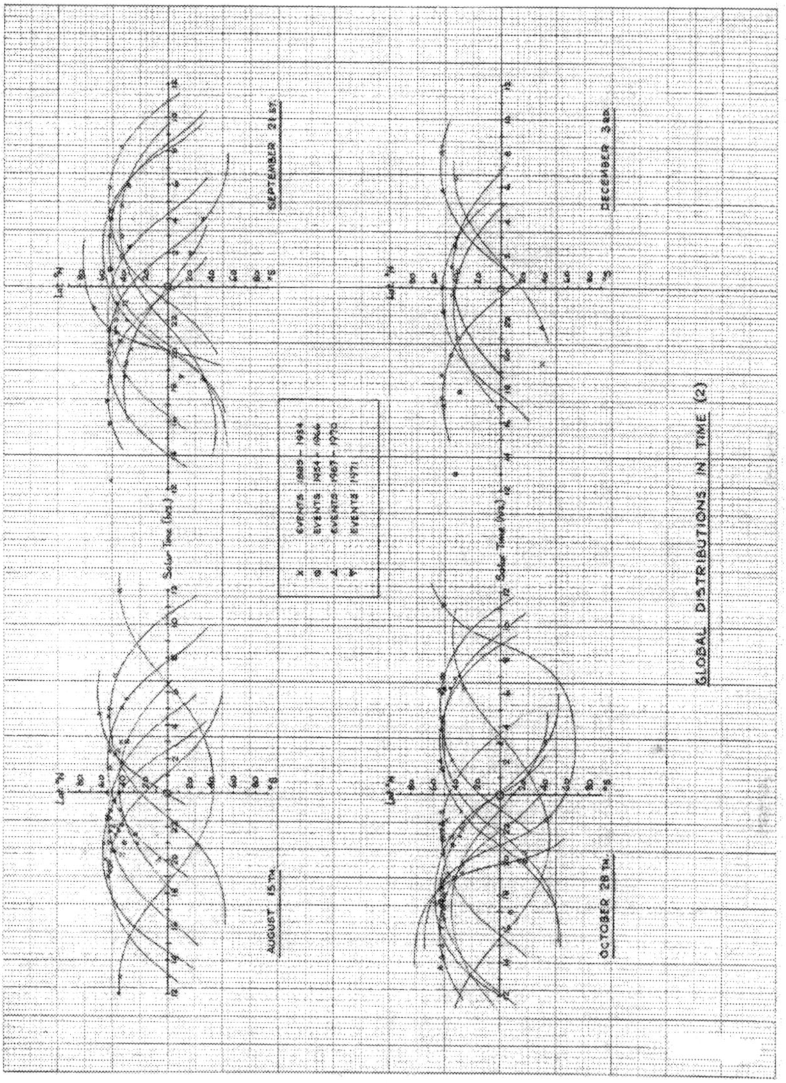


**Fig 12b**

As a means of identifying each one of the curves shown on the graphs, the Latitude (or Declination) and the Mean Solar Time of the **northern-most point** on each curve was read and listed at the date of that particular graph. The **identifier** points read off were then plotted on another graph (Fig.13 ) from which other star connections could be sought. But there was a totally unexpected bonus in store for me. As will be explained, **I**

**discovered that many of those tracks in the sky had often been linked to the position of the Sun!**

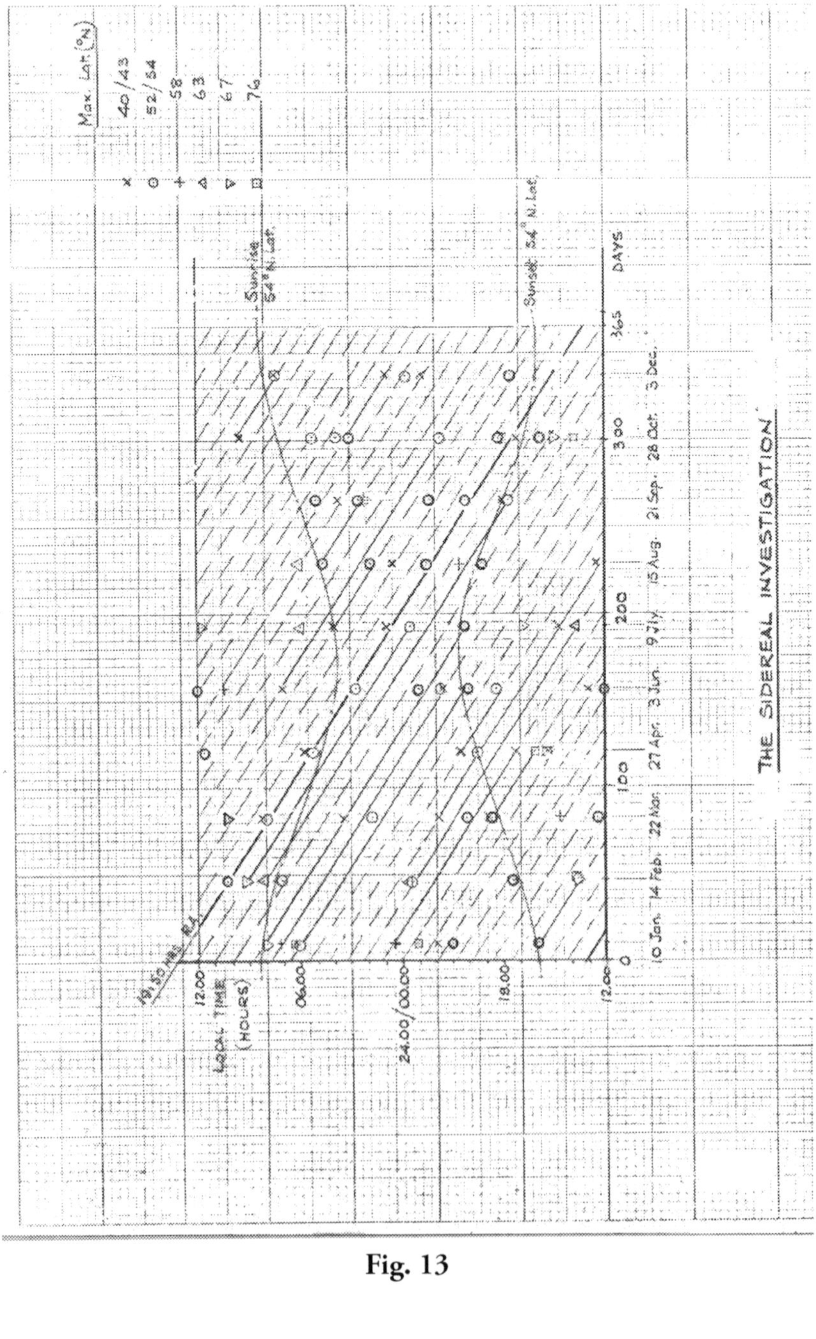


Fig. 13

The sloping straight line drawn on this Fig 13 graph of Mean Solar Time vs Days of the Year, are lines of **constant Sidereal Time** (constant star time) and they slope downwards to the right at a rate of -4 minutes per day. (As explained previously, this effect is caused by the Earth's movement in its orbit round the sun during the course of any year.) Overall, the slope is approximately -24 hours per year. The **identifier** points were superimposed onto this graph, as shown, and some interesting trends were observed. It was found that the best fit of sloping lines linking the points on this graph divided the 24-hour days into 22 equally spaced lines, the spacing between them being **65.45 minutes**. There were little clusters of points spaced in that way but it was observed that these points represented events widely separated from each other by, sometimes, periods of many decades, so clear links with particular parts of the sky or indications of orbital activity were not readily recognisable. Further close study of the pattern of points, however, suggested that paths with inclinations to the equator **between 52 and 54 degrees** had perhaps been linked to the Earth's **northern sunset terminator** at those latitudes. This observation was confirmed after a curve representing the appropriate sunset times throughout the year had been superimposed on the graph, as shown. It was not clearly evident that the northern sunrise had also been referenced in that way. Overall, the sidereal study had produced a totally unexpected result. It had demonstrated that the idea that there might be only one fixed-in-space approach path was a gross over-simplification of the situation. This discovery was not unexpected, but the observed link with the position of the sun among the stars ***seemed to suggest that the probes had been accessing the Earth from bases within the Solar System over a long period of years.***

# CHAPTER 7

## Pathways over the Earth

Up to this point in the investigation, Great Circles in space had been considered, these being supposed to represent the paths followed by the visiting spacecraft. But **when a craft is moving in orbit round the world, its path over the ground beneath it is a helical (spiralled) one**. This is the result of the relative speeds of the satellite and the rotating Earth below it. During one rotation of the satellite, the position on the equator passed over initially becomes displaced by the number of degrees of the Earth's rotation corresponding to the time taken by the satellite to complete a single orbit. A typical low altitude satellite takes about 90 minutes to complete one orbit. In that time, the Earth's surface rotates 22.5 degrees to the west of the original crossover point on the equator. This is clearly seen on NASA's and Russian projections of the Earth, which show the successive paths (ground tracks) followed over the surface of the planet as a given spacecraft (say, the Space Shuttle) continues to orbit. The time taken for one orbit is referred to as the **period** of any satellite, and this becomes greater than 90 minutes if the height of the satellite above the Earth is significantly greater than a nominal one of, say, 100 miles.

***My next step was to look for signs of spiralling lines linking the geographical locations of the reported SAC encounters.*** Considering the 90-minute satellite as the fastest likely to be encountered, I began curve matching using tracing paper with lines representing orbital ground tracks for different natural orbital periods. This turned out to be a fruitless exercise until, out of desperation, I tried **a period of 65.45 minutes**. Almost instantly, UFO-event locations on the Earth began to be linked up in significant numbers! Very soon I was able to identify 34 well-aligned sets and then went on to identify a further 32 ----- 66 in all. Table 2 lists them, in the order in which they were identified. Ground tracks numbered 5, 10, 11 15 and 27 were found to represent five of the Water Events GCs, except that they were now spiralled lines. It is also important to notice that ***some intersections with the Earth's equator were shared by tracks having different inclinations****, emphasising the importance and fixed nature of the equatorial intersections.*

| Track No. | Inclination° | Intersection (° Long.) +ve East -ve West | Track No. | Inclination° | Intersection (° Long.) +ve East -ve West |
|---|---|---|---|---|---|
| 1 | 43 | -78 | 35 | 54 | 79 |
| 2 | 43 | -60 | 36 | 67 | 146 |
| 3 | 43 | -48 | 37 | 76 | 162 |
| 4 | 43 | -5 | 38 | 43 | 32 |
| 5 | 43 | 22 | 39 | 43 | 37 |
| 6 | 43 | 72 | 40 | 43 | 48 |
| 7 | 43 | 98 | 41 | 43 | 85 |
| 8 | 43 | 153 | 42 | 43 | 114 |
| 9 | 43 | 180 | 43 | 43 | 141 |
| 10 | 54 | -78 | 44 | 54 | -41 |
| 11 | 54 | -60 | 45 | 54 | -32 |
| 12 | 54 | -48 | 46 | 54 | 37 |
| 13 | 54 | -5 | 47 | 54 | 43 |
| 14 | 54 | 19 | 48 | 54 | 46 |
| 15 | 54 | 72 | 49 | 54 | 51 |
| 16 | 54 | 98 | 50 | 54 | 59 |
| 17 | 54 | 113 | 51 | 54 | 146 |
| 18 | 54 | 154 | 52 | 54 | 162 |
| 19 | 54 | 176 | 53 | 67 | -41 |
| 20 | 67 | -93 | 54 | 67 | 21 |
| 21 | 67 | -78 | 55 | 67 | 59 |
| 22 | 67 | -59 | 56 | 67 | 79 |
| 23 | 67 | 32 | 57 | 67 | 85 |
| 24 | 67 | 37 | 58 | 67 | 141 |
| 25 | 67 | 68 | 59 | 67 | 157 |
| 26 | 67 | 112 | 60 | 67 | 162 |
| 27 | 67 | 153 | 61 | 76 | -105 |
| 28 | 76 | -93 | 62 | 76 | 43 |
| 29 | 76 | -78 | 63 | 76 | 59 |
| 30 | 76 | 21 | 64 | 76 | 79 |
| 31 | 76 | 98 | 65 | 76 | 85 |
| 32 | 76 | 114 | 66 | 76 | 180 |
| 33 | 76 | 153 | | | |
| 34 | 76 | 176 | | | |

**TABLE 2**

When a computer graphics representation of these tracks (Fig.14) was produced by professional colleagues in the Computer Services Department of HSA Ltd., Woodford, as a trainee exercise, using available 3D global software, another important feature of these discoveries became apparent. It was noticed that when the limited ground track arcs were extended to represent complete orbits, the end of one track often linked up at the equator with the beginning of another, this indicating that continuous orbiting might have been carried out.

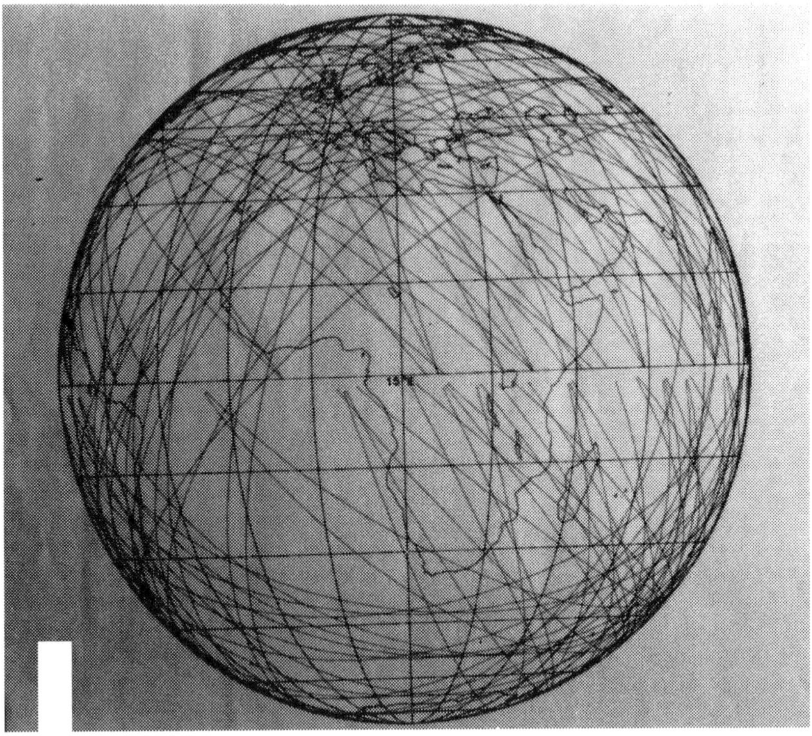

**Fig. 14**

***This being so, the spacecraft executing those operations had been literally 'out of this world'. They would have had to have held themselves in orbit by powerful and continuous thrust-vectoring directed towards the Earth's centre.*** This is a technique well beyond any foreseeable human capabilities.

*Another important characteristic of the established tracks was that they represented, without exception,* **retrograde orbiting** *of the planet; that is, the visiting craft had been moving from East-to-West, against the direction of the Earth's West-to-East rotation.* Further analysis of the intersections at the equator also showed that three distinct orbital sequences had been adopted and these are shown by Fig.15. But there were, in addition, non-conforming intersections so that the total number of intersections came out to be 101. From each one of these, 10 ground tracks with fixed inclinations to the plane of the equator could be assumed to have been generated, in a north-westerly direction, giving a total number of tracks over the inhabited areas of the Earth of 1010. ***All this represented a precisely programmed***

***surveillance activity being carried on by fully automated probes, employing technology well beyond our current understanding.***

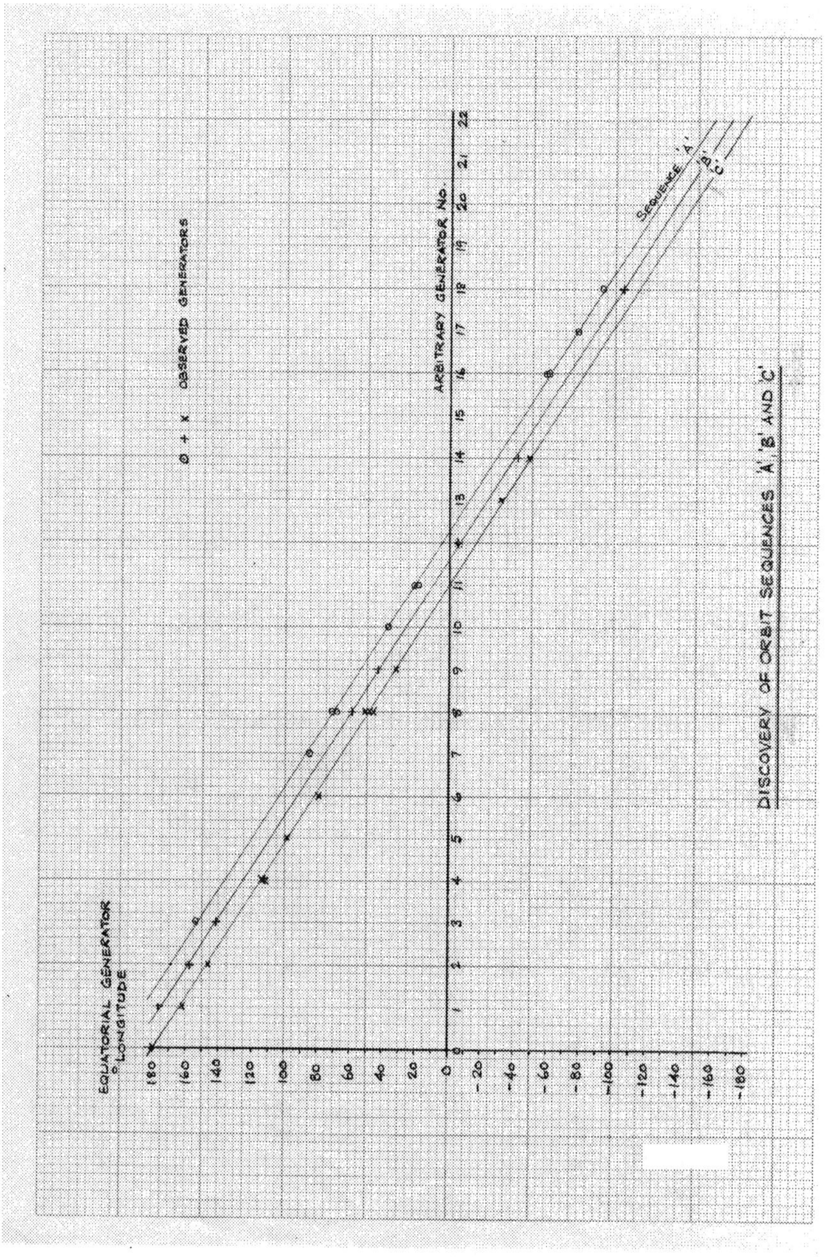


**Fig. 15**

Now, ***it is important to point out that I do not regard these fixed paths over the Earth to be followed for continuous orbiting, even though three super-orbital sequences have been identified. The evidence suggests that they are analogous to the fixed air-lanes followed by civil aircraft, except that they are followed in space. These 'space-lanes' seem to be used for rapid over-flights of targeted locations on the Earth's surface directly below. An exploration craft/probe may be launched into the atmosphere over the target area from one 'lane' and retrieved some time later by a retrieval craft following another 'lane' that passes over the target.*** This scenario will be seen to be substantiated by the encounters considered in Part 2.

## Renewed search for evidence of favoured approach paths from space

At this point in the investigation, an outstanding unknown was whether or not there had been favoured approach paths from outer space. The Fig.13 graph of Mean Solar Time vs. Days of the Year had, in a tantalising way, produced a constant sidereal time line apparently linking a series of points spanning seven of the ten chosen dates. The sidereal orientation represented by this line was **19:50 hours Right Ascension (RA) at the northern-most point on each orbital path**. It was decided to check this out by selecting 75 more recent SAC reports and, by applying to these the accumulated knowledge to that date. For each of these cases, the ground tracks passing through that location were first identified and, then, the time of the event was used to define the orientations of the corresponding orbits in space. Since more than one track and associated orbit could be linked to a given location, the number of possible RA orientations for the 75 events was 103. Fig. 16 shows how these RAs were distributed.

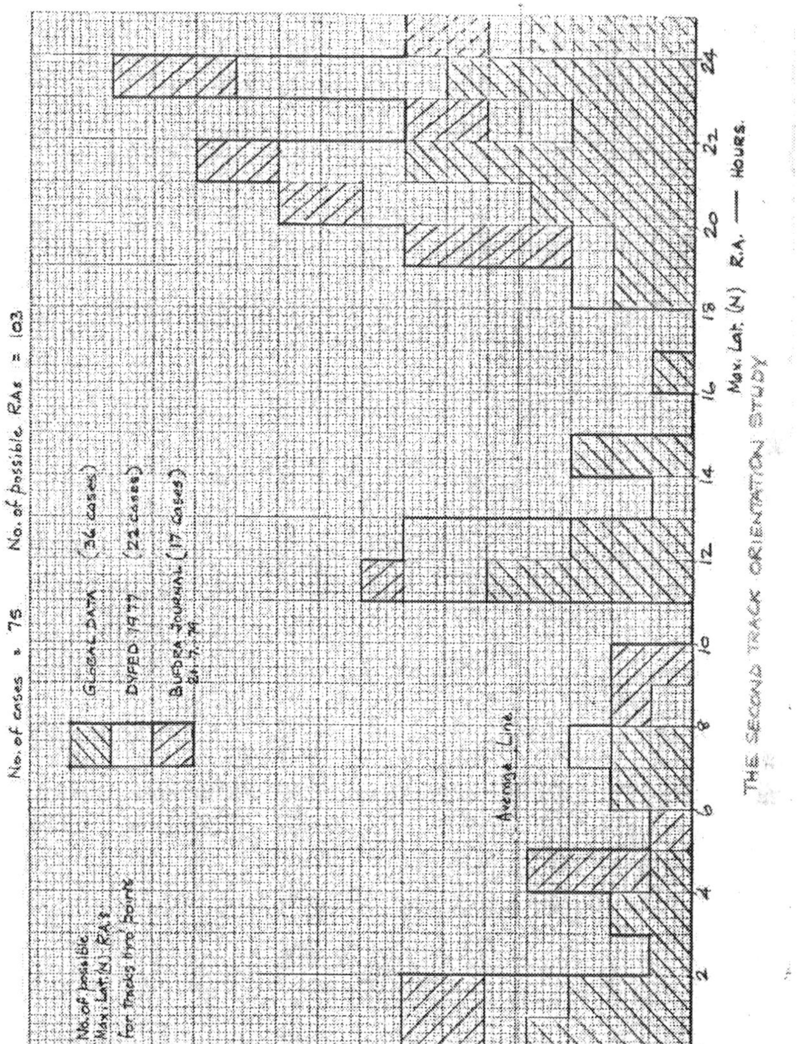


**Fig 16**

There were two most favoured areas of the sky ----- a major one in the 23:00-24:00 hours RA region and another isolated lesser peak between 11:00 and 12:00 hours RA. There was not a significant peak at 19:50 hours RA. Consequently, the puzzle remained unresolved and had to await further inputs, which came, later, from an unexpected source.

# PHASE 3:
# Contributions from Crop-Circle Studies

*Phase 3 may seem to be strangely out of place in this book, but, as will be demonstrated, it is very much part of my overall search for understanding of the SAC phenomenon. My involvement in the crop circles mystery provided new insights and resulted in an unexpected refinement of the Astronautical Theory. These contributions to my research are given here in their correct chronological sequence .*

## CHAPTER 8

## A MEETING OF MINDS

As has doubtless become apparent, my path through the UFO pilgrimage has been signposted in some very unusual and totally unexpected ways. None was more fortunate than the encounter I had on one night in May 1987.

That was the night of my lecture in a series organised by the British UFO Research Association, **BUFORA**. It was delivered at their usual venue in the London Business School. The title of that major lecture was **"We are DEFINITELY not alone"**. It took more than of one-and-a-half hours to deliver, because it contained most of the material described in the previous chapters of this book --- at least, as much of it as then existed.

After the lecture, I was approached by a group of men, who had obviously been listening very intently and with some enjoyment during the delivery. The leading person expressed his appreciation of my material and then asked if I had heard about the crop markings in Wiltshire and Hampshire. I admitted that, other than having seen a photograph of swirls in a crop on the Isle of Wight some years ago, I knew absolutely nothing about them. The leading man then told me that the kind of locations that I had associated with SAC events could equally-well be associated with the regions in which mysterious crop-circles were persistently appearing. He introduced himself as Colin Andrews (now a well-known author and researcher of the phenomenon), handed me his card and invited me to join in the research which he and his companions were pursuing. Well, it was very pleasing to get that kind of response to a long and demanding lecture ---- but those counties of Southern England seemed to be just too far away from my northern Cheshire home to make a link-up seem attractive. However, fate had other ideas on that matter. To cut a long and almost unbelievable story short, during August 1988, my wife Marion and I found ourselves driving through rainstorms to a rendezvous with Colin Andrews and others of his team in Andover, Hampshire. Where that meeting led to, over the years to the present day, will be told in broad outline with occasional detail, in the chapters of this, Phase 3, part of the story. Through the crop-circle phenomenon, everything told previously in this book has been, or is being, vindicated.

Marion and I travelled to the home of Colin Andrews and were given a very friendly welcome. Also present on that night were fellow researchers F.C. (Busty) Taylor and Don Teursley. Colin proceeded to show extraordinary video footage of crop circles and of other, possibly UFO-related, phenomena. We were suitably impressed by what we saw and heard. Busty Taylor offered to take us into one of the remaining specimens next day--- a then-unique circle surrounded by two Saturn-like rings of flattened crop.

We waited for Busty in Stockbridge, Hampshire, at lunch time, in a rainstorm. Fortunately, the rain ceased when we met and we piled into Busty's Metro for what turned out to be an exciting ride on a bumpy and lumpy old green track running into the depths of the countryside. We finally arrived alongside a large soggy field of over-ripe wheat. Therein, some metres from the edge of the field, lay the object of our attention. Wellington boots and waterproof kit were the donned by me while Busty, who was not so well equipped, stood by patiently. He and I then made our boot-clogged way into a large circular pan of flattened crop, crossing two flattened rings en route. I was taken aback by the size of this formation, in which the lay of the crop was clearly spiralled. It had lain there for some weeks and had been well trampled. I remember asking if perhaps someone with a roller had produced this specimen. Busty quickly countered that suggestion by telling me that when the formation had first been discovered the stalks of the crop had been flattened without damage. A 90 degree gentle bend had been produced at the base of each stem and the plant had not in any way been damaged. Given that information, I was clearly at a loss to understand how the formation had been produced. I looked around me and particularly noted that the field was slightly sloping uphill to our right. My next question was, "Is this formation perfectly circular Busty?" His answer was that it was not. The configuration had been measured to be 104 feet long and 100 feet wide. "Which way does the long axis lie?", I asked. Busty looked at the field carefully and then indicated that it had aligned with the slope of the field. This suggested to me that a cylinder of energy projected from vertically above could have produced such distortion on the ground. But I was baffled by that thought and wondered what had been responsible for the outward spiralling of the neatly-laid crop. As we stood pondering this remarkable happening, an army helicopter flew very low over the formation and then back again. Busty informed me that the army had a base at nearby Middle Wallop. Were we photographed? We'll probably never know. Beyond the top of the field, I learned from Busty, an ancient earthwork was located, called Danebury Rings. This seemed to provide me with a possible link with UFO activity, remembering my observations about ancient sites during my 1967 investigations in the North West. Altogether, that trip into the countryside was a memorable experience and was the beginning of a friendly association with Busty (and Colin Andrews) that has continued to this day.

## The 'White Crow' Experience.

During 1989 I participated in a joint investigative venture organised by Colin Andrews, his colleague Pat Delgado and Dr. Terence Meaden, a tornado expert and the promoter of the idea that the circles formations were being caused by natural, but rare, 'plasma vortices'. These were conceived to be fine weather whirlwinds of electrically-charged air and the concept was being warmly received and promoted by 'arm-chair' members of the scientific establishment. The 'White Crow' project had been set up in a field in Hampshire to view a low-lying field on the opposite side of the A272 road from Winchester. Formations had appeared regularly in that field for several years. The field was to be observed continually, night and day, for the period 10$^{th}$ to 18$^{th}$ June and I and two new colleagues from the North West, Michael Thomas and Harry Harris had volunteered to share the watches, though Harry Harris and I could only spare the final weekend for our participation. We had also agreed to cover the daylight hours and to return to our hotel at night.

This instrumented exercise eventually produced no evidence in the observed field, but a striking formation was created, on a bright moonlit night, in a field only half-a-mile or so from the observation site. The field involved had sturdy green wheat growing in it and the ringed formation produced in the crop had been beautifully laid down. I visited the site the following morning and there, incidentally, met and talked with Dr. Meaden for the first time. Later, I learned that Colin Andrews, Pat Delgado and several other people had walked up to an adjacent field, soon after midnight, to the site of a formation created several weeks earlier. They had been suddenly assailed by a loud twittering noise which had seemed to circle round them and then settle briefly over the next field, the field in which the new formation had been found the following morning. The twittering sound had been recorded on a cassette tape recorder and it was later analysed. Unfortunately, it resembled the twittering of Grasshopper Warbler birds and so fuelled the cynicism that inevitably followed the issued reports on the event. (On considering the information available, it had occurred to me that a flock of grasshopper warblers could have been disturbed by the commotion in the crop and had taken to the air in fright. No one else had seemed to consider that as a possibility.)

The 'White Crow' experiences provided valuable contributions to my understanding of the circles phenomenon and they were to be supplemented by other experiences during the 1990s. However, even before the 'White Crow' episode, I had been pondering the possible causes of that double-ringed circular formation of the previous year, but had found no satisfactory explanation. I had looked for analogous phenomena that might provide a model to work with. One such model arose from the sharp-edged boundaries of that Danebury Rings formation. Surely a natural atmospheric vortex could not have produced such precision. In fact, the only kind of swirling flow I could envisage was that produced in the wake of an airscrew propeller. During discussions about the modelling of such flow with my colleagues in the BAe Future Projects Dept., I discovered that one of them, Mark Metcalfe, had presented a paper on the topic **[3]**. Mark very helpfully let me have a copy of his work and I began work at home trying to understand it and to produce, mathematically and with diagrams, typical cross-sections of the flow being modelled. The results are shown by Figs. 17(a) and 17(b). However, this method turned out to be of little use to me because the results obtained were too dependent on the effects of the individual propeller blades on the flow and the distance of the cross-sections from the plane of the propeller.

## References.

[3] Metcalfe, M.P. 'On the modelling of a fully-relaxed propeller slipstream' AIAA/SAE/ASME/ASEE 21st. Joint Propulsion Conference, California, July 8th-10th, 1985.

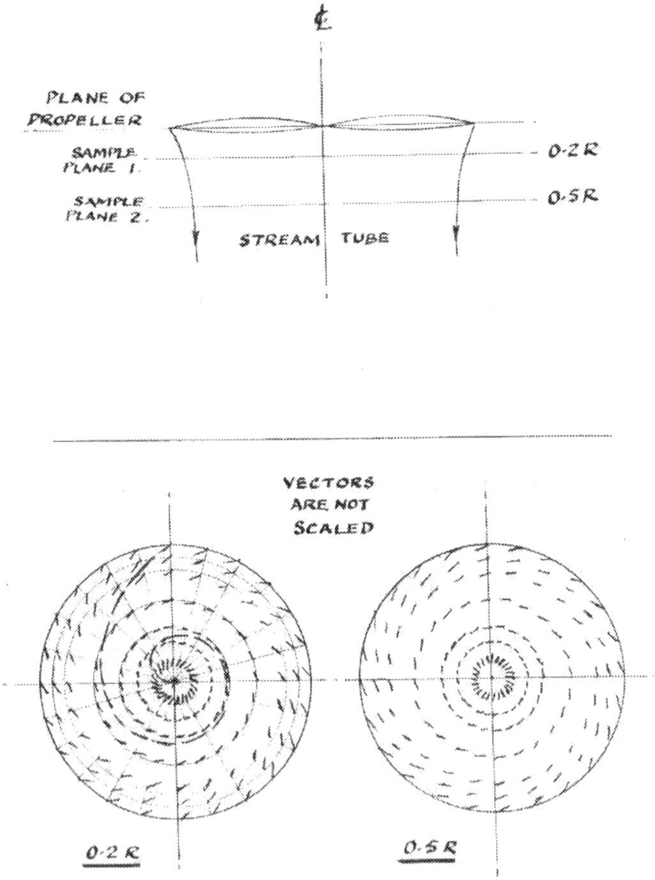


**Figs. 17(a), 17(b)**

# CHAPTER 9

## Like a circle in a spiral

*"Round, like a circle in a spiral,......."*
*SONG --- 'The Windmills of Your Mind'*

The realisation dawned that perhaps the best way to understand the processes producing the circular formations would be to study plan view photographs of several different specimens, if these could be obtained. It seemed that the obvious initial sources might be two books then recently published on the topic, one by the Delgado/Andrews duo **('Circular Evidence'** [4] ) and the other by Dr. Meaden **('The Circles Effect and its Mysteries'** [5] ). I experienced difficulty in obtaining copies of these books, but, in September 1989, received a surprise birthday gift of 'Circular Evidence' from my wife. Then I had photographs in abundance to study and I encountered problems immediately. The aerial photographs did not define the lay of the crops very clearly and there were the inevitable perspective distortions in the close-up photographs. The most important clue given by this collection was that **spiralled ridging** of the flattened crops was often evidenced and I had witnessed this myself in the 'White Crow' formation.

Each of the formations had seemed to have been unique. How could that be if they had been produced by the same processes? By then, totally bemused, I put the problem aside and awaited new inspiration.

It came during my lunch break in the office at BAe, Woodford, on Wednesday, 17th January, 1990. During my studies of the circles I had found myself often humming the once-popular song, 'Windmills of Your Mind'. Somehow the opening lines had seemed to relate to the images I'd been studying. I was musing on this when it occurred to me that perhaps I should try to define the spirals in the circles, mathematically. Spiralled flow in air or any fluid is produced by a combination of outward (or inward) flow and rotational flow, the relative strengths of these flows determining the kind of spirals produced. This principle is shown diagrammatically by Figs.18 & 19

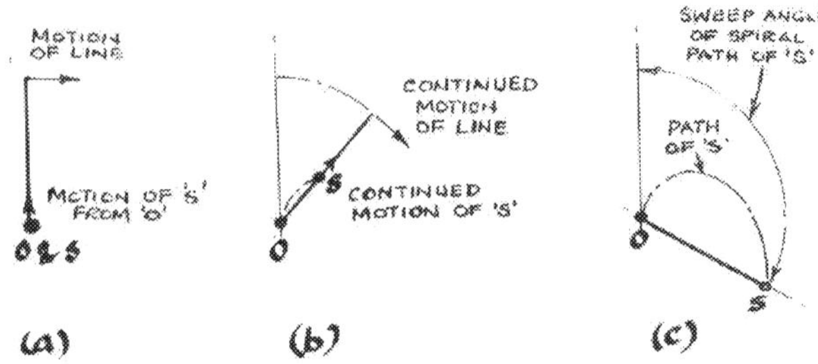


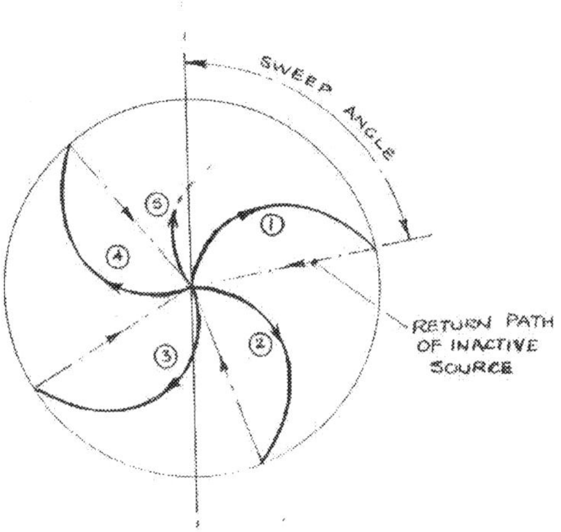


**Fig 18 and 19**

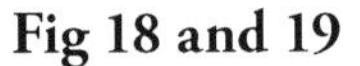

When I returned home, I paged through 'Circular Evidence' again and eventually came to a photograph (Headbourne Worthy, 1986) giving good definition of the lay of the crop without undue distortion caused by perspective effects. It was not a plan view photograph but this could be corrected by use of a simple draughting technique. From this I produced a corrected true plan view with the true shape of the spirals in the flow also reproduced. It became clear from an examination of the radial and rotational flow components at evenly spaced points on each spiral that a simple mathematical relationship linked them. It was this: **a constant outward radial flow had been combined with a steadily increasing rotational flow as the distance from the centre increased.**

To model a spiral with same characteristics, all I needed then was the angle swept through by a spiralled strip, between the centre and the periphery of the circle. Having measured this angle for several of the plan view spirals and taking an average value, I found I was able to reproduce exactly the spirals derived from the photograph.

My next idea was to write a PC graphics program to build in the features I had established manually. Having achieved this, I was then ready to try it out on other photographs. With the mathematical flow relationship built-in, all I would need would be the measured true sweep angles for the spirals in other circles. My need for more photographs or true plan views was answered by the receipt of measured drawings from Colin Andrews and several near plan-view photographs of small ('grapeshot') circles taken by Busty Taylor using his then-unique 'camera on a pole' technique. (Many other researchers were soon to follow Busty's example). I had to correct most of Busty's photographs for camera angle, but there were no appreciable perspective distortions to correct. To cut a long story short, I discovered I could reproduce the spirals very accurately in all the examples sent to me. Then I looked for a method for filling in the remaining area of each circle with spirals. The simplest way of doing this seemed to be as shown by Fig.19. As each spiral was drawn to the periphery of the defined circle, I arranged for the graphics program to start generating another spiral from the centre outwards, repeatedly, until all the circular area had been filled in. This choice of method turned out to be a remarkable piece of serendipity. As I proceeded to represent each circle modelled, I discovered something very significant. **The number of spiralled strips produced by the graphics program matched, exactly, the number of strips that could be counted in the photographs!**

It looked as if I had reproduced the method by which the spiralled circles had been formed and, furthermore, the model could be arranged to run in reverse to reproduce inward flow. The direction of rotation could also be reversed, so that clockwise and anticlockwise flattening could be produced. It then seemed that a focused, slender, beam of intense energy, programmed to perform in the same manner modelled by my computer program, could have been responsible for the observed patterns.

The problem now to be resolved was identification of the kind of energy required to heat the cells at the base of the plant stems, so as to create plasticity there and, simultaneously, to push the stems down gently in the direction of movement of the source. Nothing readily came to mind, but a picture was emerging in my mind of some sort of aerial craft equipped with laser-like projectors to project focused beams onto the crops below. The motion I had modelled was that of **a rotating line-scanner,** conceivably used for rapid scanning in that manner. It seemed to be yet more evidence that we were dealing with 'out of this world' technology and it linked with reports of UFO activity having been seen over the affected crop fields. **Furthermore, given that kind of technology, the focused energy beam could be programmed to produce complex patterns in the crops. Even the process of interweaving the crop stems would be a relatively simple activity. I envisaged that such a scanner could be linked to a computer graphics screen and could be programmed to reproduce, in the crops, whatever images had been created on the screen.**

As for that question about the mysterious energy being used, I considered its known characteristics:- heating at molecular level producing distortion of the plants' cell structure, coupled with a gentle over-pressure flattening the stems in the direction of motion of the projected beam. **(It needs to be emphasised that the idea of cell distortion by internal heating preceded, by several years, confirmation of that process by the American Dr. Levengood of BLT Laboratories).** Clearly, the beam had to be one of very high frequency energy, with wavelengths of only a few millimetres to produce the effects within the stems and it had, also, to have the capability of producing mechanical impulses. From these I conjectured that **high frequency gravitational radiation** could be involved. That immediately 'put the cat among the pigeons' for most physicists, but other evidence to be described later seemed to further validate that suggestion. When I shared my suspicions (without mentioning my guess about the nature of the energy being used) with Dr. Meaden in a crop formation one day, his response was that, surely, my solution raised more questions than it answered --- but, of course, he was unaware of the background UFO research work and probably would not have wanted to hear about it anyway. However, Dr. Meaden's new book turned out to be quite a 'god-send' later on, in that it helped to resolve the problematical 'favoured path from space' puzzle.

## References

[4] Delgado, P. Andrews, C. Circular Evidence' (book) Andrews, C. Bloomsbury Publishing, Ltd., London. 1989.

[5] Meaden, T.G. 'The Circles Effect and its Mysteries' (book) 1989 Artetech Publishing Company, Bradford-on-Avon, Wilts.

## CHAPTER 10

# CYPHERS AND SIGNS

During 1990, a few circular formations appeared in our home county of Cheshire and, generally, they appeared to be genuine ones. In June, I was asked by Colin Andrews to investigate a large ring formation in a field at Lower Peover, very close to the owner's farmhouse. This site was also only about 3 miles from the **Jodrell Bank Radio Telescope**. On site I met with the late Mrs. Jill Burton (later Hill) who had reported the ring and, as an amateur archaeologist, she was intending to carry out a survey of the site with some of her colleagues. When I arrived on site the ring was new and only Mr. Arthur Leech, the farmer, Mrs Burton and one other person had been in the field to inspect it. It didn't take long for me to recognise a genuine undamaged formation in green wheat. It measured some 22 metres (72 ft.) in diameter and was placed in a very sloping and uneven corner of the field. Whilst walking round the ring with the farmer, Marion (my wife) made a significant observation. There amidst the flattened wheat stalks she noticed two plants from the previous year's crop, laid down in the same way. They were two weedy **potato plants**. They were bent gently, without damage, at the base of the stems, and small leaves projecting from the stems had been completely unharmed by the process. These **gentle bends at the base of plants in genuine formations I began to call 'walking-stick handle' bends**.

On Monday, 2nd July, 1990, I received a call, at the office, from a reporter for BBC Look North TV, asking if I could spare a few minutes for an interview in the Lower Peover ring during my lunch break. This I managed to arrange and drove the 20 miles to the site. There I was met by Mrs Burton, reporter Steve Taylor and a video cameraman. I explained very carefully why the ring could not have been formed by an electrically-charged whirlwind and went on to share my thinking about an advanced technological explanation. Jill Burton also featured with her newly produced survey map of the formation. Then I left to drive back to the office. No fee had been asked for, nor was any paid to me for that excursion. That evening Marion and I watched the Look North programme and we were horrified by the 'send-up' we were being given. What was described as ***"the investigators' preferred, more imaginative, solution"*** was being set against a young 'armchair' physicist/astronomer, filmed impressively in front of the Jodrell Bank Telescope who, rather nonchalantly, said that scientists believed that such happenings were caused by electrified whirlwinds, which descended into the fields and produced effects such as those at Lower Peover. I leapt out of my chair to call the BBC immediately. Of course no one associated with the programme was available to talk to me.

I have related the details of this series of events to demonstrate how the British media have successfully hidden the facts about crop-circles and UFOs from the general public over the years. In fact, I am now inclined to believe they have received orders, from someone in a position of high authority, that they should not give any credibility to anything but mundane explanations of the strange things I, and others, have been seriously investigating for many years.

Throughout the 1990s, Marion and I spent at least one week of our annual holidays down in Wiltshire and Hampshire, keeping abreast of developments. We met interesting people and investigated some very challenging crop formations. Early in the 1990s the formations began to become more complicated, but some of them did not have the signs of genuineness and suggested that hoaxers had been at their mischievous work repeatedly.

## Celestial Meaden !

Soon after the Cheshire crop-ring event I read carefully through Dr. Meaden's recently published book **[5]**. I was intrigued to discover that the events he had considered were a mixture of crop-circles (7), vortices (2), sudden whirlwinds (2), glowing balls of light (5), tubes of light (2), hovering lights (1), a bell-shaped object (1) (later supplemented by a similar sighting reported by Colin Andrews), car stops (2) and a close encounter with an unlit aerial object in darkness. All these events, he claimed, could be examples of his 'plasma vortex' concept.

This was just too much for me to accept and I was inspired to check those events out against the UFO timing predictions, as the timings model, at that point of development, defined them to be.

Firstly, I created a graph of Mean Solar Time vs. Day of the Year and then superimposed the times of the Meaden events at the corresponding dates. Immediately I could see that the plotted points were running from top left to bottom right on this graph, in two, possibly three, distinct groups. So I linked them by straight lines and discovered that they could be regarded as having followed constant sidereal time lines (as defined previously). Fig. 20 shows the graph I produced, with different kinds of events being identified by different symbols. As can be seen**, a wide**

**variety of phenomena seemed to have occurred when the same stars were overhead** and it was difficult to see how naturally occurring plasma vortices could possibly have been linked celestially in that way.

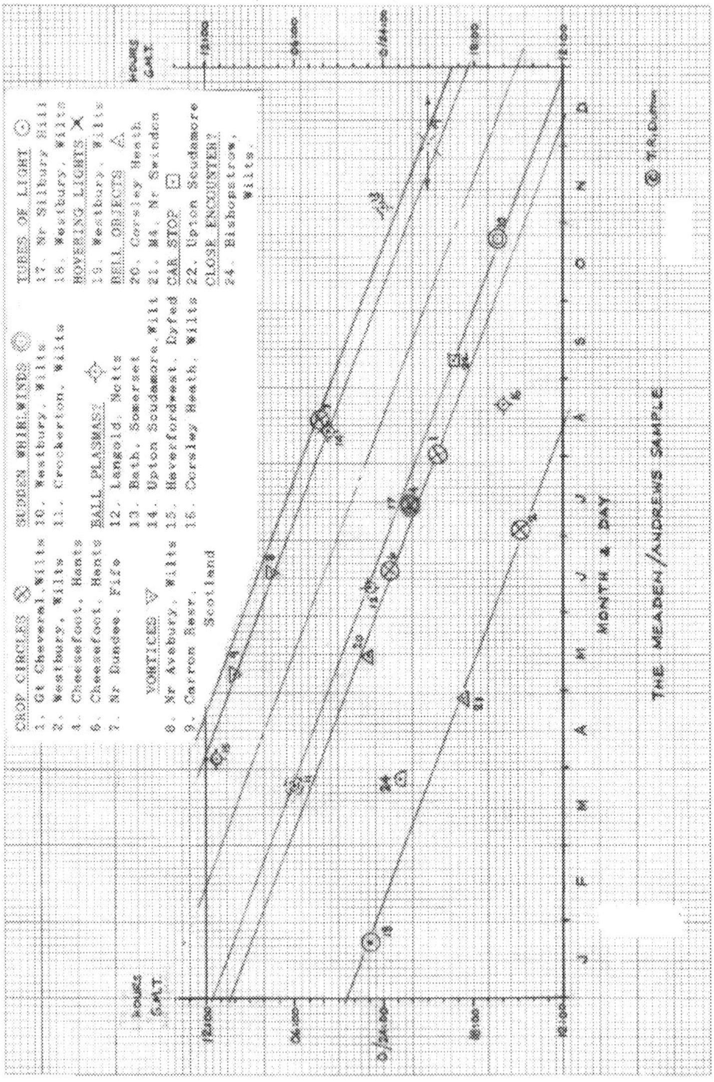


**Fig. 20**

My next step was to create a hand-drawn timing predictions graph for the Wiltshire and Hampshire area, by selecting those ground tracks from the global collection that could be considered to pass over that area.

I selected four definite tracks and two other possible ones passing over the east and west extremities, respectively. Next, the celestial orientations, as established provisionally (at the close of Chapter 7) were used, together with the sunset and sunrise times, to position the timing lines on the Mean Solar Time vs. Day of the Year graph. This graph was then superimposed over the Meaden graph. It was discovered that by adjusting the position of the predicted sloping lines vertically, they corresponded to the Meaden lines. What this meant was that with celestial orientations of **11:00 hr. RA** and **21:30 hr. RA**, the graphs were then identical. A few floating points on the Meaden graph were found to be consistent with predictions from the 11:00 hr. RA set and the sunset trends.

This was a major step forward in that **it *had demonstrated that the motley assortment of phenomena selected by Dr. Meaden had been shown to be indistinguishable from global SAC activity*** *and, also,* ***it had established two celestial orientations for the supposed approach and departure paths in space.*** After further confirmation, to be described later, these orientations were eventually used to produce a series of computer programs with the purpose of checking UFO and crop-circle reports from all over the world. Fig. 21 shows the two fixed celestial orientations superimposed on the Celestial Sphere. Only one representative path is shown for each orientation.

## Project 'Blackbird'.

I think it is worth recording at this point that a major illustrated book on my crop-circles research, which I had been encouraged to write by a leading publisher and which I'd spent much of my spare time during the autumn months of 1991 bringing to completion, was soon afterwards shelved by the publisher. This was because public interest in the topic had waned as a result of the widely trumpeted confessions of two elderly Southampton men, Douglas Bowers and David Chorley, who had claimed to have been responsible for all the formations in Southern England. The media had had a 'field day' misleading the public yet again. The claims were preposterous, as all the serious researchers knew, but no one was listening to them. One had to conclude "When ignorance is bliss, it is a folly to be wise."

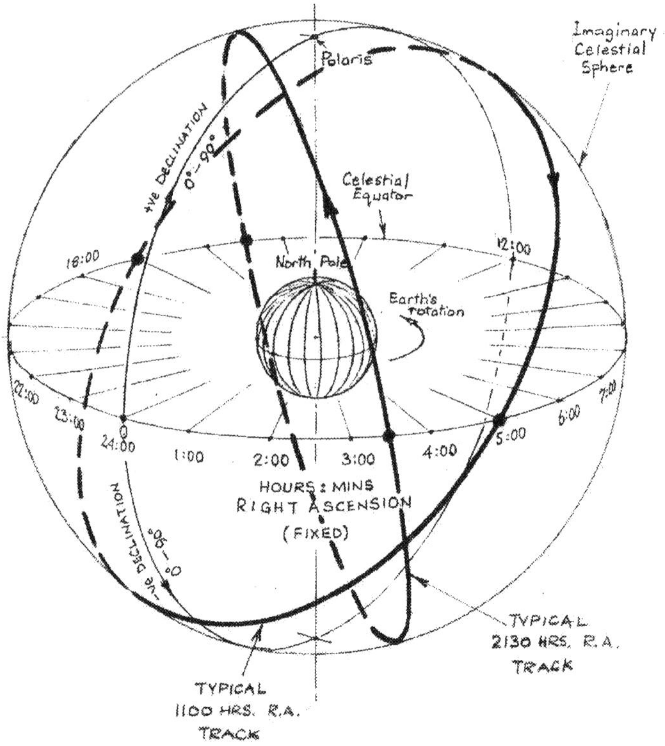


THE CELESTIAL SPHERE

AND

FAVOURED TRACK ORIENTATIONS

**Fig. 21**

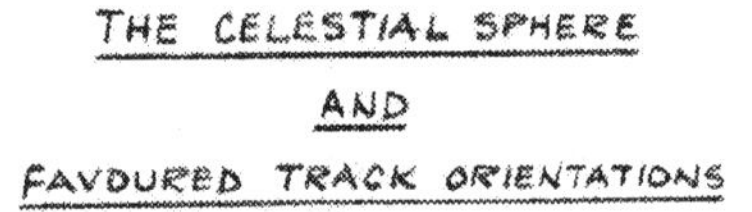

Despite such smokescreens, genuine formations continued to occur each year. Another example of material withheld from the public was that collected during another major crop watch held, during July/August 1990, from the ancient Bratton Hill Fort, Wiltshire. This **'Blackbird'** project had been jointly organised by Colin Andrews, Pat Delgado, BBC Television and Nippon Television. Marion and I had volunteered to participate during the second half of this venture, during August. Four Low Light Level TV cameras had been planted on the side of the steep embankment beyond the ditch of the hill fort. These were connected to monitor screens and recorders installed in a black caravan located on the upper edge of the ditch. Also, parked outside the caravan and similarly connected, was an IR (infrared) camera which needed to be kept covered over during daylight hours.

Prior to our arrival on site a major hoax had been perpetrated., which had been leaked to the media before the formation had been fully investigated and this had caused the project to be derided and ridiculed publicly. Fortunately, the project had been continued with the co-operation of the army, who provided two soldiers for night duty and a telescopic night sight on a tripod to reduce the chances of another hoax being undetected. We had booked in at the farmhouse situated on the other side of the road at the base of the hill fort's ditch and embankment and had been allocated a bedroom with a window facing towards the caravan. Our watch duties would be during the daylight hours.

The BBC had hired a helicopter to take aerial video footage of all new formations in the surrounding area, after Andrews and Delgado had first examined them and declared them to be genuine. Some of these formations were situated outside the Bratton area, but there were formations formed in that area, two of which occupied a field about three quarters of a mile away in the direction of a large cement works. The larger of these looked like a coiled rope and when I analysed the aerial footage later, I found my PC graphics model reproduced it exactly. But the most amazing events during our period there occurred in the early hours of the third day.

The TV set in our room was always in the standby mode, so a faint transformer hum was discernible in the quietness of the room. At about 2 am. BST I awoke to hear a transformer-like hum which was very loud and pulsating at a low frequency. I found it difficult to get off to sleep again but, I was so tired, eventually I did. I was awakened again at about 5 am. by loud twittering and chirping noises, which made me think that martins and swallows were having a big argument on the roof above. I crept past the still-sleeping Marion to the window and opened it. Immediately the noise ceased! I looked out into the greyness of the pre-dawn light. Everything seemed very peaceful and there was no sign of a flock of birds. With a shrug of the shoulders, I crept back to bed. At breakfast next morning we received a telephone call from the caravan site asking for our presence as soon as possible, because circles had been formed just beyond the area scanned by the hillside cameras and the validity of the formations needed to be checked out by the Andrews/Delgado team.

En route to the caravan we passed the field with the new formations, about half a mile from the farmhouse and close to the village of Bratton. We decided to stop and take a quick look at the formations without going into them. There were two circles, placed some distance apart and they looked as if they might be genuine ones. We then continued on our journey up the hillside to the caravan. I asked if anyone had seen anything suspicious that might have been associated with the circles. A young woman told me she had seen **two columns of mist** over that field at about 5 am.

When I talked with the two soldiers later, they told me they had returned along the ditch edge at about 4:30 am. and they'd not seen anything unusual in that field below them. So it seemed there was a circumstantial link developing between the twitterings and chirpings I'd heard at 5 am. (Grasshopper Warblers perhaps?). I consulted the hand-drawn timings graph I had produced for Wiltshire and Hampshire (as described earlier) and discovered that happenings around 2 am BST and 5:30 am BST might have been predicted for that day.

Later on, another resident at the farmhouse, a member of the public (who had been keeping a lonely watch on the hillside above the caravan) and I were interviewed by the BBC team and asked to recount our experiences of the early morning. The other resident, unable to sleep, had been watching TV at 2 am. when the picture had begun to contract and expand fairly rapidly. This had continued for several minutes. The man who had been sitting alone on the hillside had a fascinating tale to tell. At about 2 am. he had detected a low humming sound approaching from the West, apparently from an unseen source progressing at low altitude. As the source seemed to get closer, the sound level rose quickly. When the source seemed to be over the cement works it seemed to stop and the noise then became so intense that the observer felt that it was filling his head, so that it seemed to be about to burst. As panic began to overtake him, the unseen source began to move away towards the North and the noise gradually faded out. Immediately, he ran down the hill to the caravan shouting, "What was that noise?" and was astounded when he was met with blank stares and the question, "What noise?" No one in the caravan had heard it.

These interviews, and others, were recorded by the BBC, but, like the rest of the footage, were never shown in Britain. The next year when I met up again with Japanese archaeologist and journalist, Kazuo Ueno, who had interviewed me more fully on the hill-fort site, he told me he had seen the BBC interviews on Japanese television. It seemed that there had been yet another British cover-up. (As will be told in PHASE 7, these encounters in Wiltshire established a working friendship between Kazuo and myself).

## An impressive cipher.

We were told by someone on site that a huge formation, being called 'a **pictogram**', had been formed during July in a field at Alton Barnes, in the Vale of Pewsey. This was alleged to be the most complicated formation to that date. During one of our daytime periods off-duty we drove to the site and immediately saw that our informant had not been exaggerating. The farmer had set up a caravan at the entrance of the field and had been charging people £1 to enter the formation, so by then it had become much trampled. When I told the gatekeeper I was a serious researcher and involved in the 'Blackbird' project, but that I considered it was probably not going to be worth my while examining such a trampled mess, he invited me to just go in and look anyway. To show my gratitude I purchased a key ring with an attached aerial photograph of the pictogram when newly formed. As I moved into this long formation of connected circles with key-like appendages and ringed circles, I had to use that key ring photograph to help me to navigate through the maze I had then found myself in. Being in a ripe crop field on a very hot summer's day is rather like being in a desert. Some considerable time later, after a thirst-creating investigation, I got back to Marion in the welcome coolness of the car. She was delighted when I presented her with the souvenir key-ring. She treasures it to this day.

Later, we were told of several other, similar formations, which had also been formed during July. We visited some of these during our subsequent excursions. But the Alton Barnes pictogram was of particular interest to me because I felt it really was a kind of cipher of some kind and required further investigation.

In between 'White Crow' and 'Blackbird' I had noticed that several outstanding UFO Close Encounter events had occurred at times when two major planets had been aligned in conjunction in the sky. The Alton Barnes pictogram reminded me of that observation, so I decided to check it out.

The formation had been discovered during the morning of July 11$^{th}$, 1990, together with another one at nearby Stanton St. Bernard. My astronomical information told me that several major planets had been very close to conjunction during that period of July.

Referring to the pictogram, I pondered on the 'key' features of the pattern. Offset from the main alignment of circles and standing isolated in the crop were two small circles, one larger than the other, the larger one being surrounded by a thin ring. The relative sizes of these items suggested to me that perhaps they might represent the Earth and the Moon. If then I regarded the adjacent large ringed circle as representing the Sun, then, relative to the Earth, the Moon was located in what would have been its Full Moon position. My astronomical data told me that Full Moon had occurred on July 8$^{th}$, three days before the pictogram had been created. Could it be purely coincidence that the circle provisionally designated to be the Sun had a 'key' feature with three prongs pointing backwards towards the depicted Full Moon? Could this be confirming the astronomical significance of the pictogram and suggesting the means by which the rest of the formation might be interpreted?

Another large circle, connected to the assumed 'Sun' symbol by a long bar, had a 'key' feature attached to it with only two prongs. Was there a planet aligned in conjunction with the Sun two days before or after the July 11$^{th}$? The astronomical data did not seem to support this idea. Instead it showed that Jupiter had aligned in that way on July 15$^{th}$, that is, four days after the 11$^{th}$. However, that planet had been very close to conjunction with the Sun for several days prior to the accurate alignment and would have disappeared behind the Sun on the 13$^{th}$ and reappeared on the 16$^{th}$. So it is possible to regard Jupiter to have been occulted by the Sun only two days after July 11$^{th}$.

Concentrating next on the remaining part of the pictogram, it seemed to be suggesting that two bodies, one possibly being the Sun, had been aligned three days before or after July 11$^{th}$. The astronomical data showed that the Sun and Saturn had been aligned **in opposition** on July 14$^{th}$. This meant that they would have been in the sky on opposite sides of the Earth and would have been linked by a line drawn through the Earth. In the pictogram, the Earth's position between the two major bodies could have been symbolised by those two flattened strips located on either side of the connecting bar. In effect, they might be regarded as representing an opposition situation.

The two smaller circles beyond the circle with the three-pronged fork were then the only unexplained representations. They were in alignment with the major elements of the formation but , astronomically, they were difficult to identify. In fact, I had to settle for the observation that Uranus and Neptune were in the same area of the sky as Saturn, but not in conjunction alignment.

A lot of conjecture had been involved in the interpretation of the pictogram at Alton Barnes, but I felt justified in regarding this as a cipher which, perhaps, we had been challenged to decipher. Other such challenges were to arise during the following year to further substantiate that idea.

## More astronomical riddles in the fields.

During the summer of 1991, several more pictograms were produced in the fields of southern England and I felt challenged to attempt astronomical interpretation of them. Figs.22(a) and (b) represent them diagrammatically.

| Date | Pictogram | Astronomical Event Depicted |
|---|---|---|
| 12.7.90 | 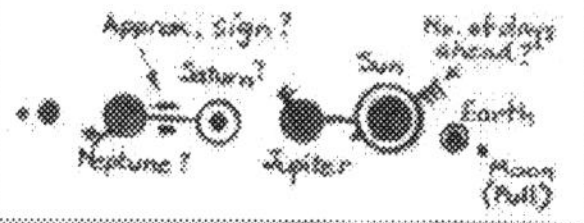 | Solar conjuction of Jupiter 3 days later: also, Saturn in opposition about that same time. |
| 20.6.91 |  | Venus/Jupiter conjuction |
| 24.6.91 | 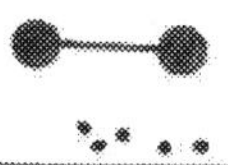 | Venus/Mars conjuction in recognisable Cancer star group. |
| 29.6.91 |  | Venus/Mars conjunction |
| 5. 7.91 |  | Uranus in solar opposition |
| 11.7.91 |  | New Moon & solar eclipse |

**Fig 22a**

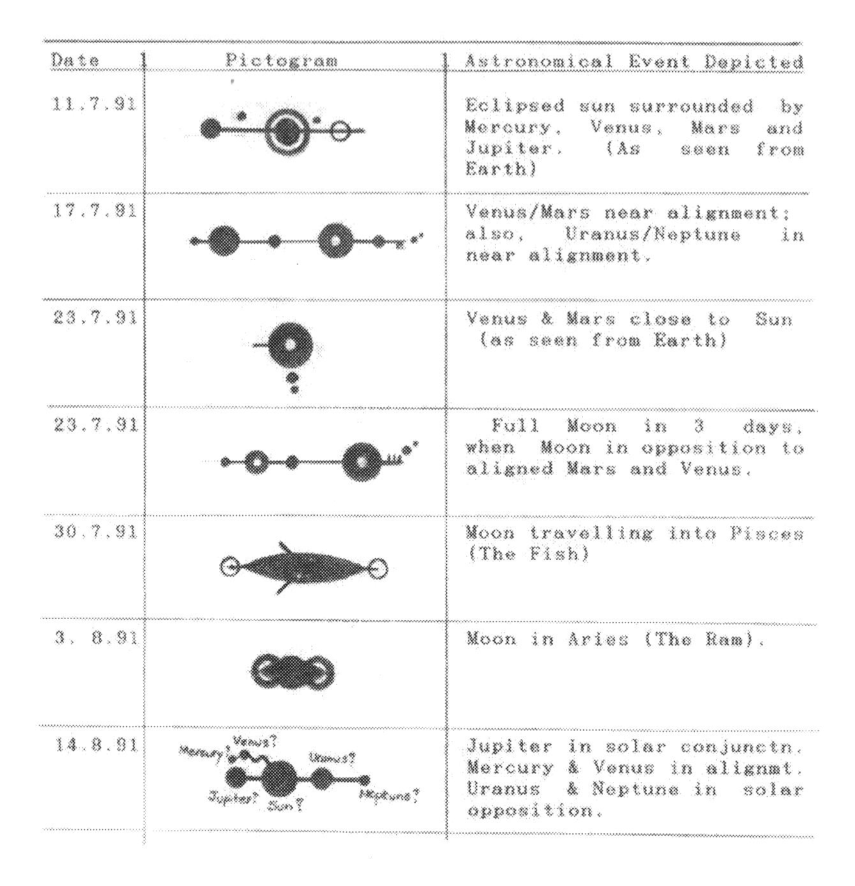

| Date | Pictogram | Astronomical Event Depicted |
|---|---|---|
| 11.7.91 | 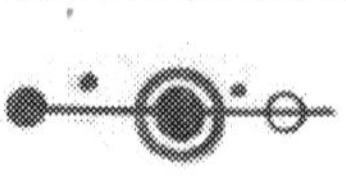 | Eclipsed sun surrounded by Mercury, Venus, Mars and Jupiter. (As seen from Earth) |
| 17.7.91 |  | Venus/Mars near alignment; also, Uranus/Neptune in near alignment. |
| 23.7.91 |  | Venus & Mars close to Sun (as seen from Earth) |
| 23.7.91 |  | Full Moon in 3 days, when Moon in opposition to aligned Mars and Venus. |
| 30.7.91 |  | Moon travelling into Pisces (The Fish) |
| 3. 8.91 |  | Moon in Aries (The Ram). |
| 14.8.91 | 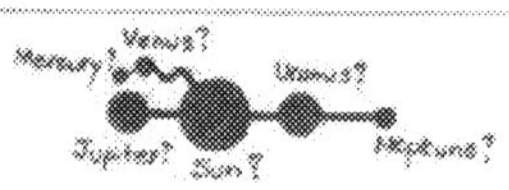 | Jupiter in solar conjunctn. Mercury & Venus in alignmt. Uranus & Neptune in solar opposition. |

**Fig 22b**

Taking them in date order, the first one considered was discovered on June 20th and consisted of a large circle linked by a bar to a smaller one. Astronomically, this could have represented the conjunction of Jupiter and Venus occurring at that time. A few days later, on June 24th, a similar pictogram had been found, but this had connected circles of a similar size together with a scattering of five small (grapeshot) circles in a cluster. This arrangement could again have depicted the Jupiter-Venus conjunction and, furthermore, the group of small circles resembled some of the stars of Cancer, the constellation providing the backdrop to the conjunction. On June 29th a pictogram possibly depicting the contemporary Venus and Mars conjunction was found. As can be seen from the diagram, this had additional features which I found to be difficult to explain.

This series of possible astronomical formations continued into July, when more complex patterns were produced, after the relatively simple one on July 5th had seemed to depict the opposition in the sky of the Sun and Uranus.

On July 11th, we were presented with something quite different. As the diagram shows, this pictogram consisted of a large ring with three straight line attachments, in the middle of which two filled-in circles, of equal size, were partially merged. Now it so happened that there was total eclipse of the Sun expected over Mexico later that day. So, it seemed to me to be a logical first step to regard the merging of the two circles as a possible representation of that forthcoming major eclipse. As in all solar eclipses, the Moon was New at that time. Astronomically, a New Moon is depicted as a simple ring and such a ring was obviously being featured, as if to confirm the nature of the event being represented. Two of the three straight lines projecting from the ring I considered might have been an attempt to represent the individual paths of the Sun and the Moon during the eclipse.

On later reflection it seemed, with hindsight, that perhaps our attention had been brought, in advance, to the importance of that ensuing event. As became very much publicised after it, Mexicans who had set up their camcorders to capture the eclipse had found themselves recording unexpected SAC (craft-likeUFOs) flitting about in the partially darkened sky. Of course, all this quickly became headline news in some parts of the world, but I don't remember such headlines being immediately seen in Britain.

The pictogram just considered was found to have a companion placed in a field some distance away. Perhaps this could have been intended to refer to the same astronomical event, depicted in a different way (see diagram), but I found it difficult to interpret.

The pictogram discovered on the 17th had a three-pronged 'key' feature and an adjacent skewed arrangement of two small circles looking as though they might represent the Earth and the Full Moon. Assuming that the rest of the pictogram referred to a date three days before or after the Full Moon, I viewed the astronomical data for the 14th and 20th July. The most notable alignments were those of conjunctions between the Moon and Mars and the Moon and Venus on the 14th. However, the pictogram did not seem to depict that situation. I encountered similar problems with the pictograms of the 23rd.

With my examination of the pictogram of July 30th, my ability to recognise meanings changed dramatically for the better. As will be seen from the diagram, this large formation on a hillside at Lockeridge, Wiltshire, resembled a stylised fish. As I was able to examine this at close quarters for myself, I can vouch for its apparent authenticity. It was beautifully laid down and featured a large swirled circle at its centre. As I considered whether the layout might have astronomical significance, it occurred to me that it might represent the constellation Pisces (the Fish). If so, could those open rings at each end represent the Moon? The astronomical data told me that the Moon was about to enter Pisces on July 31st and to leave it on August 2nd. I was pleased with that solution, especially so in view of my next success.

On August 3rd 1991, a much more compact formation was found. To me it resembled a stylised ram. Aries (the Ram) is an adjacent constellation to Pisces in the zodiac. Assuming that the circular 'horns' of the 'ram' in this picture might represent the Moon, I was overjoyed to find that the Moon had moved into Aries on August 3rd and left that constellation on August 4th.

My final attempt to interpret the pictograms of 1991 concentrated on a complicated formation discovered on August 14th. Four circles of three different sizes were aligned and linked by a bar of flattened crop. A strange zigzag of flattened crop also sprouted from the largest circle and linked it to two small circles, one larger than the other. The search for astronomical meaning seemed, at first, to be progressing well. On the 14th. Venus and Mercury were in conjunction and it was thought that these might be represented by the little circles on the end of that zigzag. Also, the Sun and Jupiter were close to conjunction and became fully aligned on the 17th. The two linked circles on the right of the diagram, however, did not seem to have any explanation, because they seemed to be placed in opposition to Jupiter ---- but there were no planets in that region of the sky, only a comet called Encke.

I don't wish to unduly extend this section. The results of my research into the crop circles phenomenon have been published over the years, mostly in **'The Circular'**, the magazine produced by the **British CCCS (Centre for Crop Circle Studies)**. I was in membership of that organisation throughout the 1990s, but my scientific approach to the phenomenon, and the proofs I produced to demonstrate that the genuine items were almost certainly being created by extraterrestrial agencies, were not shared by many members by the end of that decade. Instead, all manner of paranormal and mystical explanations were being preferred. To add further confusion, hoaxing had become developed into a fine art by teams of (I suspect, professional) jokers, probably financed by elements of the Press and officialdom, who delighted in creating ever more complicated patterns to beguile the gullible. I decided enough was enough! Even so, several friends and ex-colleagues have tried to keep me updated each year since then, and I very much appreciate their efforts.

There is one more crop circles item I want to add to this Phase 3 section. If the events described were authentic, my conclusions about how the genuine formations are produced are fully validated by them. But there is one more matter to consider before I do that ---- confirmation of the celestial path orientations suggested by the Meaden data.

## Confirmation of the Timings Model

Immediately after I had adjusted **the global entry and departure path orientations** using the Meaden data, I looked for a different selection of reports, from the same area, to check whether the changes made could also be regarded as applying to the new selection. A book written by Arthur Shuttlewood, of Warminster renown, provided me with a suitable collection of reports. From the book **'UFO Magic in Motion' [6]** published in 1979, I was able to extract 22 remarkable events from the mid-1970s and then to plot the dates and times of these onto Fig.23, the revised Wiltshire/Hampshire timings graph.

As can be seen, a very pleasing result was obtained. The lines established by the amended global model seemed to be fully supported by the manner in which the times of those additional reports conformed. The seven points standing noticeably away from the lines were displaced from them by generally less than half-an-hour. After this, I felt bold enough to incorporate the fixed celestial orientations,**11:00 hr. RA** and **21:30 hr. RA**, into the programs for computer processing of events reported from all over the world. The results obtained, when these programs were widely applied to new reports coming in to me from many sources, will be demonstrated in the Phase 4 section to follow.

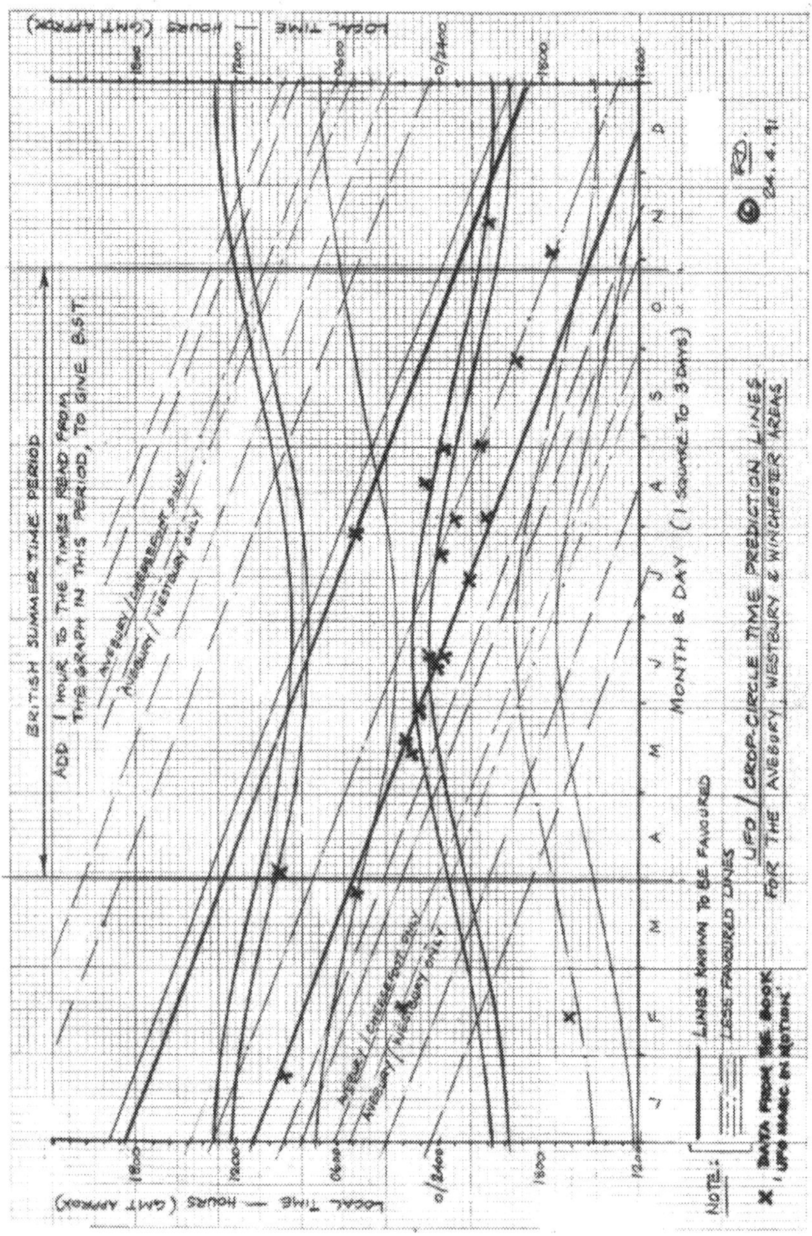
BRITISH SUMMER TIME PERIOD
ADD 1 HOUR TO THE TIMES READ FROM THIS GRAPH IN THIS PERIOD, TO GIVE BST.
MONTH & DAY (1 SQUARE TO 3 DAYS)
UFO / CROP-CIRCLE TIME PREDICTION LINES FOR THE AVEBURY, WESTBURY & WINCHESTER AREAS
LOCAL TIME — HOURS (GMT APPROX)
NOTE:
LINES KNOWN TO BE FAVOURED
LESS FAVOURED LINES

**Fig 23**

## Proof of the ET Technology Concept?

My Japanese friend and associate, Kazuo Ueno, whom I'd first met on the 'Blackbird' watch of 1990, contributed a story to 'The Circular' magazine and this was published in the autumn edition of 1994. The story seemed to be so authentic that I decided to check out the timings of the events using my global timings programs. After finding positively, I wrote an article for 'The Circular', published in the Spring edition of 1996, to share the results. The story underlying all this was quite fascinating and tells of two events witnessed in broad daylight.

On April 15$^{th}$, 1991, a Japanese boy, nine years old, who lived in the small village of Aikawa, Kanagawa Prefecture, Japan, had wandered alone away from the village. A glowing orange object had suddenly swept down from the sky and stopped only a short distance from the young witness, at an altitude estimated to be about 100 metres. The boy had seemed to be immobilised as a pillar of "transparent white steam or smoke" generated downwards beneath the hovering object. This pillar had then rotated and grown wider towards the base. When it had come into contact with the grass in the field a flattened ring had been produced in the grass, about 30 centimetres wide. During this process, the boy felt a warm wind and drops of water on his face and heard a low alternating sound like "gu-on, gu-on". Immediately after the flattening, the "**steam trumpet**" had been retracted and the object had then shot off into the sky. The startled boy had then called to a friend some distance away and asked him to come to look. While they were looking at the ring in the grass, the glowing object had suddenly returned to their location and had proceeded to repeat the performance for the benefit of two immobilised boys. They watched as two concentric rings were produced in the grass, which the 'trumpet' produced by being deployed twice. The "trumpet" (rapidly-rotating focused beam?) responsible for the inner ring had been thinner and more dense than the other. Having completed its second mission, the UFO had then disappeared back into the sky.

What a wonderful example of the kind of things I had been envisaging and which have been described in detail earlier in this Phase 3 section. It seems that the atmosphere had been very moist at the time of the events and it is probable that the conjectured rotating beam of energy had created condensation in its wake and thus produced the 'steam trumpet' effect.

As explained in my article in 'The Circular', I had investigated the timing of the two events (16:30 hr. Japanese Standard Zone Time) and had found that one of the seven computer-selected paths passing over that area had a timing associated with it, which was just over 1.5 minutes within the stated time. In the following Chapters of this book I will be providing much more evidence of similar kinds of support for the Astronautical Theory.

## References [ ]

[6] Shuttlewood, A.'UFO Magic in Motion' (paperback)
Sphere Books, Ltd. 1979.

# PART 2

## Testing the Astronautical Theory (A.T.)

# PHASE 4:
# Significant Global UFO Events

*Having established the foundation of* ***the Astronautical Theory for SAC/UFO and Crop Circle events*** *from the happenings in Southern England, it was important, next, to explore its applicability to reports gathered from all over the world. A large and varied global database was compiled, aided by my many contacts and by reference to a number of books. The following chapters demonstrate the procedures and the typical results obtained. Very well known events are considered, which are shown to be explained by the AT's scenario and predictions.*

## CHAPTER 11

## INTERNATIONAL VALIDATION OF THE A.T.

After computerisation of the Theory during the 1990s, the basic steps required to carry out investigations of reports were those of establishing the geographical co-ordinates of each location and the Standard Time Zone applying to it. Unfortunately, this was, at times, a rather difficult exercise to carry out, because some of the places referred to were too remote to be listed, even by a good world atlas. If there remained any doubts, that report just had to be excluded. Nevertheless, a large database was created (piecemeal) from ***reports never previously considered*** and the prolonged task of checking then began.

## Events in the San Luis Valley.

Through the assistance of the late Ms. Shari (Sharon) Adamiak, then the CSETI co-ordinator who lived in Colorado, I was put into contact with **Mr.** Christopher O'Brien, who was a pop-musician and a dedicated investigator of the strange happenings reported from that large valley. The San Luis Valley is a huge alpine valley extending from southern Colorado into northern New Mexico. It is unfortunately a practice area used by military aircraft which, on the face of things, might be considered to be the sources of some of the strange things reported. However, to be set against that suspicion is the familiarity of the local people with military aircraft and the exercises carried out. Chris O'Brien began by sending me reports in small batches and, later, a copy of his 1996 book, **'The Mysterious Valley'** [7] was gifted to me by Jennifer Jarvis, whose story will be told in Phase 6. O'Brien's co-operation will always be regarded with gratitude. From all the O'Brien sources I was able to extract over forty good reports for the period June 1994 to January 1995. These are shown plotted by Fig.24, on the timings graph computed for a concentrated activity area. As can be seen from the symbols used, the events also included unexplained sonic booms and UFO-linked cattle mutilations. The degree of correlation on this graph is high and it would seem that certain tracks over the area had been favoured by the visitors.

## The Australian Connections.

As the original database from which the AT had been derived had consisted of events predominantly reported from the northern hemisphere, it became very important to check out the Theory's applicability to events in the southern hemisphere. A 1975 book by Michael Hervey, with the appropriate title, **'UFOs over the Southern Hemisphere' [8]**, turned out to be a rich source of reports. I was particularly pleased to be able to form a small database of events reported from the Melbourne area of South Australia. These covered the 1950s and 60s period**.** Fig.25 is the computer-produced timings graph with extra lines drawn in to represent those which the computer had been unable to print out, owing to imposed program restrictions. As can be seen from the plot, the points correlate well with the lines and sometimes follow the programmed trends. In this way, the applicability of the AT is adequately demonstrated.

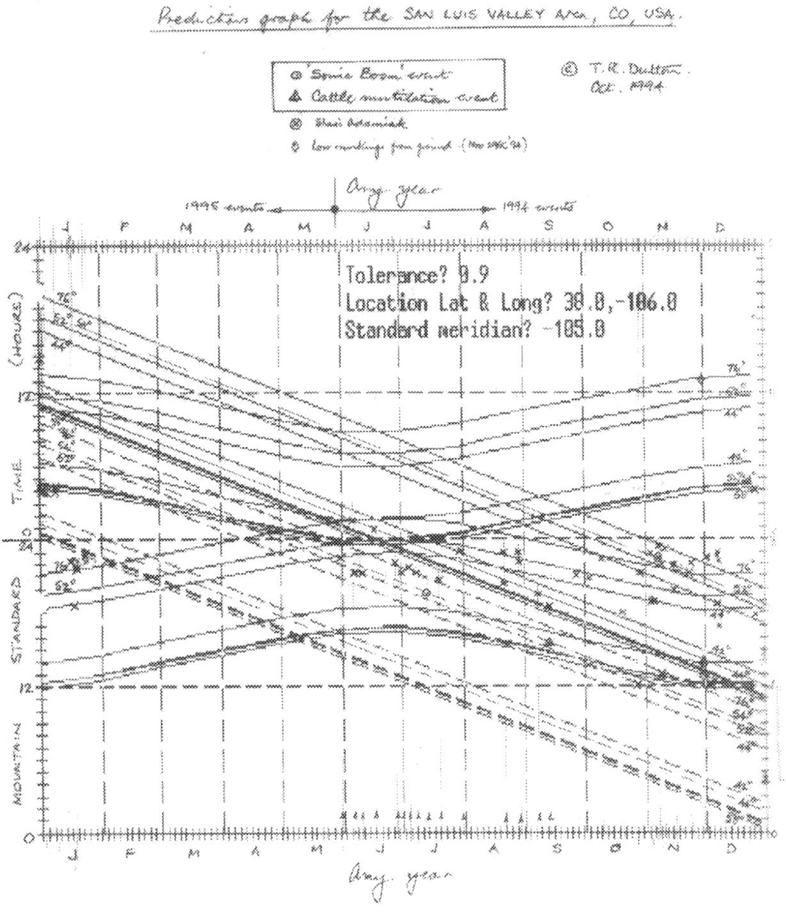


**Fig. 24**

A point on the graph which could not have been included in Michael Hervey's 1975 collection is the ringed cross plotted at October 21st. This event occurred in 1978 and the cross is marking the time of the last radio contact with Frederick Valentich, a twenty-year old flying instructor, whose Cessna 182 aircraft went missing immediately afterwards. At the time of his disappearance, 19:13 hrs. Standard Zone Time, the aircraft was over the Bass Strait with the pilot's declared intention of heading for King Island. According to the detailed account given by Timothy Good in **'Above Top Secret' [9]**, Valentich had radioed Melbourne Control at 19:06 hrs to query the presence of a large aircraft flying below him. The response to that was a negative one. The unknown aircraft was displaying four bright lights and within a minute had climbed above the Cessna. Still, Melbourne Control had no explanation. Within another minute or so the SAC headed back towards Valentich from the East and then it proceeded to fly over him two or three times, at speed. After having given his flight level as being 4500 ft. at 19:09 hrs., Valentich declared that the object was not an aircraft. Twenty four seconds later he described it as a "long shape" as it flew past. At just after 19:10 hrs, Melbourne received the information that the object was then stationary and that Valentich was circling beneath it. The object was now displaying a green light and it seemed to have a shiny metallic surface. Less than half a minute later, "It's just vanished!" Melbourne Control were obviously dumbfounded by that and at 19:11 hrs. they wanted to know if the 'aircraft' was still with the Cessna. A few seconds later, the response was that it was then approaching from the south-west. Shortly before 19:12 hrs the Cessna'a engine began rough-idling. Valentich then declared his intention to go to King Island and added that the strange 'aircraft' was hovering above him, but it was not an aircraft. Soon afterwards he opened his microphone to speak but said nothing. The microphone remained open for 17 seconds.

From the timings graph's evidence, it seems reasonable to speculate that the SAC had an urgent rendezvous to keep in space and could have taken Valentich and his aircraft with it, as no trace of either has been found.

## References

[7] O'Brian, C.'The Mysterious Valley' (book) St. Martin's Press, N.Y. September 1996

[8] Hervey, M.'UFOs over the Southern Hemisphere' (book) Robert Hale & Co., London. 1975

[9] Good, Timothy 'Above Top Secret' (book) Sidgwick & Jackson, London, 1987 Grafton Books, London, 1989.

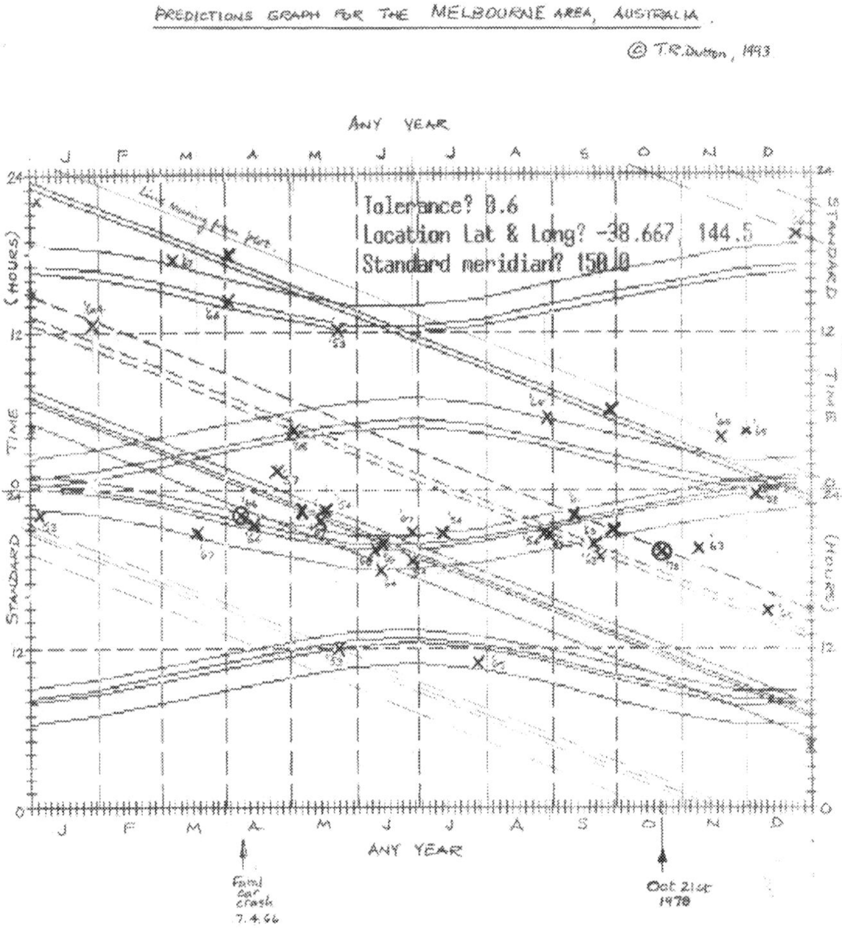
PREDICTIONS GRAPH FOR THE MELBOURNE AREA, AUSTRALIA.
© T.R.Dutton, 1993
ANY YEAR
Tolerance? 0.6
Location Lat & Long? -38.667, 144.5
Standard meridian? 150.0
STANDARD TIME (HOURS)
ANY YEAR
Fatal car crash 7.4.66
Oct 21st 1978

**Fig. 25**

# CHAPTER 12

## MILITARY CLOSE ENCOUNTERS.

There have been numerous reports of Close Encounters with SAC experienced, over the years, by military personnel and civilian police. I propose to consider only a few examples in this chapter to demonstrate why I believe the accounts to be true and validated by the Astronautical Theory.

It has been quite clear that, from time to time, orders were issued to Service personnel and police, by members of the security organisations, to place such encounters under confidentiality wraps. During the Cold War period I suppose it was their concern that the public should not be unduly panicked by such stories, when the threats of infiltration and conquest by a foreign power were very real. However, in Britain, those rules seemed to have been relaxed from time to time. The degree of secrecy applied to events seems to have depended on the sensitivity of the military site and on whether to admit to having failed to apprehend the SAC would reveal certain weaknesses to the potential enemy. Despite all that, news has occasionally leaked out that military bases had been visited. During the late 1940s and the 1950s, a number of events were made public knowledge and I will deal firstly with my investigation of one of those. After that, I will consider a few high profile happenings from the following years.

Apart from the over-sensationalised Roswell events, for which it is now difficult to separate the facts from all the moneymaking fictions, the case with the highest profile during the late 1940s was probably the one resulting in the tragic death of the American reservist flyer, Capt. Thomas Mantell, Jnr. The events have been reported by several writers but my definitive view of the them has been derived from the late Major Donald E. Keyhoe's book of 1953, **'Flying Saucers from Outer Space' [10]**. Major Keyhoe was an ex-Marine who had friends in the Pentagon and other high places within the U.S. military establishment**. Timothy Good's 'Above Top Secret' [9]** supplemented Keyhoe's account by giving details of the official summary of the incident.

## The Mantell Encounter --- 1948

On the afternoon of January 7[th], 1948, hundreds of people in Madisonville, Kentucky, spotted a huge round glowing object overhead. As they watched, the SAC moved over to **Godman Air Force Base**, close to Fort Knox, when the State Police, in their message to Fort Knox, estimated the object's size to be at least 250 feet in diameter. Capt. Mantell and three other **F51 Mustang** pilots were on a training mission and flying through that area when Godman AFB called on them to investigate the intruder. After climbing through broken cloud, Mantell reported he could see the object, that it looked metallic and that it was huge. He reported that the SAC was beginning to climb. It was above him and making his speed or faster. (This invalidates the suggestion that the object might have been a high altitude research balloon). He said he was going to climb to 20,000 ft. and would abandon the chase if he seemed then to be no closer.

Meanwhile, one of his companions decided to continue on, to complete his original mission. Two others accompanied Mantell up to 22,000 ft. and then had to abandon the chase because they were low on oxygen. Mantell, however, continued to climb.

Shortly afterwards, his aircraft crashed and was seen by an eyewitness to disintegrate in the air. It seemed likely that Mantell had suffered oxygen starvation, blacked out and lost control of the aircraft, which may then have entered a powered nose dive. The location of the crashed aircraft was given as being 2 miles south-west of Franklin, Kentucky, and the approximate time of the crash was given as 16:45 hr. Central Standard Time.

I checked out the details of this terrible accident using my programs. A computer-generated timings graph for that area of Kentucky was produced and then the date and the time of Mantell's crash was plotted onto it. (The crash time was assumed to be also the approximate time of the SAC's departure). As can be seen from **Fig.26**, that time fell upon a predicted line, 21:30 hr RA orientated, which represented the last opportunity (for about 18 hours) for that supposed automated SAC to rendezvous with its retrieval ship in Space. A glance at the little Latitude/Longitude grid map of the area, always provided above the (later) computed timings graphs, shows that the associated track over the ground (No.4) of the ship in Space would have placed it in a good position, directly overhead, to retrieve the exploration craft. In the light of all this, it seems particularly unfortunate that Mantell chased the SAC just as it was forced to depart skywards in a hurry.

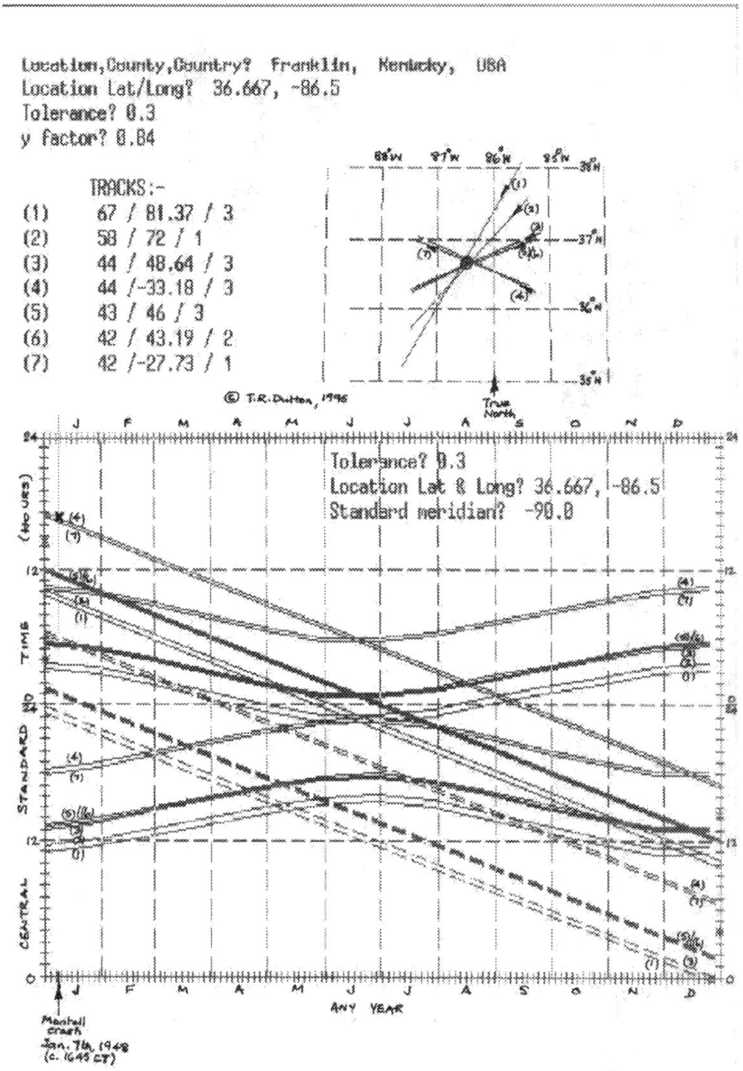


**Fig. 26**

## The Bethune Encounter --- 1951

During the dark, clear, night of February 9th/10th, 1951, a U.S Navy **C-54 (DC4) transport aircraft** droned along on its lonely way from Keflavik, Iceland to Argentia, Newfoundland, following the Great Circle route linking those two places. It was carrying cargo and its passengers were military personnel. At the controls, in captain's seat on the flight deck, was the young Lt. Com. Graham Bethune. The flight was proceeding at 10,000 feet, in an uneventful way up to the point when the navigator confirmed their course and estimated the distance to Argentia to be only 200 miles. What happened shortly afterwards would almost certainly have changed the lives of everyone onboard that aircraft, not least, Lt. Com. Bethune's. (A full account of the ensuing events is given by Michael Hesemann's book **'UFOs – The Secret Evidence' [11]).**

Ahead and off to the right, in the 12:30 position (the 12 o'clock position would have been directly ahead), an underwater glow was spotted from the flight deck. It seemed to be of considerable size and an estimated 30 miles away. Five minutes later, the glow suddenly disappeared and was replaced by large ring of white lights, located in the same position, but about 20 miles away. Shortly afterwards the relief crew members were called to the cockpit. The lights were then at the 1 o'clock position. Suddenly, the underwater lights were extinguished and a smaller ring of light, more like a yellow halo, appeared above the surface of the water about 15 miles ahead. This ring, becoming oval and growing larger as it did so, suddenly zoomed upwards towards the C-54 with unbelievable acceleration. Startled by this and fearing a collision was imminent, the pilot disengaged the autopilot and initiated an evasive nose down manoeuvre. However, that turned out not to be necessary. The ring resolved itself into a huge disc-shaped craft with a large upper-surface dome and it stopped suddenly ahead of and below the aircraft. Bethune estimated it had a diameter of about 200 feet. It moved to the starboard side and seemed to be examining the C54 before it accelerated away into the night. (It became known after the event that the departing disc had been tracked on radar and its measured speed had been 1,800 mph.)

Within the aircraft, the sudden evasive manoeuvre had caused the standby crew members behind the flight deck to be thrown about and two of them had been injured. Bethune then left the co-pilot at the controls and entered the passenger cabin to check on the well-being of his passengers. The first man he encountered was an important Navy psychiatrist. He asked, "Did you see that?" The man responded, *"Yes, it was a flying saucer --- but I didn't see it because I don't believe in them"*. Bethune, guessing there was an official cover-up policy prevailing, rushed back into the cockpit saying, "Whatever you do, don't tell anyone about this!" The co-pilot replied, "Too late. I've already contacted Gander Control".

Some years ago, the Michael Hesemann video, "UFOs --- The Secret Evidence" had introduced me to this happening. In an interview with

Hesemann, Commander (rtd.) Bethune related the exciting elements of the encounter. He appeared to be a practical and unexcitable man, even when he described the excitement and shock of the experience. He was an excellent witness.

Then, during 2006, whilst watching through an on-line **'Disclosure Project'** video, I heard Com. Bethune tell his story again in the company of other retired US service men, most of them high ranking, all of whom had experienced UFO encounters during their years of service. One of those on the platform was retired (USAF) Major George Filer. This man had been a member the crew of a US patrol aircraft based in the U.K. during the 1960s. One night, whilst returning from a patrol over the North Sea, his aircraft had been instructed to seek out and investigate a UFO located somewhere west of London. They had succeeded in this mission, but the UFO had then left the scene with amazing acceleration that left the onlookers open-mouthed. In retirement, Major Filer eventually went on to create a MUFON-related web site called **'Filer's Files'**. These files, updated every week, reported the latest UFO events reported world-wide and could be accessed by subscription.

My second viewing of Com. Bethune inspired me to investigate his event in detail. As this book has already revealed, by then, I had created a means of checking UFO events involving Strange Aerial Craft (SAC), using the computerised astronautical theory (AT). This tool had proved its worth during the processing of well over 1000 new global events collected subsequent to its creation during the late 1980s.

I needed accurate positioning and event times to apply this work to the Bethune encounter. Major Filer seemed to be an obvious source and link with the Commander, so I contacted him by e-mail. He seemed very happy to help me and even gave me details of his own encounter. He supplied me with Com. Bethune's telephone number, with the rider that they hadn't been in touch for more than a year.

When I rang the New Jersey number, immediately I found myself talking with Graham Bethune. After my introduction, he answered my first enquiry about his wellbeing by telling me he was confined to bed. When I expressed a hope that it was nothing too serious, he brushed my concern aside by saying that the medics these days make a 'big deal' of a broken toe-nail --- and then kindly proceeded to answer my questions. When I told him I suspected he had encountered an **underwater base**, he surprised me by saying, "Yes, it was". He had no doubt about the presence on this Earth and in our skies of extraterrestrials. He referred me to the extensive account he had given to Michael Hesemann, which had been reproduced in Hesemann's book. **[11]**

With that, I decided not to prolong the conversation with a sick man, thanked him and wished him a speedy recovery. (Regretfully, a few months later, I saw it announced that Commander Graham Bethune had died. I felt very privileged to have been able to talk with him before he departed from this world.)

Using the new information, I set about analysing the elements of the C-54's encounter. First, I established the path the aircraft had been following and the location over the ocean from which the glow had first been seen. The graphics programs were then run and the graph and map produced were printed out. On the latitude/longitude grid map, the assumed (input) location of the C-54 had been marked by the program, in the usual way. Next, I superimposed the path of the aircraft. Its two additional positions along the flight path were estimated and added to this. Using the viewing angles of the lights and the time lapses between them, I then projected lines from the each location of the aircraft and placed marks on those lines at the estimated distances of the underwater lights. From this exercise I determined the probable position of those lights. This marked up diagram (Fig. 27(a)) is shown below:

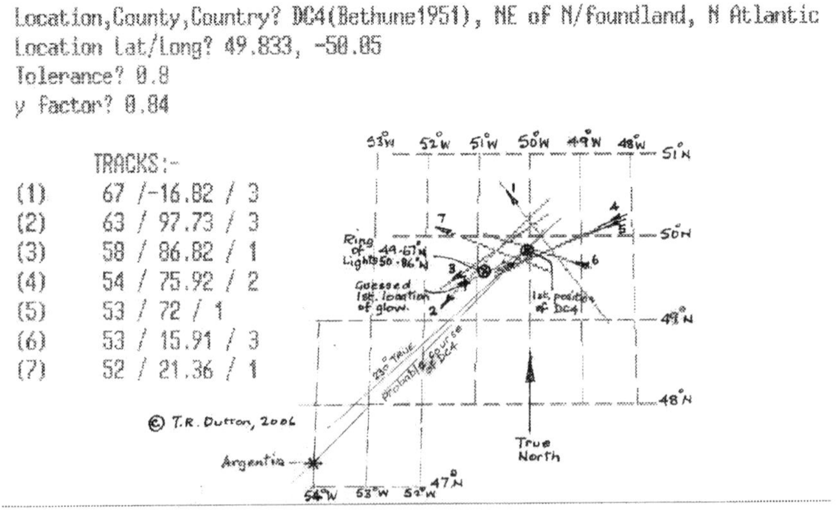


**Fig. 27(a)**

In this way having tentatively determined the position of the **submerged lights** as being 49°40' N / 50°52' W, I ran the programs again for that location. The map output was as given below (Fig. 27(b)):

```
Location,County,Country? UFO base?, NE of Newfoundland, N Atlantic
Location Lat/Long? 49.67, -50.86
Tolerance? 0.6
y factor? 0.84

         TRACKS:-
(1)      76 /-32 / 3
(2)      63 / 97.73 / 3
(3)      63 /-11.36 / 1
(4)      58 / 86.82 / 1
(5)      58 /-0.45 / 3
(6)      53 / 72 / 1
(7)      52 / 19 / 1
```

©T.R. Dutton, 2006

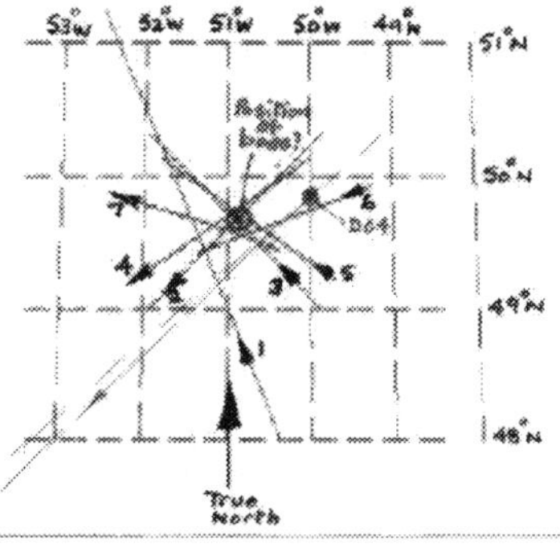


**Fig. 27(b)**

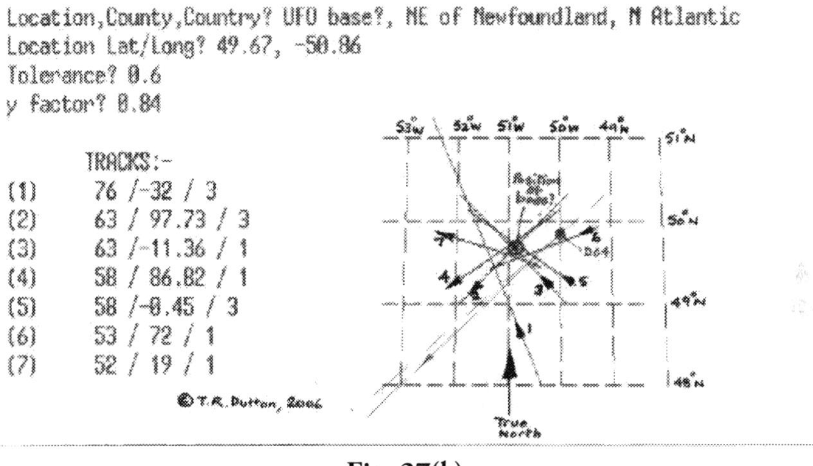

The output revealed the estimated location to be an intersection area for the computed tracks associated with it. This was something of a triumph, because it meant that the lights' location could be accessed by ET exploration craft released from any of those paths in space. In other words, it was an ideal location for a base. The large craft witnessed by all those onboard the C-54 could have been departing or had just previously arrived from space. The underwater illuminations had been essential to one of those routine operations, which had been unexpectedly interrupted by the approach of the C-54.

***There are many locations all over the world where such track intersections occur. Some of them are known to have been associated with similar unexplained underwater activities.***

Com. Bethune had remembered the time of the encounter as being probably just after midnight, local Zone Time. To produce the timings graph, I used the Time Zone meridian of Newfoundland. The result is shown by the output graph (Fig. 27(c)) below:

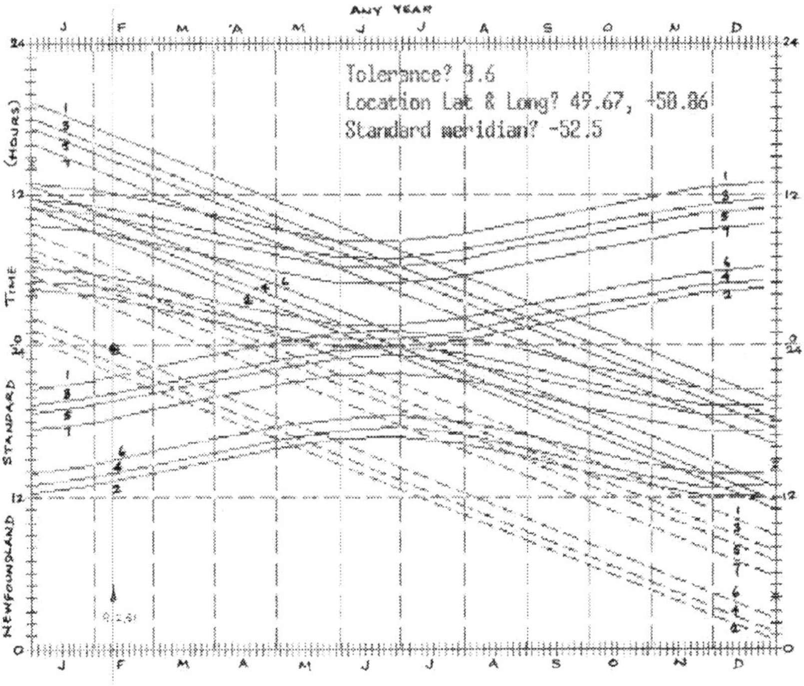


**Fig. 27(c)**

The nearest prediction to midnight on 9th/10th February is 23:50 hrs. and this relates to star-related track No.6 . That time could have been the actual time of arrival from space of that disc/dome craft, or a programmed departure time, which was subsequently delayed by the interruption.

As will be continued to be demonstrated, the work described in this book sheds light on many other SAC encounters and establishes, without reasonable doubt, that **programmed on-going surveillance** and exploration of our planet is being carried on by **extraterrestrial agencies**.

## Intruders at Woodbridge and Bentwaters, U.K.

One of the most publicised happenings involving military bases and aircraft to come to public attention in Britain occurred during the summer of 1956. During the night of August 13th, UFO activity broke out and continued for some hours over East Anglia. The RAF and USAF were based in that area and had radar controllers on constant watch for

Soviet intruders. At 10:55 pm BST, an American Ground Controlled Approach radar operator at RAF Lakenheath detected a fast-moving target, approaching from the east at an estimated distance of 30 miles. As it came in over the sea its speed had seemed to be between 2000 and 4000 mph. The object quickly overflew RAF Bentwaters and was lost on the radar scope when it was about 30 miles away. The physical nature of this object was visually confirmed by the air traffic controller on the control tower, but to him it was travelling so fast that its shape was blurred. Interestingly, no sonic boom was heard. The UFO had headed for the neighbouring base of RAF Woodbridge. Observers there saw the same UFO (or a similar one) slow down and hover before it shot off back towards the east. A Venom night fighter was then scrambled from RAF Waterbeach and this was vectored towards the radar target still visible on the screens. The fighter pilot made visual contact and locked his gun-sight onto it as he flew closer to the object. Suddenly, the target disappeared from view and from the fighter's radar scope and the surprised pilot asked the controllers where it had gone. When they informed him that they had seen the target very suddenly flip backwards and that it was now following him, the RAF pilot then began rapid manoeuvres in an attempt to reverse that situation. Alas for him, the object was seen by the controllers to follow his every manoeuvre precisely, as if it was physically attached to his aircraft. At this point, running low on fuel, the pilot broke off the engagement and returned to his base. As he did so, another Venom fighter was scrambled to take up the chase. However, that attempt to intercept the intruder also failed.

My various sources of information on this invasion of East Anglian airspace had informed me that UFO activity continued until well into the early hours of August $14^{th}$, until about 4:30 am BST.

This authenticated series of happenings was of particular interest to me for analysis. As will be shown later, the timings graph for that area, Fig. 28, produced predictions that clearly marked both the start and the alleged finish times of that prolonged activity. The reason for delaying revealing the results obtained, at this point in the account, lies in the fact that the military bases at Bentwater and Woodbridge came into the public spotlight again during 1981.

## The Rendlesham Forest Events.

Probably most people in Britain now know that a senior serving USAF officer, Lt. Colonel Halt, in a memorandum to the RAF (MoD), reported that UFO activity and strange happenings had occurred in the forest adjacent to his Woodbridge airbase during the closing days of 1980. This report was leaked to a popular Sunday newspaper and became headline news. The woodland being referred to is Rendlesham Forest, an area which has since become a focal point for UFO investigators and documentary film makers. At first, I had only Lt. Col. Halt's report to investigate. It became clear that he himself had been involved in some of the forays into the forest. Thereafter the situation became thoroughly confused by claims and counterclaims made by people who said they had witnessed strange events in Rendlesham Forest during that post-Christmas period. To investigate the incidents I needed good solid facts, so nothing could be processed until I was able to be satisfied about the authenticity of the information.

During 1992 I was contacted by Brenda Butler, a resident in that area of Suffolk, who had investigated the events personally. In fact, she was a co-author of the 1984 book **'Sky Crash'** which dealt with the investigations that Brenda, Dot Street and Jenny Randles had carried out soon after the news of the Halt memorandum became public knowledge. Brenda had heard about my analytical work and wanted to know more about it. It was arranged that she would travel up to Cheshire to familiarise herself with the work and to discuss Rendlesham matters with me. I shared my findings with Brenda and then produced the graph I had created for the Woodbridge area. Brenda had brought along seven recent UFO reports from the area. I think she was impressed when they were found to correspond well with the lines on the timings graph. This graph is now shown as Fig.28. Brenda Butler's reports are shown as ringed points. The 1956 radar/visuals, dealt with previously, are depicted by two diagonal crosses linked by a dashed vertical line. The lower cross represents the time when a UFO was first detected approaching from the sea and the upper one corresponds to the time when the aerial activity seemed to cease. Both these points lie on predicted lines and the dashed line linking them suggests that other programmed delivery/retrieval paths could have been utilised during that activity period of some four-and-a-half hours.

During the years following my meeting with Brenda Butler, more definitive information about the Rendlesham happenings became available to me and these reports were processed as I received them. It seemed that strange happenings had occurred between December 27th and December 30th, 1980. Altogether, I judged that there had been six key timings, two on the night of 27th/28th and the remainder on the night of 29th/30th. I plotted these times appropriately and got the results which are shown as diagonal crosses at the extreme right of the Fig.28 graph. The events seem to have been linked to two delivery/retrieval paths from space, Nos. 1 and 2, and these had been celestially linked to the position of the sun and to one of the fixed orientations (11:00 hr. RA.) relative to the stars. I felt that this result was a very significant one and validated Lt. Col. Halt's accounts. In this way, ***the Rendlesham happenings have been demonstrated to have been perfectly consistent with the recognised pattern of programmed SAC happenings on the global scene.*** The strangeness of the events was also consistent. The first step towards understanding the phenomenon is to concede, as I did in the late 1960s, that ***the technology being displayed to us is definitely not of this world.*** The burning question is why so many people seem to expect it to be?

## References

[10] Major Keyhoe, Donald. E. 'Flying Saucers from Outer Space' (book) Henry Holt & Company, N.Y. 1953.

[11] Hesemann, Michael 'UFOs – The Secret Evidence' (Book, 1998) Marlowe & Company, 841 Broadway, Fourth Floor. New York, NY 10003

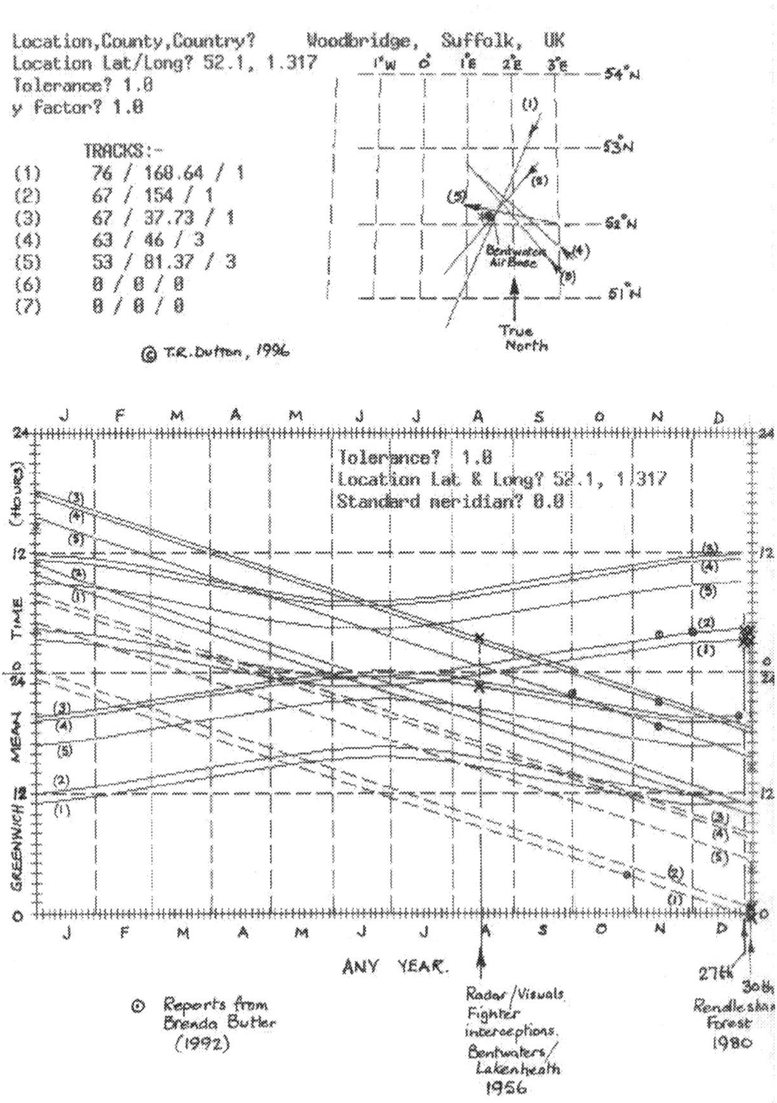

Location,County,Country? Woodbridge, Suffolk, UK
Location Lat/Long? 52.1, 1.317
Tolerance? 1.0
y factor? 1.0
TRACKS:-
(1) 76 / 168.64 / 1
(2) 67 / 154 / 1
(3) 67 / 37.73 / 1
(4) 63 / 46 / 3
(5) 53 / 81.37 / 3
(6) 0 / 0 / 0
(7) 0 / 0 / 0
© T.R. Dutton, 1996
1°W 0° 1°E 2°E 3°E
54°N
53°N
52°N
51°N
True North
Tolerance? 1.0
Location Lat & Long? 52.1, 1.317
Standard meridian? 0.0
J F M A M J J A S O N D
GREENWICH MEAN TIME (HOURS)
ANY YEAR.
Reports from Brenda Butler (1992)
Radar/Visuals Fighter interceptions. Bentwaters/Lakenheath 1956
27th
30th
Forest
1980


**Fig. 28**

# CHAPTER 13.

# Civilian Encounters with Automated Craft.

*Judging from the evidence collected over more than thirty years, it is likely that most encounters with SACs* ***involve fully automated exploration craft without occupants****. These craft seem to employ* ***terrain-following*** *navigation techniques during their mission time within the atmosphere and have built-in* ***artificial intelligence****, which enables them to investigate interesting events and specimens during their clandestine activities. They also seem able to respond to human attempts to attract attention by the use of light signals. Members of the American* ***Center for the Study of Extraterrestrial Intelligence*** *(****CSETI****), with whom the author and his wife had personal contact during the crop circle investigations of the 1990s, appear to have had considerable success in initiating light signal dialogues with strange aerial craft, sometimes at close quarters.*

## Car Stops.

Similar craft seem to have been responsible for stopping motorcars on numerous occasions, by neutralising the cars' electrical systems. Typically, during such car-stops, the occupants have found their cars surrounded by a powerful beam of light, which was being projected from a dark object hovering directly above them. During that period of immobility, all the electrical equipment was put out of action, this including radio links used by police patrolmen. But that operational mode was not the only one adopted by the visiting craft. Another commonly used technique involved blocking the road ahead of the vehicle to force the driver to brake, before disabling the car's electrical equipment. Two examples of the deployment of this technique will be given later. Both events were followed by periods of amnesia experienced by the occupants of the cars, and the timings of these events provide evidence in favour of the witnesses' suspicions that they may have been abducted for periods of, typically, one hour.

Before leaving this topic, it seems relevant to describe a very atypical car stop investigated by the author during 1978. A full and detailed report was sent to BUFORA, which was later published by the Association in a 1979 compilation of such events with the title **'Vehicle Interference Project'**. The main features of the case will be summarised here for readers' consideration.

A local (Bramhall, Cheshire) resident had reported seeing a bright light in the night sky. This light had seemed to be moving above broken cloud. The author visited the witness at his home and received a first-hand account of the sighting. It became evident that the bright light had been the planet Venus, which had been made to appear to be moving by the movement of the broken cloud layer. This is a common problem encountered by UFO investigators. The witness seemed to be satisfied by that explanation, but then went on to tell of a night, some ten years earlier, when he had experienced a car-stop event and he related the story of that happening.

It seemed that the event had occurred on March 4$^{th}$, 1968, at about 9 pm in the evening, when he had been driving his 1967 soft-top Triumph Spitfire sports car along a narrow country lane, on a hillside on one side of the Longendale Valley, between the towns of Marple and Glossop. The night had been dry but very dark. He had been travelling in the direction of Glossop with the high ground to his right and the valley to his left and had been approaching a T-junction. Suddenly a small golden object had appeared ahead of him, travelling at high speed, from over the ridge to his right. At virtually the same instant all the car's electrics had failed. The light from the headlamps had been extinguished, the radio had ceased to play and the car had lurched abruptly to a halt as the engine stopped. The object in the sky had appeared to be elliptical as it sped at phenomenal speed across the valley. Its trajectory had been low and flat and it had seemed to climb in order to overfly the higher land on the other side of the valley. The witness estimated it's time in view as being about ***two seconds***. As soon as the object had traversed the windscreen, the headlights had blazed out again and, after the startled driver had collected himself enough, the engine had been re-started without problems. However, the radio had been completely silenced and had never played again.

In the company of the witness, I travelled out to view the site of this amazing series of events. From compass bearings, the measurements of angles indicated from the author's car by the witness and the distance between the high ground on both sides of the valley, the following conclusions were arrived at:-

1. The flight path bearing (measured some distance from the car) had been 345° Mag.
2. The object had been approximately 0.6 mile ahead of the observer;
3. Directly ahead of the car's position, its flight altitude had been 730 ft. above the level of the observer (ie. 1,480 ft. above sea level)
4. It had cleared the high ground to the observer's right by about 1000 ft. and the higher ground across the valley by about 700 ft. and had seemed to have curved downwards before disappearing from sight. From these indicators, the object had appeared to have been ***terrain-following.***
5. The speed of the object across the valley was calculated to have been about 7000 mph., or Mach 9 at low altitudes, but there had been no shock wave heard or felt.

(The witness had repeatedly emphasised that he had never witnessed such transit speeds, even during air show high-speed, low altitude, passes by fighter aircraft.)

(6) From indications given by the witness, using the 'fingers-at-arm's-length'method, and from the estimates of distance from the observer, the length of the object was able to estimated to have been about 30 ft.

The witness had stored the unserviceable radio after the event, hoping to get it repaired, but had not succeeded in this. On our return to his home he handed it to me for further investigation. Two members of the Avionics Department of Hawker Siddeley Aviation, Ltd., Woodford, (our employer at that time) had always shown interest in my UFO investigations and very obligingly consented to check out that radio. Some weeks later they demonstrated that they had found the fault and repaired it. Two PNP transistors had been overloaded and had simply been replaced, but they were puzzled by the fact that the damaged items had been on the aerial side of the circuit and they were unable to explain that. When the full circumstances were explained to them, they then became opened to the idea that an external powerful electrical impulse, through the aerial, might possibly have been responsible.

This possibility set me thinking that a similar pulse effecting the flow of electrons through the car's bodywork (the common earth) might have also accounted for the other electrical failures. That hypothesis is still the best one the author has yet come across to explain the car-stopping phenomenon in all its forms. But the stopping of a car by a distant and high-speed object was then (and perhaps still is) a unique situation. In view of the very transient duration of the headlights' failure, it seemed that any external electrical impulse of the sort suggested would have had to have had the nature of an electrical shock-wave, analogous to the acoustic shock-wave one would have normally expected, but which was not evidenced. This aroused the suspicion that the SAC may be capable of controlling the acoustic pressure waves and converting the energy within them using powerful electrical fields. (Not a stupid suggestion if such fields could be generated within an aircraft.)

As ten years had passed between the event and the investigation, the witness had been unable to give an accurate time for the happening, so it has not been possible to conduct a correlation study using the UFO timings software. Even so, it seems that the features of this sighting reveal a lot of information about the SAC operating all over the world. *Their ability to fly silently through the atmosphere at supersonic speeds, one way in which they can induce sudden power failures, both in motor vehicles and power distribution lines, and their ability to fly fast and low, using terrain-following techniques, have been adequately demonstrated.* Wherever they were contrived, we can be sure that it was not in any manufacturing facility set up by humans.

## The Livingston Encounter.

A bizarre encounter with a hovering SAC occurred during 1979. It became headline news in **Scotland** and led to a full investigation of the event by a leading **BUFORA** investigator, astronomer Steuart Campbell. Campbell subsequently produced a very full report for BUFORA **[12]**. Being a member of the Association at that time, I was able to obtain a copy of that document, to which I have referred, to ensure accuracy in the summary presented here.

It seems that on the morning of November 9th, 1979, a 61 years' old foreman forester, Robert Taylor, had driven out to inspect young trees in a plantation situated alongside the M8 motorway. (This links Edinburgh and Glasgow.) He had taken his dog along with him and the two of them had left the vehicle to walk the narrower path that led to the site to be inspected. As they rounded a bend in the path leading into a clearing, at about 10:15 am GMT, they were confronted by a surprising sight. There, hovering low over the ground, was a large hemispherical craft, the base of which was surrounded by a narrow rim or brim. This rim had a number of strange antenna projecting from it. The shape of the craft below the rim could not be discerned clearly, perhaps because it was heavily shadowed. The object's colour was dark grey and the surface texture seemed to be rough, but becoming smooth in patches, in a random fashion, over different parts of the surface. The craft's diameter was guessed to have been about 20 feet (6m.) with an overall height of about 13 feet (4m.). Suddenly, two dark grey spheres, each about 2 to 3 feet in diameter and equipped with about six (a number not clearly stated) long spiky legs with expanded ends, rolled swiftly from beneath the hovering craft. Rolling on their spikes, they moved rapidly towards the witness with plopping noises as the legs contacted the soft grassy ground. After arriving one on either side of him, the spheres each attached a spike to a trouser leg, just below the pocket, and began to drag the witness towards the large craft. He tried to resist but was overcome by a choking odour. He lost consciousness and fell forwards.

When the witness had come back into consciousness, still lying on the ground, his first remembrance was of his barking dog, madly chasing round him. He had been unable to speak to her because he had lost his voice and, at first, he was unable to stand. This had meant crawling, on all fours, almost all the way back to his parked truck. Being still unable to speak he had not been able to contact his employers by radio and had then failed to turn the truck round when it sank into soft soil during reversing. Feeling stronger, he had decided to set out to walk back to his home, located more than a mile away across woods and fields. He had arrived home at about 11:15 am.GMT with a headache and feeling very thirsty. His thirst had lasted for several days and he had felt nauseated by the taste of the overpowering odour still lingering in his mouth.

Campbell's excellent report details all the subsequent developments. From all the known details, he estimated that Mr. Taylor had been unconscious for about 20 minutes. A local doctor had conducted a physical examination, the Head of the Forestry Department had visited the site and arranged for it to be fenced off. On seeing strange ground markings, he had informed the **Police** who then went to the site and took measurements and photographs. They later visited the witness and took away items of clothing for examination. Very quickly news of the happening reached the Scottish and then the national Press. In fact, Campbell had become involved on the night of the event as a result of a telephone call he had received from a reporter employed by **The Glasgow Herald.** Many tests were carried out on Robert Taylor, but none gave any cause to doubt the man's integrity or fitness. Examination of the torn trouser legs showed that the tears were consistent with the circumstances described by the witness.

Back in the forest, strange markings impressed into the grassy surface seemed to link with the position of the hovering object and, furthermore, these marks were surrounded by forty deep (10 cm.) circular holes impressed in a sloping manner into the soft surface. No other damage to the surface was in evidence. Even soil sampling revealed nothing about the nature of the reason for the observed damage.

Campbell (a 'reductionist' scientist) examined the possible nature of the craft described by the witness. Creditably, he dismissed the ideas of those who wanted to believe that it was some sort of secret man-made device and went on to consider the possibility of its having been from an extraterrestrial source. He argued (correctly) that the craft's strangeness was not in itself proof of extraterrestrial origins and no one had witnessed its arrival or departure. However, he claimed to be willing to reconsider that possibility should any further evidence come to light in favour of it.

[Campbell's subsequent attempts to explain away the event as a temporary epileptic seizure suffered by the witness (with incidental ground marks caused by forestry equipment), or as a mirage, or as a form of ball lightning, seemed to indicate that he had closed his mind to any such ET evidence. That is the kind of trap that ensnares those who try to find mundane solutions for all extraordinary phenomena. Unfortunately, today's scientific establishment is too well populated by such individuals, as the author knows only too well!]

If Stueart Campbell's mind is still receptive, then the evidence he claimed to be willing to consider is about to be presented here. After processing the details of the Livingston case through the AT's computer programs, it was found to have complied admirably with the global behaviour patterns established by hypothetical ET craft over a period of more than a century. The scenario presented by the AT is one in which exploration craft are delivered into the Earth's atmosphere from space, in accordance with the rules of a long-established programme, and are subsequently retrieved after a predetermined mission time in the atmosphere. During its mission such a craft may have to hide itself, to await the time of rendezvous in space and there is much evidence, worldwide, that this often occurs. Many stationary craft have been encountered in woodland clearings, disused quarries and entering or exiting shallow reservoirs, lakes and seas. Fig.29 presents the timings graph for Livingston. A vertical line is drawn to represent November 9$^{th}$. Mr. Taylor's estimate of the time he had encountered the craft was 10:15 am. GMT. Two horizontal lines have been drawn to represent this time in both sections of the two-section timing scale for the same day. It can be seen that these two lines intersect the vertical line very close to two of the predicted timing lines. One intersection is almost exactly superimposed on the curved sunrise-orientated No.3 track's timing line and the other lies just below the sloping 1100h RA No.3 line.

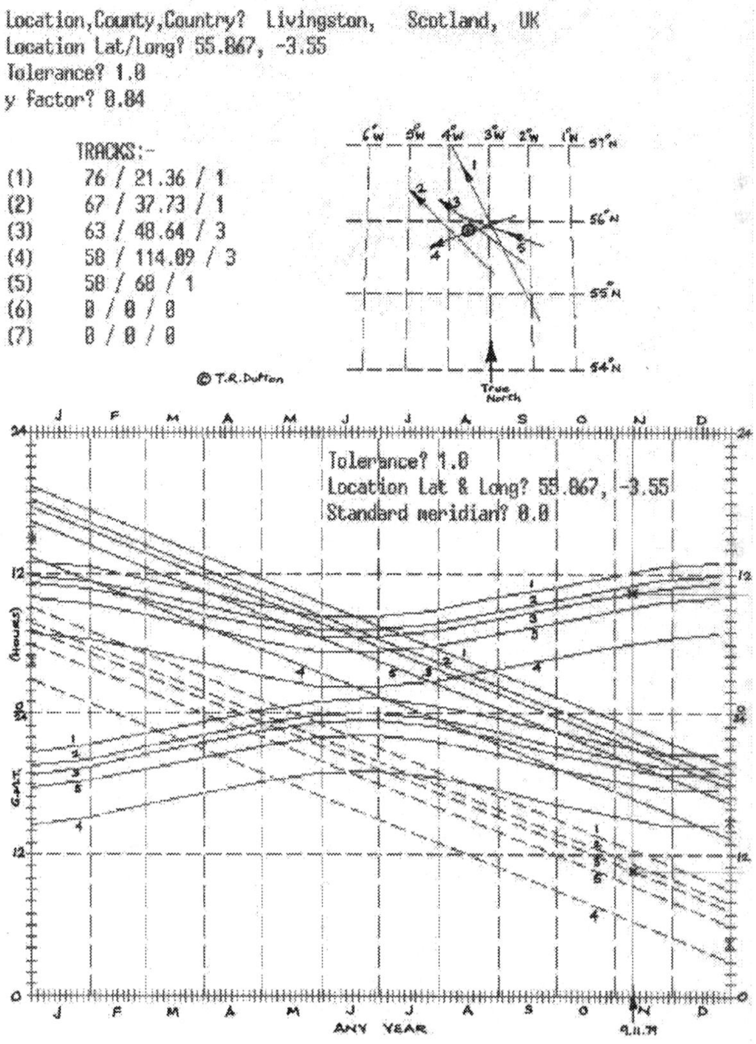


**Fig. 29**

This is an excellent result to explain the happening in that plantation. The indications are that the craft had arrived on site just prior to Mr. Taylor's arrival there and it had taken action to detain him, unconscious, until it's mission period expired. What that mission was, cannot be determined, but within half an hour it was due for retrieval by a larger craft travelling along the No.3, 1100h RA, path in space.

## References

[12] Campbell, S. 'Close Encounter at Livingston' (report) British UFO Research Association (BUFORA Ltd.) July 1982.

## Close Encounter in Wiltshire.

During the 1990s period when my wife, Marion, and I were spending time in Wiltshire and Hampshire investigating the crop formations there, we were brought into contact with a small American group led by **Dr. Steven Greer M.D.**, a paediatrician, then working within a hospital in South Carolina. The occasion for that meeting was a Crop Circles symposium in Glastonbury, Somerset, at which I was an invited speaker. During the afternoon tea break on July 26$^{th}$, 1992, Busty Taylor came over to our table and introduced us to the group of people accompanying him. Their spokesman was introduced as Dr. Steven Greer, who then explained his interest in the work I had been doing on the UFO enigma. Having become convinced that some of the SACs reported worldwide were intelligently controlled ET craft, he had created an organisation called **Center for the Study of Extraterrestrial Intelligence (CSETI)**, with rapid response teams enabled to go to any part of the world in which UFO activity was taking place. Their objective would be to make contact with the intelligence guiding the craft by following a protocol of meditation and signalling methods, including the use of powerful hand lamps. The co-ordinator for these activities, the late Ms. Sharon (Shari) Adamiak, was then introduced to us along with the other two members of that group. Shari became a very helpful link with the group for several years later, until her untimely death. Another member of the group, artist Ron Russell, has also kept in touch from time to time.

Dr. Greer then went on to explain that several nights before, whilst on a skywatch in the crop fields of Wiltshire, he and his team had witnessed a spectacular SAC overflight of their location before midnight and he wondered if this would have been predicted by my graph for that area. I took the **hand-produced graph** (the computer programs had not been created at that time) from my documents case and, together, we checked out date and time in the usual way. We were all very happy to find that the observed event had occurred exactly as predicted by the graph. I was then asked if I could give them predictions for the coming night of the 26th/27th. I read off three likely times and soon afterwards they left to return to their Wiltshire base, with the assurance that they would let me know how they had fared. About three months later I received a letter from Shari Adamiak, informing me that events had occurred at all three predicted times and that one of them had been an interactive close encounter with a large SAC. Steven Greer had labelled such encounters, **'Close Encounters of the 5th Kind ' (CE5)**. Later still I received a copy of a full report of that CE5 event written by Dr. Greer. (During our subsequent videoing of the crop circles scene for a trilogy of videos, produced later, we were fortunate to be able to record a first-hand account, on location, from Dr. Greer with Shari Adamiak in attendance.)

I will try now to summarise the very long, but interesting, story.

According to Dr. Greer's account, I had given the team three BST predictions: 10:30 pm. 12:30 am. and 1:30 am. After a brief break for dinner, they had proceeded to the chosen research site at Alton Barnes, Wiltshire. Clouds were gathering when two unidentified lights had risen up above the cloud bank in the East and moved towards the group, being last seen heading south. This had occurred at 10:25 pm. At about 11:00 pm, the clouds above the group were seen to be illuminated from above by a rotating cartwheel of brilliant lights, spinning anti-clockwise. This display lasted for about ten minutes. The cloudbank in the north-east had grown larger when they observed four elliptical objects detach from the western edge of the cloud and then disappear into its eastern edge. Soon afterwards it began to rain and this quickly became a torrential downpour. They retreated to their cars, being very much aware that their equipment was in danger of getting soaked. At this point, some members of the group decided to catch up on lost sleep by returning to their accommodation, leaving four stalwarts to carry on with the watch. These remaining people had shared two cars, one being Dr. Greer's, and they had decided to drive onto a hard-paved road on the adjacent farm. They'd sat and hoped that the rain would subside. At around 12:30 am., as the rain eased, one of the team in the other car came hammering on Dr. Greer's window yelling that there was a spaceship coming across the adjacent field, only a few hundred feet away from them. They all then saw the object and what they saw astonished them. Dr. Greer described it as "a large, disk-shaped craft with brilliant lights rotating counter-clockwise along its base. The object had a high dome or cone on which sat three amber lights on top of this structure. The report goes on to state that it was hovering no more than thirty feet (10m) above the ground and was located about four hundred yards (metres) from their position. They could see metallic structure between the rotating blue, green, amber and white lights and at times believed they could hear a humming sound. The team's video recorder had been inoperable so that Dr. Greer had to give a commentary into his micro-cassette tape recorder. Using the fingers-at arms-length method, the diameter of the craft was assessed to be somewhere between 80 and 150 feet, this being dependent on the assumed distance away. It manoeuvred about the field as they watched and at one point seemed to flip up as if to reveal its underside, when the rotating lights gave it the appearance of an illuminated Christmas tree. Subsequently, one of the amber coloured lights on the dome had suddenly detached itself and flown off into the mist. A new Swedish compass with a life-time guarantee had been indicating all manner of directions for North during this encounter, slowly moving in a counter-clockwise direction, like the lights on the craft. The high-powered signal lamps had then been used to flash light signals towards the craft and, to their satisfaction, the craft had flashed back in the same sequences. Contact had been made! Afterwards, it seems that the craft had moved away into the mist and had been lost to view. What happened to it then is something of a mystery.

A large amber-coloured UFO was seen in the south-west at about 1:30 am. This had seemed to wobble as it hovered and was thought to be about one or two miles away. When flashed at by the signal lamps, it had obligingly responded in the same sequences before disappearing from view below the tree-line. It had been quite a night for the CSETI team!

**After being told of all this I had remarked that, whilst the SETI scientists search the skies for radio signals from alien civilisations, a group of amateurs equipped only with signal lamps seemed to have**

**already achieved that much sought-for contact**. I added that they seemed to have had a far better success rate than I'd ever had myself!

Scientists are generally disinclined to even consider the possibility that technological aliens may have been surveying this planet, perhaps for millennia. For this reason, most of them cannot conceive or concede that my Astronautical Theory can have any validity whatsoever.

So, *it needs to be emphasised, again and again, that we are not dealing with our kind of astronautics or space hardware. The technology being displayed is beyond our current understanding of physics. The craft manifest as solid objects when they are interacting with us, but are capable of metamorphosis when in operational mode; for example, whenever they move from one location in space-time to another. We can have no idea how they are able to perform in that way. Their technology is literally 'out of this world'. But as my colleague and science author Edward Ashpole has repeatedly pointed out, that is exactly as we should expect it to be. It can only be regarded as being extremely fortunate that a programmed strategy, understandable in our terms, has been adopted by the visiting craft to facilitate their surveillance of this planet and that this has been able to be approximated by the AT.*

The strangeness of the technology being used was demonstrated at close quarters to a retired engineer (who was once **Head of Engine Development with Rolls-Royce Engines, Derby),** during the winter of 1998. Prior to his established period with Rolls-Royce, he had flown in Lancaster bombers during WW2 and survived. The story follows.

## The Lewis-Goodwin Happening.

After moving to Devon, I had been pleased to discover that there was a very active Torbay Astronomical Society in Torquay and that it had meetings open to interested members of the public once every fortnight. One of these was a public talk held at Torquay Town Hall each month. Having been interested in astronomy from schooldays, I decided to attend some of the 'open' meetings. This went on for several years and, even though I never became a member of the Society, I received a copy of each year's programme through the post. During early 1998 I had asked the Secretary if the members would like a lecture on my scientific UFO research, which had considerable astronomical content. He replied that he'd have to confer on that and then come back to me with the answer. I was very pleased to be told later that the committee had consented to the idea, but I was warned that I could expect a lot of outspoken scepticism. That did not deter me and the date was set to be September 17th, 1998, the first lecture in the programmed series.

I had prepared a lot of material and went on to deliver it in the same manner as for a serious technical lecture (which was appreciated by some of the visiting members of my audience), but the meeting turned out to be a fiasco. The hostility of the Chairman was clearly evident at outset and he continued in the same vein during the question-time period. When some of his colleagues displayed the same mind-set, it became clear to me that I had been talking to many closed minds. It was a very disappointing outcome, because I'd looked forward to getting involved in an intelligent discussion after the talk.

However, something quite unexpectedly wonderful did result from that meeting. As I packed away my belongings, a man and a woman came over to me with some very interesting news. They were, then, the owners of Torbay Holiday Motel and the adjoining permanent chalet and caravan park, Beechdown Park, located between Paignton and Totnes. A resident of the Park had claimed to have had a weird experience involving the sudden appearance of an object on the adjoining hillside one night, when walking his dog. He had written a full report of the incident for MoD, but had received a dismissive answer. Would I like to investigate this event? Of course, I would! My contact, Mr. Graham Booth, posted to me a copy of the report, first sent to MoD, after consulting with the witness. After reading the very detailed report, I got in touch with the witness, the late Mr. Fred Lewis-Goodwin. From the time of our first meeting, Fred and I got on very well and his sudden death in 2001 was a great blow.

On the evening of February 11th, 1998, Fred L-G, a resident of Beechdown Park, had set out to walk his dog shortly after 8 pm GMT. He had walked some 150 yards (metres) to the wire fence separating the Motel grounds from the adjacent grassy hillside to his left. Further along the fence was a style, which had to be climbed over to get into the field beyond. As he had reached the fence, his greyhound-lurcher dog had stopped abruptly and 'pointed' with its nose towards a location on the hillside. It then sat upright with ears cocked, staring intently at something Fred couldn't see. It was a bright moonlit night and, with the moon situated above the hillside to the left of the witness's position, the silvery grey dog had looked very strange in that lighting and in that upright position. Then, in a matter of a few seconds, an object had materialised where the dog was pointing, in a position about 45 degrees to the left and about 40 feet (12m) away. Fred had watched in disbelief as he'd seen dark lines drawn quickly in the air to form the framework of a large rectangular box. In the next instant 'blinds' had seemed to be drawn across the framework to form the base, top and sides of the box. The estimated dimensions of the box were 10' (3m) long, 4' (1.2m) wide and 3' (1m) deep and it had seemed to be skewed about 45 degrees to the viewing line. It had had a dark grey (charcoal) colour in the moonlight and, as the still unbelieving witness watched, two circular apertures, each about 6" in diameter, had formed in the long side visible to him. These had been centrally located. The lateral spacing between them had been about 6 feet (2 metres). The apertures had glowed brightly orange-yellow and they began pulsing in unison. The object had then raised itself to a height of about 2 feet above the sloping ground, as the changed moon shadow had confirmed. The rate of pulsation then increased, from about 90 to something like 300 cycles per minute, causing Fred to drag his still-mesmerised dog further along the fence to a safer distance from the object. Wave-like small disturbances had been visible in the air above each corner of the box throughout this early stage. Another change of the witness' position along the fence had been made when the object had begun to climb steadily uphill, maintaining its orientation and height above the ground as it climbed.

A patchy hedge ran across the hillside about 100' (30m) above the witness' position and a power line, feeding a homestead at the right-hand edge of the field, also crossed the field just below the hedge-line. This power line was mounted some 20-30 feet (6-10m) above the ground on wooden posts. As the object had moved beneath the power line, Fred had noticed that a flood-light mounted on the residence to his right had suddenly flared and then dimmed --- and he'd thought it was a "goner". But the light had been restored as the box moved on, still climbing the hill at the same pace. On reaching the embankment on which the hedge was growing, the object had then climbed over the embankment by passing through a large gap in the hedge. It had then moved sideways following the line of the hedge until it had reached an open gateway, next to another hedge. This was roughly at right-angles to the original one and providing the boundary of the field beyond the gateway where the box had become stationary. With the two apertures still pulsing out bright light, the box had then remained motionless for about 15 minutes before a very dramatic change had occurred. Suddenly, it had become a brilliant ball of white light. This had then climbed uphill from the gateway just a short distance and had then hopped over the adjacent hedge before moving very rapidly to the right, to be lost to the witness' view.

This experience had, to say the least, made a big impression on the worldly-wise Fred Lewis Goodwin. When I met him during September of that year, I found him reading up on Quantum Mechanics in the hope that he might find a solution to the problem therein. He was as sure as I was that he had seen a demonstration of advanced technology. It had seemed almost as if this 'thing' had been drawn out in the air by some sort of computer graphics system, but his overwhelming impression had been that it had materialised "from another set of dimensions". So, he had come to terms with the idea that he had witnessed other-worldly technology. He had been particularly incensed by the MoD's 'brush-off' and had written a scathing letter in reply. He had noted that the site of the event had been beneath major civil air routes and the object, whilst sitting in that gateway, could have been monitoring the radio and TV emissions from major masts mounted on the hillside above Marldon, situated only a few miles away. The masts were in clear view from that hillside. I had checked out the timings of the event as given by Fred L-G and found them to be consistent with the AT's predictions for that day. A marked-up graph for the Torbay area, centred on Torquay, is given as Fig. 30, below.

This had impressed Fred and he'd wanted to know more about the work I'd done. I passed on copies of papers I'd written and he soon became conversant with them. In fact, he was keen to engage in regular skywatches from his site and promised to call me by 'phone if he ever saw anything again. From where I live on a hillside in Torquay, situated at 500' above sea level, I was able to see the top of Fred's hill beyond another, intervening, hill crest. On several occasions thereafter he called me to check out lights in the sky he had been watching outside in the Park. Unfortunately, I can't remember ever having been able to confirm anything. Then, in August 2000 he told me he had been sitting out on a bench adjacent to the Motel, during a late dog-walk, when he had seen two pulsating lights, side by side, move steadily up the same hillside, hop over the hedge embankment, as before, and then to continue climbing until they were lost to view behind the curved surface of the hill crest. The time of this event was again congruent with my predictions. On another occasion, in daylight, he had seen a long, black, javelin-proportioned, pipe-like object hurtle overhead, beneath the overcast, heading towards the sea.

During filming for Part 3 of my **video trilogy 'We are DEFINITELY not Alone'**, I introduced Fred to **Roy Rowlands**, my friend and video co-producer, and he consented to be interviewed at the location. In view of his death in 2001, I consider myself very fortunate to have been able to record Fred Lewis-Goodwin's personal account of the 1998 happening for posterity's sake. I imagine Fred also became widely known in America as a result of a coast-to-coast radio broadcast he shared with me on August $12^{th}$, 1999 --- but more about that later, in PHASE 6.

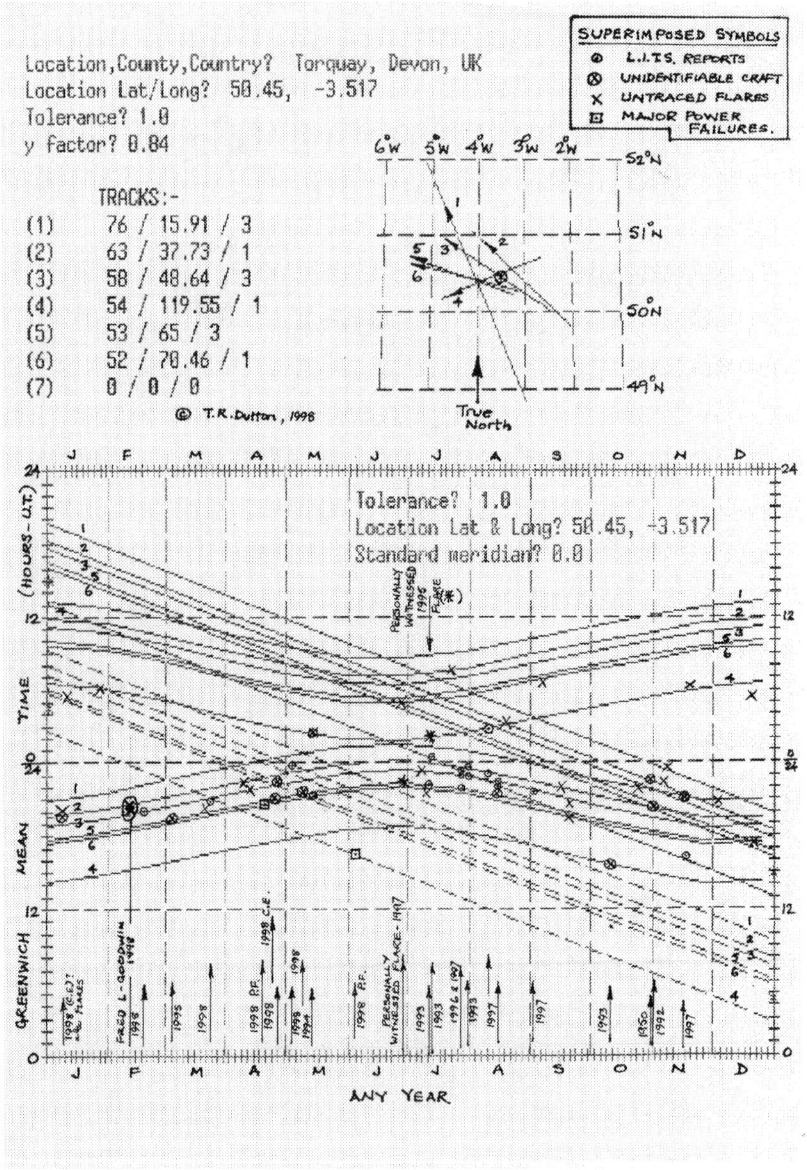
SUPERIMPOSED SYMBOLS
L.I.T.S. REPORTS
UNIDENTIFIABLE CRAFT
UNTRACED FLARES
MAJOR POWER FAILURES.
Location,County,Country? Torquay, Devon, UK
Location Lat/Long? 50.45, -3.517
Tolerance? 1.0
y factor? 0.84
TRACKS:-
(1) 76 / 15.91 / 3
(2) 63 / 37.73 / 1
(3) 58 / 48.64 / 3
(4) 54 / 119.55 / 1
(5) 53 / 65 / 3
(6) 52 / 70.46 / 1
(7) 0 / 0 / 0
© T.R.Dutton, 1998
6°W 5°W 4°W 3°W 2°W
52°N
51°N
50°N
49°N
True North
J F M A M J J A S O N D
Tolerance? 1.0
Location Lat & Long? 50.45, -3.517
Standard meridian? 0.0
PERSONALLY WITNESSED 1995 FLARE (*)
GREENWICH MEAN TIME (HOURS - U.T.)
ANY YEAR

**Fig 30**

# CHAPTER 14

# ENCOUNTERS WITH OCCUPIED CRAFT (CE3S)

*Before going on to share the results of the following studies, it is important to explain that during initial selection of the data from which the Astronautical Theory was eventually derived, no events involving encounters with occupied craft were considered. In fact, they seemed at that time to be very rare occurrences, and accounts like that of George* ***Adamski*** *were thought to be highly imaginative. But this attitude was forced to change when such CE3/4 cases came to be considered in detail after formulation of the Theory. The selected cases to be considered in this chapter will be examples of processed* ***Close Encounters of the Third Kind (CE3)****, when witnesses have found themselves to be observing and interacting with not only strange aerial craft, but also their occupants. The scenario presented by such accounts is one in which mutual interaction has occurred between humans and extraterrestrials.*

*As stated previously, over the years, the author has successfully processed more than 1000 selected UFO reports through the AT software. Relatively few of those cases were of the CE3/CE4 varieties, so it has been a difficult task to select a small number of them for inclusion in this book. Those in the database did not include George Adamski's reports, which have been studied independently and the surprising outcome of those investigations will be presented later.*

## Encounter in South-West Wales.

During the late 1970s the region of **Dyfed** in South Wales became a targeted region for strange UFO activity. As well as I have been able to ascertain, many strange happenings were reported during the period November 1976 and September 1977. ( A prolonged visit was made to the area in the summer of 1978, when all activity seemed to have ceased.) Two books were published during 1979, from which I was able to obtain more information. One was the **Pugh and Holiday book, 'The Dyfed Enigma' [13]** and the other was a book with the title **'The Welsh Triangle' by Peter Paget [14]**. Both these sources gave details of the event about to be described and analysed.

At about 9 pm. GMT on the night of March 13$^{th}$, 1977, a young man called Stephen Taylor was returning to his home, walking alongside the

NATO installation at Brawdy Airfield. It was a fine night and he noticed a strange light with an orange halo in the sky. He had called on some friends and they'd all laughed about the light. He had then continued his journey home still walking alongside the perimeter fence of the airfield. Suddenly, he was startled when a dog raced out of the darkness towards him. Then he noticed that the lights of the base and of a farm he had usually been able to see from that position were not visible. It seemed to him that a large black dome-shape with a dim glow at its base was responsible for this blocking of his view. The object seemed to be occupying part of a field in a position opposite the entrance to the oceanographic research station installed within that secure base. As he leant on the entrance gate to light up a cigarette, the witness heard the sound of someone standing close beside him and turned to look. Presumably by the light of his match or lighter, he could make out a skinny human-like figure, about six feet tall. The figure was clad in an all-over silver suit, but had no helmet, only a small box over its mouth with a tube running rearwards from it. The creature had large human-like eyes and high cheek bones. To escape, the witness remembered taking a wild swing at the visitor and running furiously until he arrived at his home, about three miles (five km) away. Fig. **31** shows that the timing of the events, on that date, fitted the AT predictions for that area very well.

The features of this happening are similar to others reported in various parts of the world. Silver-clad, tall, human-like creatures seem to represent one variety of the visiting explorers from space. But, as will be shown, even though the varieties differ, they all seem to be working to the same game-plan when interacting with us. Generally, the AT is able to demonstrate that the strategy used by all the different kinds of entities is common to all their activities. It seems that each entity represents a contingent of some kind of **'united galactic study group'** and the human species is being given special attention. Therein may lie the reason underlying the alleged encounters with human-like ETs. **Perhaps humans have been bred by the other species for the purpose of making contact with Earth-dwellers less traumatic for the people being investigated?** Given such circumstances, it could be that other galactic species have been similarly bred by a 'master race' species for interactions elsewhere in our galaxy. The predominance of one kind of ET would nicely explain why the AT seems to model all the activities taking place.

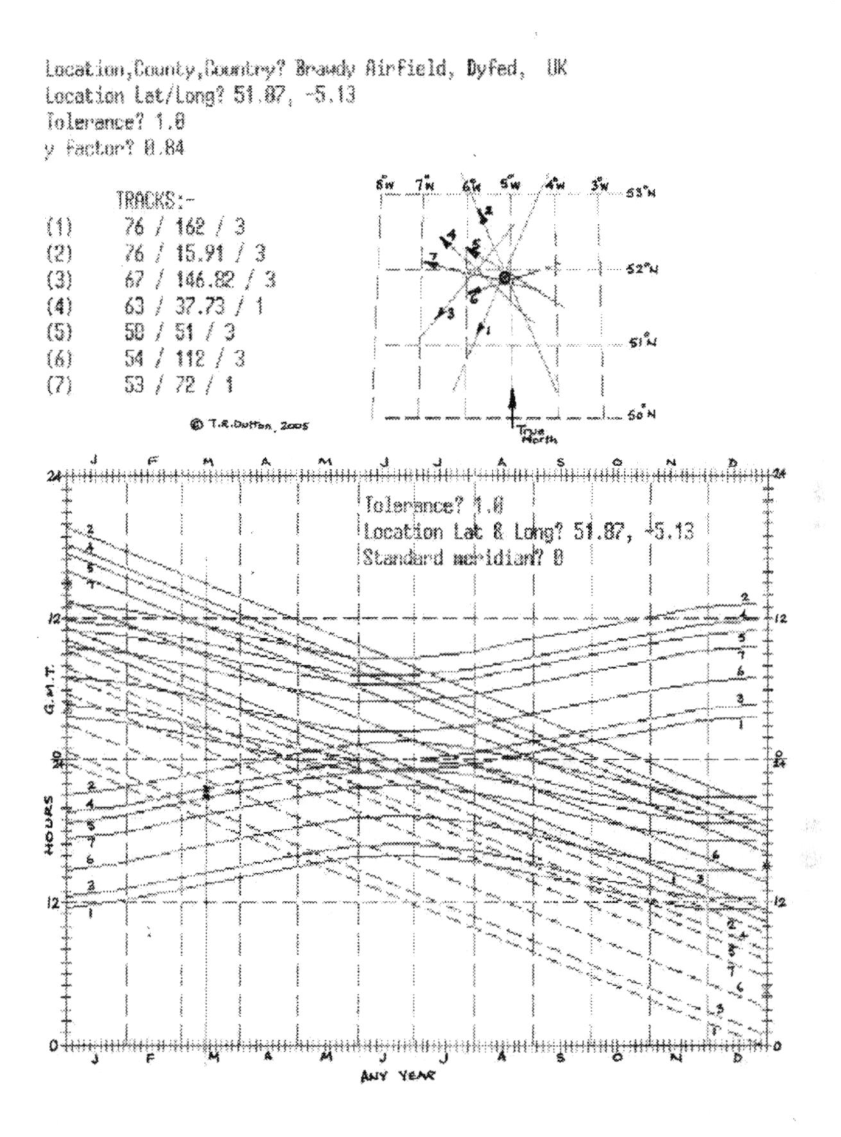

```
Location,County,Country? Brawdy Airfield, Dyfed,  UK
Location Lat/Long? 51.87, -5.13
Tolerance? 1.0
y Factor? 0.84

       TRACKS:-
(1)     76 / 162 / 3
(2)     76 / 15.91 / 3
(3)     67 / 146.82 / 3
(4)     63 / 37.73 / 1
(5)     58 / 51 / 3
(6)     54 / 112 / 3
(7)     53 / 72 / 1
```

© T.R.Dutton 2005

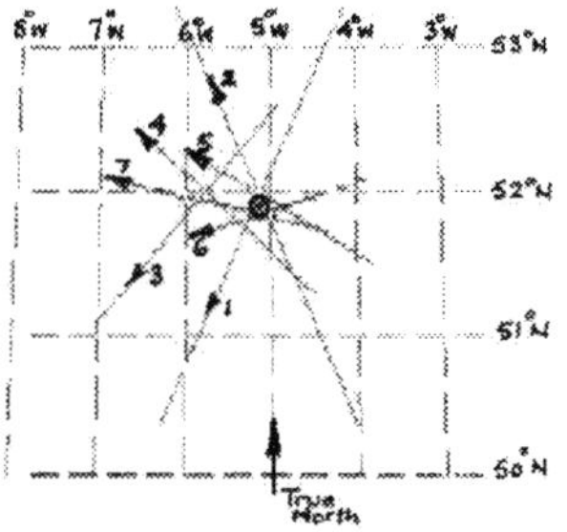


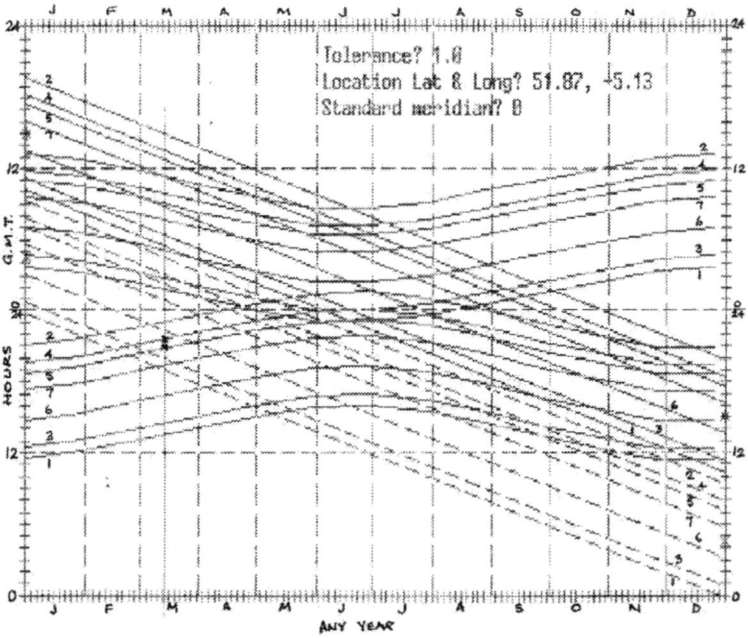


**Fig. 31**

## The Aldershot Encounter, 1983

A little-known but fascinating CE3 is related by Timothy Good in his 1987 book, **'Above Top Secret' [9]**. A detailed account of the happening is given by the book. The story goes that in the early hours of August 12th, 1983, an elderly angler, Alfred Burtoo, set off from home at 12:15 am. BST, to go fishing in the local Basingstoke Canal, accompanied by his dog. Having set up his fishing tackle and attached the dog's lead to the detached base of his large umbrella, spiked into the ground, he settled down to fish. Just prior to this, he had heard a local clock strike 1 o'clock. At about 1:15 am BST he decided to pour himself a cup of tea from his flask and stood up to drink it. A brilliant light then approached him from the south. It seemed to be at very low altitude and wavered in its motion as it passed over a railway line. It then seemed to settle and the light went out. Mr. Burtoo decided he needed a cigarette. He put down his cup and lit up. Then the dog growled and he looked up to see two 'forms' coming towards him. When they were within five feet of him they stopped and he and they looked at each other. The dog's growling had been silenced by his master's command. (Mr. Burtoo's description of the entities is a bit puzzling given the lack of natural light.) They were about 4 feet tall, wore pale green coveralls and helmets equipped with black visors. By arm gestures they persuaded the witness to follow them and he obliged. With one of the entities walking ahead of him and one trailing behind, the trio walked along the towpath, climbed a fence beside a canal bridge, crossed a road and then continued to progress along the towpath. Then, the startled witness saw a large object, 40 – 45 feet (12 – 14m) in width, standing ahead, on the towpath. Portholes were set into the 'hull' and it seemed to be resting on two ski-like runners. The account then goes on to relate that Mr. Burtoo followed the leading entity (or 'form') up a set of steps into the craft and found himself in a black octagonal room with a central column. The other 'form' stood behind him. He was asked to stand beneath a ceiling-mounted amber light, which he did for about five minutes. A voice asked for his age. He replied that he would be seventy-eight next birthday. After another five-minute time lapse the same 'sing-songy' voice then told him that he was too old and infirm for their purpose and that he could leave. He left via the steps and then walked towards the canal bridge. Halfway there he turned and saw the top of the domed object rotating like a chimney cowl. He continued walking until he reached his fishing pitch where, presumably, his dog was patiently awaiting him. The first thing he did was to finish off his now cold cup of tea. Then there was a whining noise. The object lifted off and the brilliant light from it lit up the scene brightly. The object left at high speed over a military cemetery to the west and Mr. Burtoo last saw the light moving out of sight at about 2 am BST. Apparently unperturbed by all that excitement, he continued fishing until 12:20 pm, because fishing he had set out to do and fishing he had been determined to do! The only after-effects were a temporary loss of appetite and weight loss, sleeplessness and a disinclination to go out. It seems that Mr. Burtoo had not taken UFO reports seriously prior to his encounter, but his experience had radically changed his mind. Timothy Good followed up on the story after the witness' death in August 1986. Mrs. Burtoo confirmed that her late husband had never changed his story and had not been the kind of man to have fantasies. He had been very 'down-to earth'. It seems almost needless to say that **the AT's timings graph plot supported the details of his case**, as shown by Fig. 32.

## The George Adamski Controversy.

This Californian's story of his face-to-face encounter with a human-like being, in association with a hovering craft of the disc/dome variety (see PHASE 1), became the object of wonder and ridicule after he had written about it in his co-authored book, **'Flying Saucers have landed' [15]**. The alleged happening had occurred sometime between 12:15 and 1:00 pm PST on November 20th, 1952. George and two friends had set out into the desert in the hope that the strange aerial craft Adamski had observed and photographed from his home on the slopes of **Mount Palomar** would return to the area. At a location in the wilderness close to **Desert Center** they had first seen a cigar-shaped craft, high in the sky, at about **12 noon**. Within about a quarter of an hour, they had then observed a silvery disc-like object drop down from the sky and appear to land, out of sight, behind rocks some distance away. George had set out alone to investigate this craft, leaving his companions behind to monitor the scene through binoculars. As he approached the top of a rise he was confronted by a fair-haired man in a one-piece suit. The craft was hovering, noiselessly, over the rocky terrain behind the visitor. The observers later confirmed they had seen this meeting even though the craft had still been out of sight to them. They went on to sign legal affidavits to confirm their sincerity. George claimed he had conversed with the visitor using sign language and symbols drawn in the sand. He'd wanted to know where in space the visitor had come from and drew a picture of the orbits, round the sun, of the planets of the solar system. The visitor pointed to the orbit of **Venus**. For George Adamski that had meant that he was meeting a Venusian.

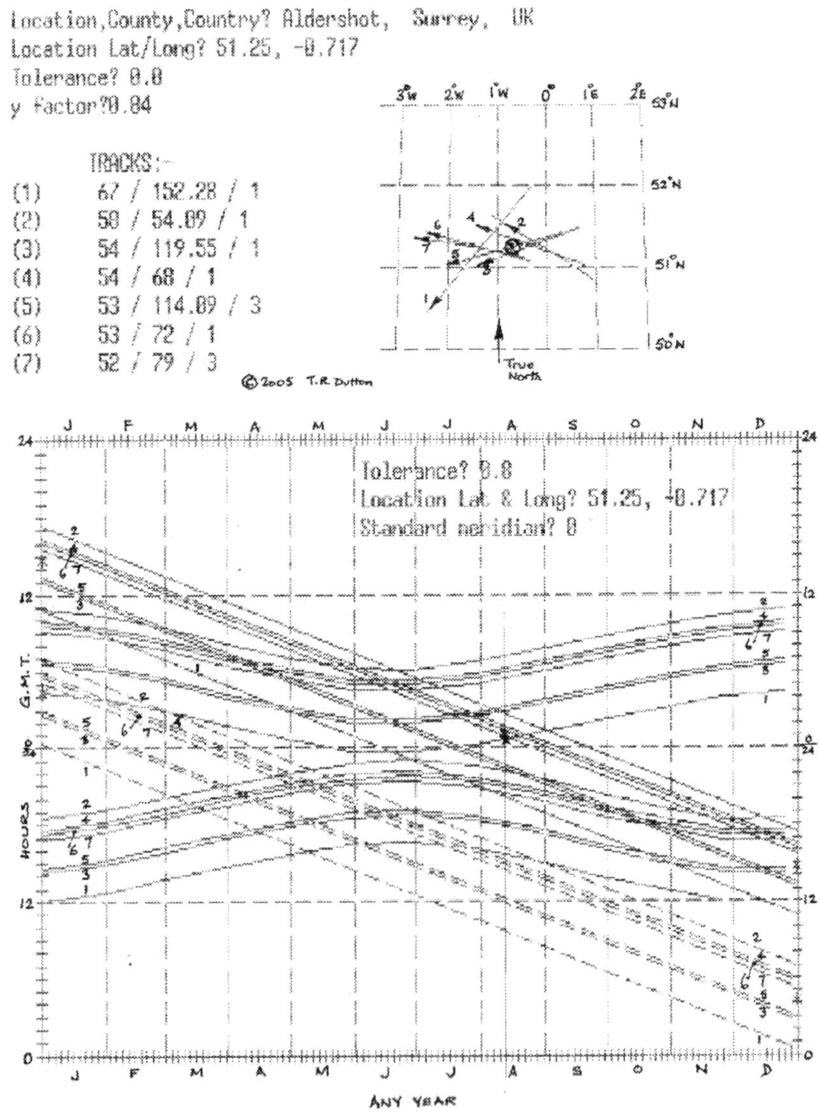


**Fig 32**

The craft and its occupant had come from Venus, he concluded. As the visitor returned to his delivery craft, George trekked back to his friends. Then they all witnessed the take-off of the disc and saw it disappear into the bright sky. From the accounts, this departure probably occurred at about 1:00 pm. PST.

Of course, while people without astronomical knowledge were prepared to wonder and marvel at Adamski's assertions, scientists and especially astronomers, knowing that Venus is extremely hostile to human-like life, either howled in derision or shook their heads in pity. How could any informed person in his right mind believe such a tale? I had to agree.

But aspects of the happening challenged me for years. In PHASE 1 I discussed possible practical features of the Adamski craft as revealed in his close-up photographs of a similar craft. Given an advanced propulsion system, such a craft could have been operable. When I checked out Adamski's times, I had another surprise. Fig. 33 shows the results I obtained

The witnessed arrival of a cigar-shaped delivery craft at 12 noon is shown by the lower cross in the right-hand top corner of this graph. That time corresponds to an arrival using the No.3, Sunrise track option. The departure at approximately 1:00pm (13:00 hrs.) PST is seen to relate either to the upper cross on the No.1, Sunrise track option; or to the No.7, 21:30hr. RA option as shown by another cross in the lower right-hand part of the graph. This degree of correlation was, in itself, very challenging, but an exercise I carried out during Autumn 1999 led to even more interesting discoveries.

I had noticed that **the AT's predicted approach and departure tracks, as suggested by the timings of individual events, had seemed to be quite frequently aligned with bodies in the solar system.** The major planets featured most of all, but they were often rivalled by certain comets and minor planets (major asteroids). In October 1999, I produced a report summarising this study, with the title, **'Astro-navigational links with correlated tracks ---- A Pilot Study'.**

Just over a year later, I created an Appendix to that report, with the title, **'Close Examination of all processed CE3 and CE4 cases in the database at November, 2000'.** The Appendix was longer than the original report. (The entire report will be given as **Addendum 1** at the end of this book.) In the Appendix, twenty eight cases in all were analysed, with the Adamski happening heading the list. The results of the alignments investigation for each event were tabulated on the left-hand side of each page and a marked-up diagram depicting the Earth at the centre of the Ecliptic Plane was presented on the other side of the page. The meridians of Right Ascension were radiating out from the centre of the Earth with spacing adjusted for the 23.5 degree tilt of the Ecliptic to the plane of the sidereal equator. The position of each aligned body was then superimposed on the basic diagram. Fig. 34 is an extract from the Appendix of the report and shows the results obtained for the Adamski happening together with three other documented CE3s.

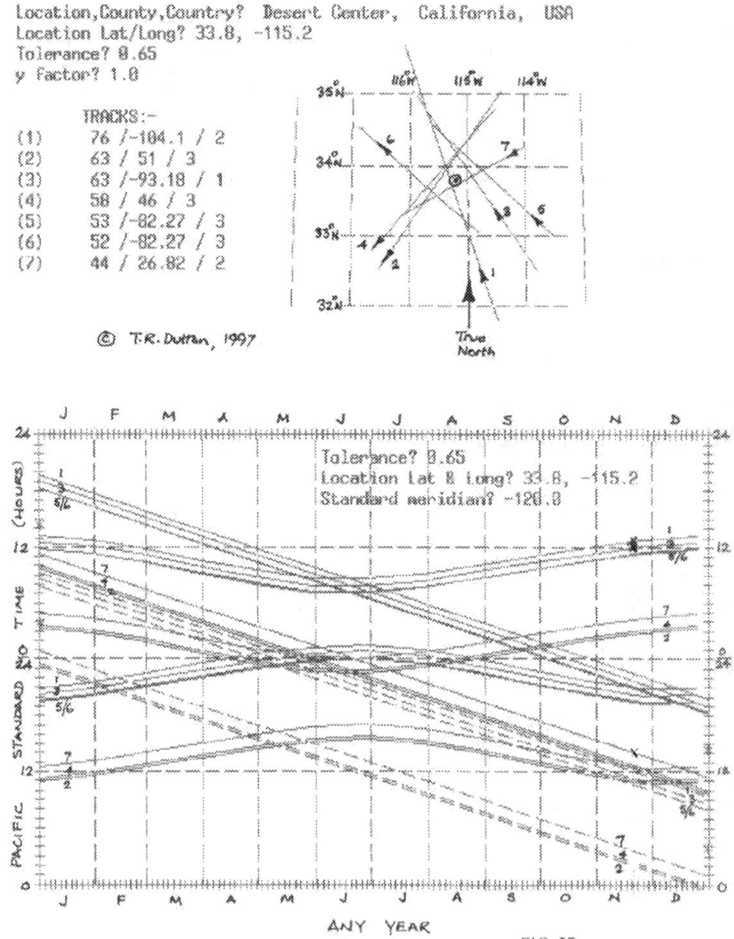


FIG. 27

**Fig 33**

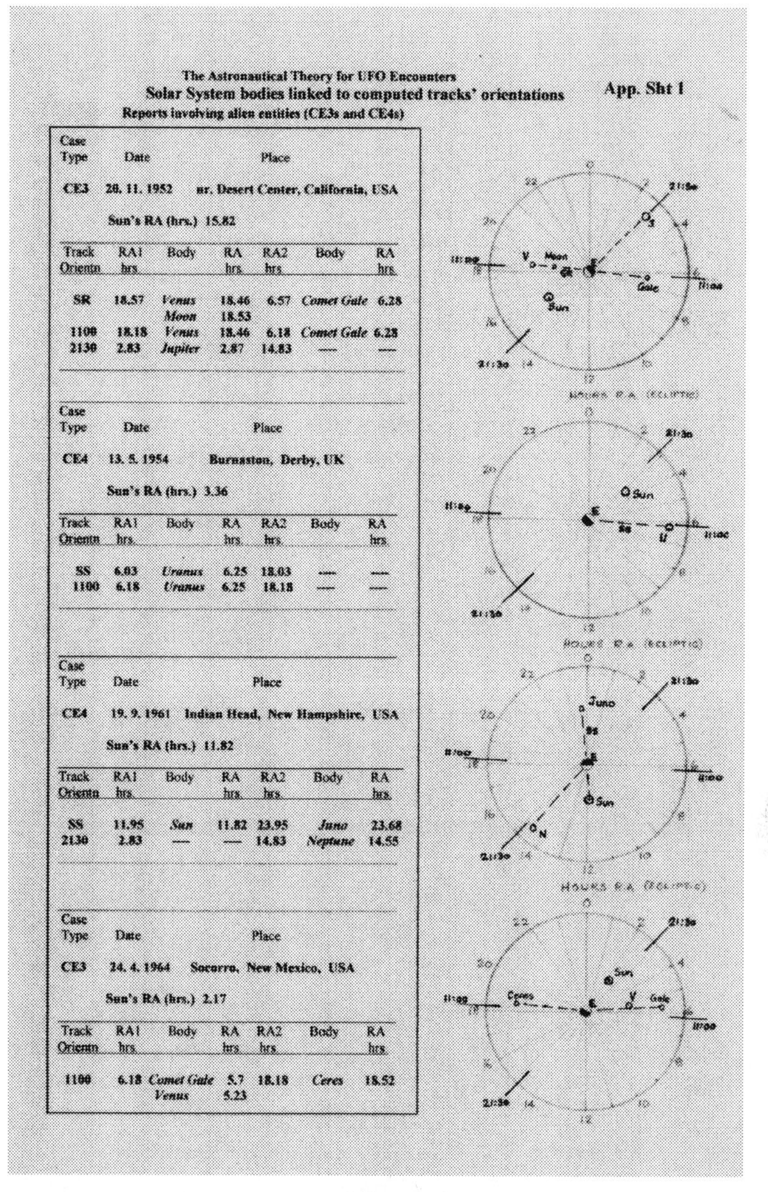

The Astronautical Theory for UFO Encounters

**Solar System bodies linked to computed tracks' orientations** **App. Sht 1**

Reports involving alien entities (CE3s and CE4s)

| Case Type | Date | Place |
|---|---|---|
| CE3 | 28. 11. 1952 | nr. Desert Center, California, USA |

Sun's RA (hrs.) 15.82

| Track Orientn | RA1 hrs | Body | RA hrs | RA2 hrs | Body | RA hrs |
|---|---|---|---|---|---|---|
| SR | 18.57 | *Venus* | 18.46 | 6.57 | *Comet Gale* | 6.28 |
| | | *Moon* | 18.53 | | | |
| 1100 | 18.18 | *Venus* | 18.46 | 6.18 | *Comet Gale* | 6.28 |
| 2130 | 2.83 | *Jupiter* | 2.87 | 14.83 | — | — |

| Case Type | Date | Place |
|---|---|---|
| CE4 | 13. 5. 1954 | Burnaston, Derby, UK |

Sun's RA (hrs.) 3.36

| Track Orientn | RA1 hrs | Body | RA hrs | RA2 hrs | Body | RA hrs |
|---|---|---|---|---|---|---|
| SS | 6.03 | *Uranus* | 6.25 | 18.03 | — | — |
| 1100 | 6.18 | *Uranus* | 6.25 | 18.18 | — | — |

| Case Type | Date | Place |
|---|---|---|
| CE4 | 19. 9. 1961 | Indian Head, New Hampshire, USA |

Sun's RA (hrs.) 11.82

| Track Orientn | RA1 hrs | Body | RA hrs | RA2 hrs | Body | RA hrs |
|---|---|---|---|---|---|---|
| SS | 11.95 | *Sun* | 11.82 | 23.95 | *Juno* | 23.68 |
| 2130 | 2.83 | — | — | 14.83 | *Neptune* | 14.55 |

| Case Type | Date | Place |
|---|---|---|
| CE3 | 24. 4. 1964 | Socorro, New Mexico, USA |

Sun's RA (hrs.) 2.17

| Track Orientn | RA1 hrs | Body | RA hrs | RA2 hrs | Body | RA hrs |
|---|---|---|---|---|---|---|
| 1100 | 6.18 | *Comet Gale* | 5.7 | 18.18 | *Ceres* | 18.52 |
| | | *Venus* | 5.23 | | | |

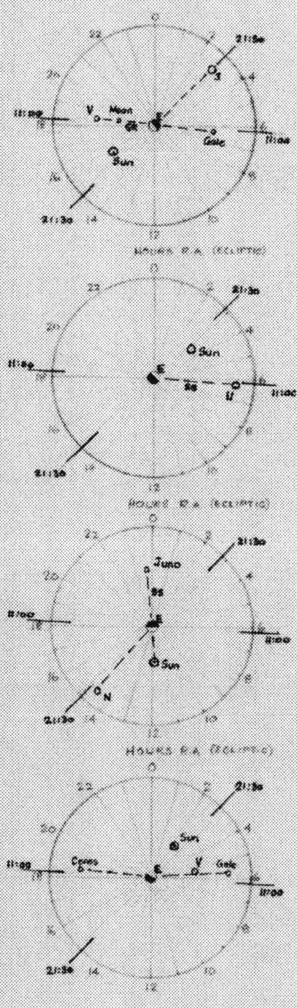

**Fig. 34**

The Adamski event is dealt with at the top of this extracted page. Quite significantly, ***the Sunrise track options were found to be aligned with Venus and the Moon, in close conjunction. The 21:30 hrs RA (fixed star option) path was aligned perfectly with Jupiter.*** As I had expressed in the main report, I had suspected that solar system bodies might be being used as sighting markers to aid navigation of the probes within the planetary system. The result obtained for the Adamski event perhaps now sheds new light on the visitor's selection of the orbit representing that of Venus. He could have been implying that he had travelled to the Earth via Venus, or with the aid of Venus as a marker. If so, it seems that his retrieval craft had had the option of returning via Venus or via Jupiter, which was situated in a different part of the sky.

This degree of accurate alignment applied to the majority of the cases considered by that Appendix, and this is demonstrated by the results for the other three cases shown by Fig. 34. I feel there has to be further research into this aspect of my studies, but until the astronomers take the rest of the work seriously, it is unlikely to be done. As far as Adamski's credibility is concerned, he made even more incredible claims throughout his lifetime, many of which seemed to me to have been delusions, but I cannot now deny that his original claims seem to have had substance.

## Eduard Meier and his Pleiadians/Plejarens.

A man who over the past thirty years has created even greater controversy than George Adamski is the Swiss backwoodsman, **Eduard (Billy) Meier**. I had largely discounted all I had heard about this man's outlandish claims until 2004, when I began to receive information about him by mail from a promoter, **Mr. Lawrence (Larry) Driscoll**, an American. On April 30th I received a package of information from Mr. Driscoll with a request for comments on the material enclosed. I was intrigued on several counts. Why had this unsolicited material been sent to me? How did Larry Driscoll get my details? Anyway, putting those questions aside, I decided to look through the material. It contained far more information about Meier and his encounters than I had been aware of before, but it all seemed still to be OTT (over-the -top).

I responded by e-mail within a few days and received a response from a rather surprised Driscoll. He'd forgotten how he had come to send the material to me or, even, where he'd found my mailing address. However, he'd been pleasantly surprised that I had bothered to read the material and had replied to it. Having a mind that's always open to new information, I consented to read through more of the material on offer. So began frequent e-mail dialogues between us, interspersed by more extracts through the post.

These are some of the things I discovered. Eduard Meier had been born in 1937, in Switzerland. He'd claimed to have been contacted throughout the period 1942 to 1953 by an extraterrestrial from the Pleiades with the name, Sfath. From 1953 to 1964 the contacts had continued, but, then, with another character from the DAL (?) universe, a female called Asket. But these contacts with extraterrestrials really 'took-off' on January 28th, 1975. Up to August 20th, 2004, it was being claimed that 664 personal contacts had occurred and that these had been supplemented by 854 telepathic contacts. Meier had obtained many daylight photographs of the **disc/dome** 'beamships' allegedly carrying his contacts and also had captured some of them on 8 mm. cinefilm. (A few years ago I had watched through a video recording of a Japanese film crew trying, unsuccessfully, to fault Meier's 8 mm. film records). Meier's colour photographs and films were all the more remarkable because he has only one arm, having lost the other in an accident at work some years earlier. It was being claimed that he had met more than 70 different ET contacts and a list of some of their names was supplied to me.

But Larry Driscoll's objective in sending all this material to me had been to obtain scientifically-informed comment on some of the 'revelations' Meier had received from his ETs. The first set of extracts were alleged dialogues between Meier and a young-looking Pleiadean woman, who gave her name as Semjase. The dialogues were presented as in the script of a play or a film. Semjase gave information about how the solar system had been transformed by a huge 'Destroyer' comet, allegedly responsible for, among other things, establishing our Moon and Venus in their respective orbits. This same comet had also been responsible for major cataclysms on the Earth in times past. She claimed that her people had succeeded in deflecting the comet into another orbit so that it was no longer a threat to Earth. The years and the nature of the cataclysms were given and even a value for the density of the 'Destroyer'.

On consideration of all this impressive information, I had to conclude that very little of it could be checked out by me and that some of it seemed implausible. Given the length of the dialogues, I couldn't help wondering how they were supposedly written down and asked Larry Driscoll for an explanation. The answer was that, after returning to his home, Meier had been able to recall them, verbatim. That, I just could not accept, in view of their complexity. But add to that the claim that Meier had been contacted frequently by telepathic means, the information put out by him just had to be suspect, in my view. Against this, Meier claimed that he had often been summoned to go out into the forest, by an inner impulse, when a rendezvous with a 'beamship' was in the offing and it seemed, from the amount of photographic evidence, he had usually gone to the right place at the right time. This thought prompted me to want to investigate such encounters using my AT techniques. I e-mailed Larry Driscoll and asked if he could supply me with a list of the contact events, together with the date and time of each, explaining my intentions. The initial response I received was that such an analysis might not be successful, because Meier had been informed that the Pleiadeans had created a number of Earth bases, one of which was alleged to be situated near Mont Blanc. Activity stemming from those bases would probably not be following the AT's rules. Nevertheless, Larry promised to try to obtain the requested information from his main source of information, someone associated with **F.I.G.U**, the founders of the **'Semjase-Silver-Star-Center', Schmidruti, Switzerland.** Soon afterwards I received another package in the post with information that included a list of 219 dated and timed encounters for the periods 1975 to 1984, 1989 to 1995 and three events from 2004. I had previously been supplied with maps showing the locations of the landing sites.

During the processing of these data, one item had to be eliminated because the date given for it was stated to be uncertain, so the total number of events processed was 218. The numerical processing, which compared predicted time with actual time for an event to within 1 minute of discrepancy, showed that 131 encounters (60%) had occurred within 20 minutes of predictions for arrivals from, and departures into, space, 163 (75%) were within 40 minutes, and of the remaining 56 events, only 19 (8.7%) were shown to have occurred greater than one hour removed from the predictions. Of the 17 listed photographic sessions, 12 (70%) had occurred within 20 minutes of predictions, 2 were within 40 minutes and for the remaining 3 the discrepancies were greater than one hour. In view of my reservations about how accurately Meier had recorded the times of his encounters, whether they had corresponded to the beginning, end or middle of his prolonged sessions and whether they had been listed correctly, this was quite an impressive result. A marked-up timing graph, for the area in which the encounters were alleged to have occurred, confirmed these results visually. This is shown by Fig. 35.

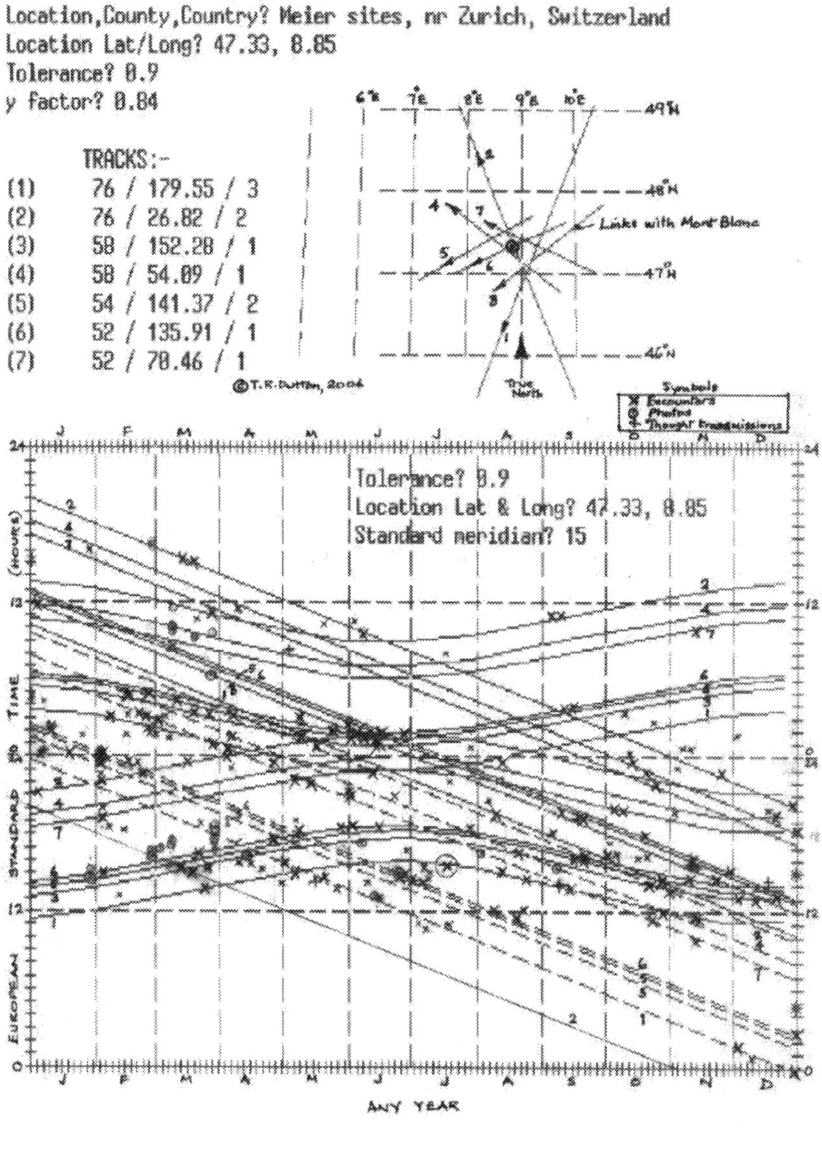

Location,County,Country? Meier sites, nr Zurich, Switzerland
Location Lat/Long? 47.33, 8.85
Tolerance? 0.9
y factor? 0.84
TRACKS:-
(1) 76 / 179.55 / 3
(2) 76 / 26.82 / 2
(3) 58 / 152.28 / 1
(4) 58 / 54.09 / 1
(5) 54 / 141.37 / 2
(6) 52 / 135.91 / 1
(7) 52 / 78.46 / 1
Tolerance? 0.9
Location Lat & Long? 47.33, 8.85
Standard meridian? 15
ANY YEAR


**Fig. 35**

I was now faced with a dilemma similar to the one I'd encountered when considering the Adamski claims. I had found the 'revelations' of Meier to be generally without checkable valid substance, but then had discovered very strong links with the AT. How was I to explain this? One thing I could do was to check whether the AT indicated there might be a link between the Hinwil area of Switzerland and Mont Blanc. Sure enough, Track No. 3 serving the Meier sites also served Mont Blanc and, some 27 miles (44 km.) south-west of the mountain peak, formed one track of a four-tracks intersection. Could that intersection be the location of the alleged base? Perhaps someone ought to investigate that by surveying the land in that area? And, although I have received no UFO reports from that region, perhaps the local people know it to be a hot-spot?

Meier also alleged that Semjase had told him more about her home planet (Erra, some 80 light years beyond the Pleiades cluster) and the means by which she and her people travelled to Earth. The original technique, which had since been superseded, entailed accelerating to near light speed for seven Earth hours. At that point in the journey, they were then able to transfer instantly to any other chosen location. The 'beamships' witnessed by Meier were also able to transfer in the blink of an eye from one place to another. If Meier's film records are genuine, this ability was demonstrated to him on several occasions. Semjase is also claimed to have told Meier that the seven hour acceleration phase on the outward and return trips from and to Erra were no longer necessary. In view of my earlier observations that we are dealing with technology far in advance of our own, all that may or may not be true.

Here is another mind-boggling claim. Semjase told Meier that they are **humans** like ourselves and that the other contacts from the DAL universe (which seems to be portrayed as a parallel universe, co-existing with our own) are also of the same stock. It is alleged that we humans are the results of occupation of this planet by her people aeons ago. (Larry Driscoll accepts this as an unquestionable fact and will not allow the use of 'humanoids' or 'human-like' to describe Meier's contacts to go unchallenged). So, now, readers of this book must judge for themselves whether there could be any truth in Eduard Meier's CE3s. For me, despite the impressive AT results, the jury is still out.

## References

[13] Pugh, R. Jones 'The Dyfed Enigma' (book) Holliday, F.W. Faber & Faber, 1979 / Coronet Books, London. 1981.

[14] Paget, Peter'The Welsh Triangle' (book) Panther Books, 1979. (Granada Publishing, Ltd), London

[15] Leslie, D Adamski, G. 'Flying Saucers have Landed' (book)Neville Spearman, Ltd., 1953, 1970 Futura Publications, Ltd., London. 1977, 1978.

# PHASE 5:
# Studies of Lost Time Events

*In this phase of the study I took a special interest in alien abduction (CE4) stories I had not taken very seriously until the late 1980s. Why I regard them very seriously, now, will be revealed in the following chapter.*

## CHAPTER 15

## CLOSE ENCOUNTERS OF THE $4^{TH}$ KIND

Until my BUFORA lecture in London during May 1987, I had paid little attention to stories of alleged abductions by ETs. Only the Barney and Betty Hill story had made any imprint on my mind, as far as I can recall. But my thinking came to be very much changed when the late Mr. Ken Phillips, the Chairman during my lecture and my overnight host, brought to my attention such an event in my area of the country. I was preparing to leave for my train back to Cheshire when Ken mentioned this to me. He asked how the date and time details might fit in with the AT's predictions for East Didsbury, Manchester. As I happened to have a hand-produced graph for the Manchester area among my documents, I brought it out onto the table and we studied it. I told Ken that the time he had given me was about 1 hour earlier than the indicated nearest time on the graph. As I left him, Ken said he would check out the time with the witness and let me know if he'd got it wrong. On arriving back home several hours later, I was just in the process of removing my coat when the telephone rang. The caller gave her name as Mrs. Linda Jones and then, obviously excited, she asked, "How did you know the time Ken gave you was wrong? It was an hour later!" I tried to give a quick explanation over the 'phone but Mrs. Jones wanted to know more and invited me to visit her and her family at home, so a time was arranged. When I arrived, I was met at the door by Mrs. Jones and three huge dogs. Following a cordial welcome, the dogs meekly backed off and then sat quietly as I sat down in the lounge. Mrs. Jones marvelled at the dogs' acceptance of me, because they usually went mad at first when strangers called. The story then told to me by Mrs. Jones was impressive. The event had occurred on the evening of August 19th, 1979, and had involved herself, her young teenage daughter and her (nearly) five years' old son. This is the story of the event that had left them traumatised and intrigued.

## Encounter and Lost Time in East Didsbury, Manchester.

On that fine summer's evening in 1979, Linda Jones and her children had gone on a nature walk, before sunset, alongside the newly canalised River Mersey behind their home. The disturbed river bank area had been in-filled and had become a flourishing natural grassland and meadow. The path they had walked is set below the meadow, so, it would have been separated from them by a high grassy embankment. Linda had met a friend on their walk and they had continued walking and talking. When the friend crossed a bridge to return home, the little party then, also, turned for home. But, by then, the sun had set and twilight had descended upon them. Suddenly, they looked up to see a large elliptical object, glowing flame orange-red, hurtling out of the sky towards them from across the river. They had all expected to be killed by this huge missile and threw themselves into the long grass on the embankment in the desperate hope of escaping it. Next, they were surprised by the fact they were still alive and by the lack of an explosion. Looking up the embankment they had discovered why. The object was by then hovering over the meadow above them. As they watched, it slowly descended out of sight. Gathering courage, they had scrambled up the embankment to view it and were met with a very strange sight. Hovering only a few feet from the ground was a shape that reminded Mrs. Jones of a gondola. It had a curved 'basketwork' underside, glowing red, which had an upward-mounted pole on the right-hand end. Surmounting the pole was a flashing amber light. The upper part of the craft they were unable to see because a brilliant white light obscured it. Linda's curiosity caused her to want to get closer but the frightened children were keen only to go home. As they'd turned for home, some even stranger happenings had begun. The long grass was being blown about and they'd been confronted by a group of men in blue coveralls. These men had seemed to be floating over the ground. The next thing they recalled was that they were all running home in darkness. As they reached the gate of the meadow, some ten minutes later, they turned towards the encounter site and saw a brilliant light lift off skywards. After first tagging on behind an aircraft approaching Manchester Airport, the light hurtled off into the sky at high speed. When they arrived back home about two minutes later, they found that the time was by then 10:15 pm BST. A concerned Trevor (Linda's husband), who had just arrived back home himself, asked them why they were home so late. After telling their incredible story, they realised that they had no recollection of where they had been for an hour or so. In fact, they had appeared to have suffered the kind of amnesia claimed by the Hills in America, though, at that time, UFO reports had made no impact on them and they'd had no knowledge of that kind.

I was taken along the footpath they had walked and shown the location where the object had been seen hovering. Its approach path seemed to have been over a long, recently built, motorway bridge, the construction of which had necessitated the re-channelling of the River Mersey in that area. In my view, the area had many of the features favoured for UFO activity. (See Phase 1)

When I showed the timings graph for the Greater Manchester area to Mrs. Jones and explained that their event had seemed to fit in with the global pattern of UFO events, she was overjoyed. She told me that, among the many investigators she'd had to relate the story to, I was the only one who had been able to offer any sort of explanation. Some had just dismissed the story out of hand. In effect, all I had done was to give some confirmation of their observed UFO event, but I could not, at that time, explain the lost hour. That was done later, when I produced a computerised graph targeted on East Didsbury. This is shown as Fig.36 .

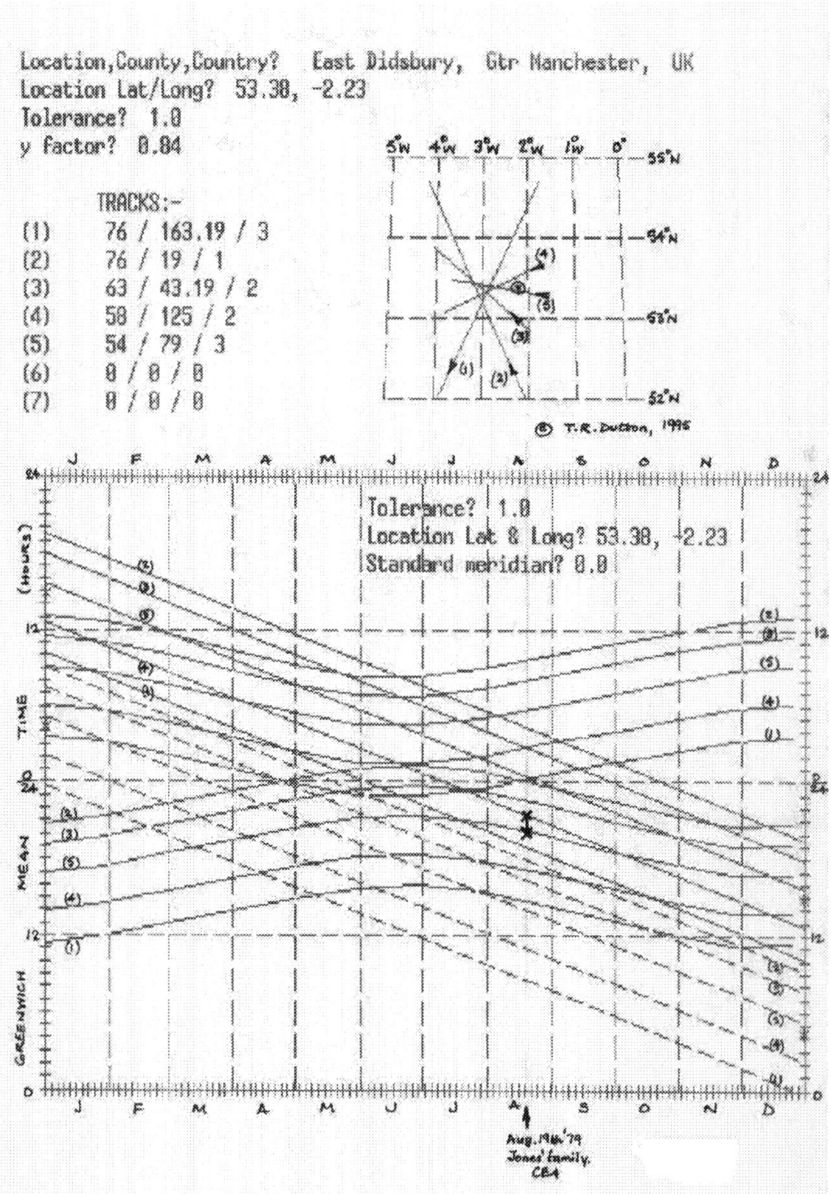


**Fig. 36**

The two diagonal crosses on the timings graph mark the significant times associated with the Jones event. From the time of sunset on that evening in August and the fact that the family group had begun to return home in the deepening twilight suggested to me that the object had descended from the sky at about 9 o'clock BST, that is, 20:00hrs GMT, as depicted by the lower cross. They had seen a brilliant light lift off from the location and head skywards at just before 10:15 pm. BST, say at 21:13 hrs. GMT, and this time is marked by the upper cross. Well, Well, Well!!!

The computed graph (produced by the globally-derived AT) just happened to have produced two lines passing through those points. The interpretation suggested is that an exploration craft, delivered from a delivery spacecraft travelling temporarily along sunset-orientated path No.5, had landed in that East Didsbury meadow and had remained there for over an hour to await the next retrieval opportunity. This had occurred when a retrieval spacecraft had overflown the area on the star-orientated (2130hrs RA) path No.4. If reference is made to the Latitude/Longitude grid map above the timings graph, we can see that paths Nos. 4 and 5 both provide ready access to that East Didsbury location. That particular arrangement of path (or track) lines was of great interest to me, because *I had noticed that Close Encounter events of all kinds had seemed to favour (but not exclusively) a combination of sun-orientated and star-orientated paths to determine the time allowed for a given mission.*

When Mrs. Jones was shown this later graph, she was doubly excited by it, because not only was her account being validated, but an explanation for the family's lost time had been suggested. It looked as though I had provided tangible indication that they had been detained by the occupants of that craft, even though Linda was personally inclined to disbelieve that suggestion. Others had put forward the same idea and she had been very disturbed by it. As I was soon to discover, one of those 'others' had been a Manchester solicitor (lawyer) Mr. Harry Harris, whom I was invited to meet at the Jones' home one evening. I was told that Mr. Harris had been arranging hypnotic regression sessions with medically-qualified hypnotists for UFO witnesses who had had similar 'memory-loss' experiences.

At that first meeting with Harry Harris at the Jones' home, he was accompanied by fellow-researcher Mrs. Linda Taylor. During the introductions, I learned that Mrs. Taylor had had a UFO-related time-loss encounter during 1982 and that she had consented to hypnotic regression arranged by Harry Harris. She had been so impressed that she had wanted to participate in Harry's future investigations. Both she and Harry had been intrigued when Mrs. Jones had informed them of my participation in the investigation of her experiences and they had requested a meeting with me. The discussions we had then revealed that Linda Jones and her daughter had both been hypnotically-regressed, on Harry's suggestion, (the boy being considered to be too young for that process). As usual, Harry had videoed the sessions. The outcome had been traumatising for both Linda and her daughter. Under hypnosis they had, individually, given similar accounts of all that had happened after their initial encounter with the SAC, including physical examinations. They had been disinclined to believe their recorded accounts, but, as I discovered later, they had been given other reasons to believe them. It seemed, also, that my discoveries had inadvertently reinforced their worst fears. (Linda's daughter was not present on that evening, nor at any other meeting with the Jones'. In fact, I have never met her.) The outcome of that meeting was that I was invited to join in Harry's activities..

Through this link-up I learned, first of all, the details of Mrs. Linda Taylor's encounter and subsequent regression and, later, of several other similar events.

## An Interrupted Journey on the A580 'East-Lancs' Road.

On January 10th, 1982, Mrs. Linda Taylor and her mother were returning from a visit to a relative living in Southport, a seaside resort on the west coast of Lancashire. The night was cold and moonlit (there was a full moon) and snow lay on the fields. Their route home towards Manchester took them onto the A580 trunk road at a junction near St. Helens. This was to have been a journey they had made many times before, but that particular journey was to be anything but routine. The first indication had been the sighting of a bright light apparently dodging through the trees on one side of the road. Then this light had suddenly disappeared. In the vicinity of a road junction to the west of Leigh, Linda's Ford Cortina car had begun to behave oddly. It had seemed to be losing power and resisting all Linda's attempts to compensate for this. Speed dropped off, the car seemed to be bouncing up and down and all the electrics began to go berserk. Suddenly, the rear end of a large black limousine of a 1930s variety had appeared directly ahead of the Cortina, so close that it seemed that a collision would be inevitable. Even Linda's mother sitting in the rear seat shouted out in alarm and, having no means of controlling her car's behaviour, Linda tried unsuccessfully to sound the horn. By then frantic, she had opened the driver's window to shout at the car in front and in so doing had seen a huge craft ("as big as a double-decker bus on its side") directly above her car. Its underside displayed large coloured lights and the glow from them lit up structure above them. Linda had quickly pulled her head back into the car and, to add to her utter confusion, the car in front had disappeared. As the Cortina had then seemed to be behaving normally again, it was driven frantically to the haven of a filling station, a short distance away. On arriving there, Linda and her mother saw a brightly lit craft circle the station and then hover over a tall tree on the opposite side of the road. It suddenly shuddered before accelerating vertically upwards at an unbelievable rate and disappearing from sight. Unfortunately, there was no one else to witness the departure of the SAC. After arriving at her mother's home they both realised that the journey had taken about one and half hours longer than usual. On arriving back at her home in north-east Cheshire, the door had been opened to the still-shaking Linda by her husband, who thought she must have been involved in an accident. He had noticed also that she was not wearing her winter coat and had asked where it was. Linda had been unable to give an answer. She'd had no recollection of having taken it off !

What a story! It involved a bright light moving through trees, a series of car problems like those encountered with car-stop events, a close encounter with a structured craft, a phantom car dating back to the days when the A580 had first been opened and time loss. Mrs. Taylor had not been impressed by the outcome of her hypnotic regression session. She felt it seemed more like a dream sequence than a real series of events. I offered to try to process the facts to see if there had been any connections with the established global patterns for SAC encounters.

Two essential facts needed to be established ---- the location and the approximate time of the happening. As the location appeared to have been within about 3 miles from the town of Leigh, Lancashire, the location of Leigh would be close enough for the initial investigation. As the journey had been made many times before, Mrs. Taylor was able to give an estimate for the initial time of the event based on the time of departure from Southport. She estimated that 7:30 pm. GMT would have been about right.

Given these approximations I was able to proceed. Fig.37 shows the computer- produced output, targeted on Leigh. Superimposed on the timings graph at January 10$^{th}$ is a cross marked at 7:30 pm. It lies very close to the timing line for a No.4 sunset track. If it is now assumed that 1.5 hours were lost during the encounter, this creates another time of 9 pm. for the close of the happening. As can be seen, this cross lies on the timing line for sunset track No.2, which represents the next possible retrieval track (or path) option in space. Once again, the indications are that the time allocated for the SAC's mission had been determined in that way. From the latitude/longitude grid map it can be seen that No.2 and No.4 paths are well placed for this happening to the west of Leigh.

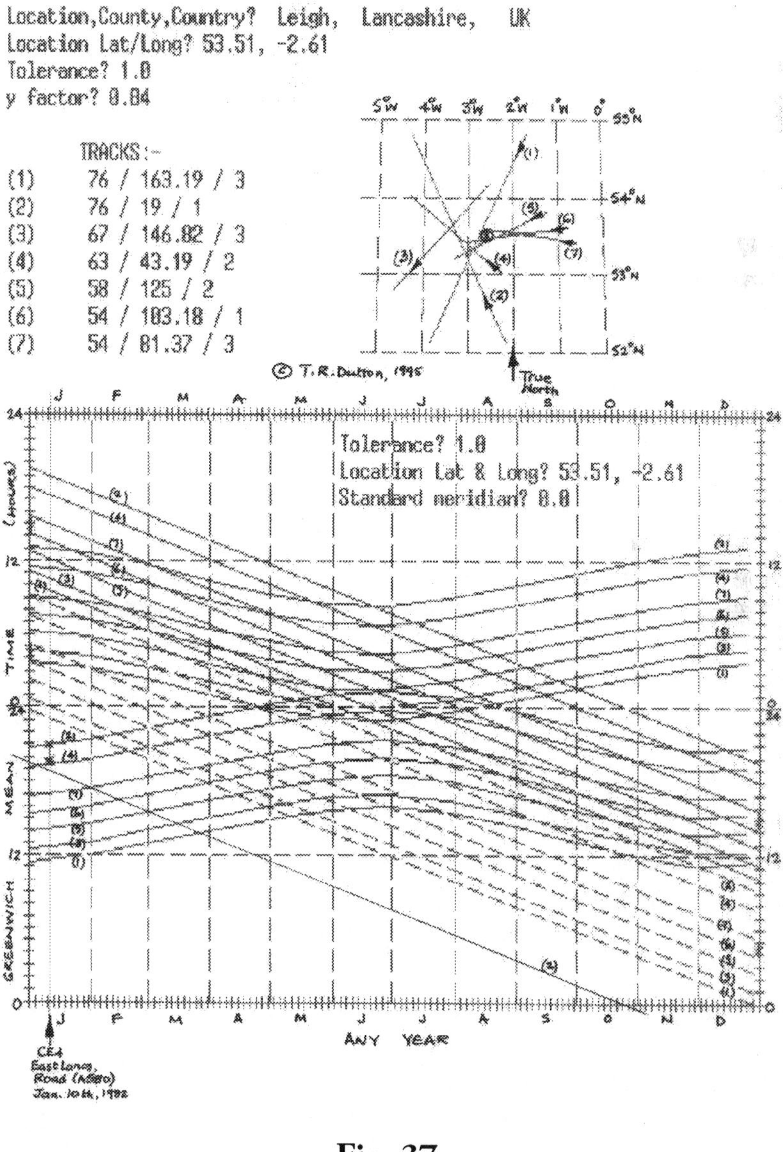


**Fig. 37**

Through my continued association with Harry Harris, I learned about other CE4s he had investigated using hypnotic regression. The first one, if I remember correctly, had involved PC Alan Godfrey, a police patrol car driver who had encountered a large hovering object blocking the road ahead when he was on his way to investigate a report of a herd of cows wandering round a housing estate in Todmorden, a small Yorkshire town on the border with Lancashire. He braked, brought the car to a halt and tried to contact his HQ by radio. He found he could not even radio to his colleagues. As he sat watching the SAC, he decided to try to sketch it on his notepad. The next thing he'd been able to remember was driving further along the same road towards the Police Station. On his arrival there he reported the incident and noticed he'd lost track of about half-an-hour. This event had been, subsequently, widely reported by the local press and the publicity had resulted in an invitation to PC Godfrey from Harry Harris to participate in a hypnotic regression session. (Harry told me, later, that he had been inspired to do this because his own wife had been regressed by a dentist friend, and she had been able to remember incidents in her childhood she had consciously forgotten. These forgotten incidents had then been confirmed by her mother.) The revelations given by Alan Godfrey were to become familiar aspects of subsequent ones given by other CE4 participants whilst in hypnotic trance. Typically, they were a bright light, a floating sensation, finding themselves in a strange room containing a surgical bed or couch, meeting with humanoid creatures (some of them very human-looking), a voice in the head telling them to lie on the bed, initial refusal then meek compliance, then being attached by cables to electronic equipment with arrays of coloured lights, being physically examined and, finally, finding themselves wherever they were when the encounter began. (In Britain, the creatures encountered have been generally described as being human-like, in contrast with those described by many American participants.) My processing of Alan Godfrey's event was able to confirm the object's departure time. The arrival time was uncertain because the object seemed to have been the one which police officers had been chasing round the adjoining moorland for several hours and which had last been seen dropping down towards Todmorden just prior to the Godfrey encounter.

After trust had been built up with Harry Harris he invited me to meet two other participants who, with a third person, had experienced a CE4 during the summer of 1981. These three young women had been regressed and had revealed some remarkable new aspects of the abduction phenomenon. One of them had since emigrated. Harry had contacted the remaining two, suggesting that another regression using a different technique might be helpful. They had responded and requested a meeting at Atcham, Shropshire. Harry drove Linda Taylor and I to meet these young women, Valerie and Rosemary, at a public house (inn) near to the scene of the event. I was told about the circumstances of the happening, which had occurred just a few miles further along the road. Then, we all went to the site. En route I noticed we were passing a restored Roman settlement called Wroxeter and commented about it. It reminded me of the link I had discovered between UFO sightings and ancient sites. Valerie said that when she had first sighted the lights in the sky they would have probably been over Wroxeter. The full story, as well as I now know it, follows.

## An Unforgettable Night Out in Shropshire.

Valerie, Rosemary and their friend Vivienne had developed a routine of visiting a night club in Shrewsbury every Wednesday night for light relief from their daily work chores. On that particular night of July $15^{th}/16^{th}$ , 1981, they had not found the entertainment to be very much to their liking and had decided to leave a bit earlier than usual. They had not been drinking alcohol. Prior to their leaving, they had been approached by two young men wearing white clothes and shoes, who had seemed to want to become better acquainted, but these 'cricketers' had received no encouragement. (The women had thought the men were dressed very inappropriately and later wondered if they had been linked in some way with the subsequent CE4 event.)

It was a fine night and the moon was almost full. As they drove back towards their hometown of Telford in the early hours, with Vivienne driving, Rosemary in the front passenger seat and Valerie in the rear seat, Valerie said she could hear a high-pitched whistling noise. Looking out of the side window, she saw a group of bright lights in the sky which seemed to moving in their direction. She brought these lights to the attention of the others then, suddenly, the lights disappeared. On rounding the next bend in the road they found their way blocked by a craft of some kind and the car came to a halt. The next thing they remembered consciously was that the craft had gone and Vivienne had then driven them to the first Police Station they'd encountered. As they looked at the clock, on walking into the Police Station, they realised that they were now running half an hour or so later than usual, since they knew the time it usually took for them travel from Shrewsbury to that location. They told their story to the policeman on duty and then headed for home.

After the event they were all troubled, especially during sleep. Eventually, Harry Harris came to know of their encounter and time loss and offered his services. The women had agreed to regression therapy and were regressed individually by a medically-qualified hypnotist. As usual, Harry Harris had paid the substantial fees for these sessions. His rewards would be in helping the CE4 victims to resolve their difficulties and in the creation of a collection of unique video recordings of the proceedings.

Having acquainted me with the two young women, Harry then invited me to be an observer during their next regression session. Of course, I was only too keen to accept that opportunity.

## Hypnotic Regression Sessions.

### No.1

The background story I was given told me that, during the early regression sessions, Valerie had been able to give a lucid and coherent account of events that followed the car-stop event, whereas Rosemary had become very stressed as she re-lived the early stages of the happening. The hypnotherapist had wisely told her to sleep and had ceased to question her. As related earlier, at the time when I became involved, Harry had contacted the two women and suggested that, perhaps, it might help Rosemary if the two would consent to be regressed together. It was being suggested that, perhaps, Rosemary could be persuaded to live through her remembered trauma if she were to be able to sit next to Valerie in a dual hypnosis session. Harry contacted a GP doctor hypnotist and the doctor was happy to attempt this unusual arrangement. I was invited to be present as an observer.

The session took place in the Harris' lounge one Sunday afternoon. The doctor asked to be left alone with the two subjects as this would facilitate the achievement of rapid trance-states. We observers, including Harry, retired to the adjacent dining room and waited for our call. When called, we moved quietly to our seats facing the settee on which Valerie and Rosemary sat with eyes closed and looking very relaxed. Harry took up his position beside the tripod-mounted camcorder and the doctor sat to our left, close to the settee, with a stethoscope hanging from his neck.

Before the session, Harry told me that he had told the doctor that the two women had suffered a traumatic experience some years ago, when they were in their friend's car, and that they could not recall all that happened, and that Rosemary had panicked during an earlier attempt to regress her. The doctor had not been told anything about the nature of the trauma, so that the manner of his questioning would be similar to that he would use when providing therapy for, say, a terrible car crash.

The session opened with exploratory questions from the doctor about the circumstances of their trip in the car that night in July, 1981. Gradually, the story of a craft blocking the road ahead unfolded. Then, we had accounts from both women of small creatures approaching the immobilised car and how they had approached the nearside passenger door. Then it was revealed that the door had been opened from the outside and something or someone was trying to drag Rosemary from the car. Rosemary began to get extremely excited and began resisting these attempts. She became so agitated that the doctor walked over to her, took her pulse and decided it was not safe to continue. He ordered Rosemary to sleep and asked her no further questions. He resumed his questioning of Valerie.

According to Valerie's account, driver Vivienne sat there staring straight ahead and motionless. Then, in the moonlight, she could see small human-like creatures coming towards the car and approaching the passenger's door. The door was opened and Rosemary screamed that they were trying to pull her out and was resisting furiously. Then a black mist enveloped her and, in the blink of Valerie's eye, she was gone!

Valerie then remembered climbing out of the car, leaving Vivienne still comatose, and walking through an open gate into a nearby field, feeling completely bewildered. She felt a pair of hands placed on her shoulders from the rear and a strong masculine monotone voice said to her, "Don't be afraid". (Valerie imitated that voice very convincingly). A remarkable description of the effects of high vertical acceleration and deceleration then followed. She said that everything had gone black but had then turned red before she had found herself standing in a well-lit room with two human-like creatures dressed in green. One seemed to be a woman, being smaller than the other, who seemed to be more like a man. She described how the 'woman' began hobbling around in Valerie's high-heeled shoes as if to try to amuse her. When asked what happened next she said she was on a bed.

If I recall correctly, this is when the doctor decided to terminate Valerie's regression, because she was becoming tired. (Harry Harris's video recording, which I have no access to, will be able to validate my account.) Before the doctor left, I had a quiet word with him and asked him for his assessment of all he'd heard. His response was that undoubtedly these two young women had suffered great trauma, but he could not try to guess the nature of it.

So ended the first hypnotic regression session I had ever attended. Given the stated objectives, the outcome had been disappointing, but it had served to demonstrate to me that these participants had been genuinely re-living a real event in their lives. There had been no leading questions from the doctor. He had allowed the women to guide him to the next question each time they had answered. It had occurred to me that the doctor might very well have experienced a profound 'culture-shock' himself, when he had realised the nature of the trauma he was delving into.

Rosemary was extremely disappointed that she had not been allowed to proceed further into her remembrance of the event, but accepted it had been medically inadvisable.

Some time later, I received a call from Harry Harris telling me that the women had requested another dual session with their original hypnotherapist, because they felt they'd be more at ease with him than they'd been with the methodical and clinical approach of the doctor. Would I be interested in sharing the cost of such a session with him? After he'd told me what my share of the cost would be, I asked him to proceed with the arrangements, provided I could view the video afterwards, for analysis.

**No.2**

The next session was arranged for the afternoon of March 24th, 1991, and the general procedure adopted was as it had been with the doctor. However, this therapist had already gained an awareness of the nature of the event and, also, experience of Rosemary's problem. As previously, the therapist put the two women under trance in private and the observers were then invited in when this had been accomplished. The layout of the room was as before, with the therapist sitting close to the settee and on our left. The first questions were asked by the hypnotist and thereafter by Elaine, Rosemary's sister, who continued with the questioning throughout the session. Harry Harris was again operating the camcorder. The account of the No. 2 session, given below, has been taken from my word-for-word written record of the contents of Harry's video recording of that session, of which, later, I received a copy.

At the point in the story when Rosemary was resisting attempts to extract her from the car, she and Valerie became very agitated. The therapist, in a reassuring and commanding voice, assured them both that what they were remembering was not happening now and that they must imagine themselves looking at it all from the safety of thick coverings of warm lead. The effect of this command on Rosemary's demeanour was dramatic. She became calm and collected immediately and began to open up her mind to us. After going through the fight with the entities trying to remove her from the car and being enveloped by a black mist, she said she felt very strange, couldn't breathe and didn't know where she was. The next thing she remembered was finding herself in a strange white room, with white lights, opening into a double (? inaudible) and there was just a table in the room.

Elaine then switched her attention to Valerie and asked her to tell her story of the events. After re-telling the story of the events in the car and her experience in the field, she then gave a more informative account of the being lifted at high speed and everything going red and black. She described the experience as being "sucked up a vacuum". She couldn't get her breath. It was like being in a lift. Then she found herself in a white wedge-shaped room. There were no objects in the room and she couldn't see any doors. The walls were like white metal. She could hear shuffling. A man and a woman were then in the room, just looking at her. She asked them where Rosemary was and they replied that Rosemary was being looked at and that Vivienne was being examined. Then we heard, "They're now touching my clothes and my hair". They hadn't told her why Vivienne was being examined, but the woman was by then hobbling around in her (Valerie's) shoes, as if trying to make her laugh.

At this point, Elaine again switched her questioning back to Rosemary and asked her to continue her story of being in a white room. Rosemary said she was very frightened. She was standing by a wall, not daring to move. Then she told us that she was getting onto a table, because "I just know in my head, that I have to. I don't want to, but I must". She could hear a shuffling sound. Round heads with no faces --- like robots --- appeared all around her. There were six of them. They were about three feet tall and were on little rollers or castors. They were looking at her and she was looking at them. She was not frightened because she knew they were not going to hurt her. Next, she was getting off the table and had to go with them ----- through the door and down a long white corridor with bright lights. There were doors and openings along the corridor. She could see strange things --- machinery. She mustn't describe this ,because it would betray them. She then entered a small room, narrow, with doors and windows, in which there were two beds and she thought these were empty. In an oblong-shaped room, she could see Valerie. [Rosemary told us shortly afterwards that she had floated along that corridor, horizontally, in the company of the six little robots. She had then been taken into the oblong room and placed on a bed, next to Valerie, who was asleep.]

The questioning was then directed at Valerie. After the shoes incident, Valerie told us she was being told to go to sleep. Then ---- she was with Rosemary, but she didn't know which room she was in. Rosemary took up the story. After the robots left the room, she felt immobilised, but realised that Valerie was conscious again and seemed to know she was there. This made her feel better, but she was concerned there was no sign of Vivienne.

The two women, between them and in accord, then proceeded to describe the subsequent happenings. A tall dark man with dark eyes and long hair, wearing a long green gown, had walked into the room. He had placed one hand on Rosemary's left arm and the other on Valerie's right arm. Then began an account of a process of telepathic communication being established between the two women, accompanied by an overwhelming sense of wellbeing. They felt 'at one' with this man. When asked why they thought this was happening to them, they replied that he was trying to demonstrate that this form of communication was possible. Rosemary described the experience as a kind of ecstasy. Asked why Vivienne was not being included in all this, Valerie responded, "It's part of his plan". Rosemary added that Vivienne wasn't strong enough to cope with it. *They were not the only ones this had happened to and the purpose was one of learning. Valerie interjected, "That we can all be one". Rosemary went on to say she thought it was a two-way learning process.*

Then followed a most unexpected and, for me, the most exciting revelation of the session. When asked if the creatures had given them any information about where they had come from, the first response from Rosemary was a 'brush-off', "You wouldn't understand". Pressed further by Elaine, she went on to say they had been shown something like a map, but she couldn't draw it. Not like a road map. It was difficult to describe, but it was like a black window with shapes on it. Valerie again interjected, "They're teaching us". [At this point the hypnotherapist suddenly interrupted their flow by commanding the women to remember every detail of that map so that they would be able to draw it for us afterwards.]

Still in trance, after telling us that this was not the first time they had been taken and that they felt they were falling behind in the learning process, the two women then described how they had been floated from their beds and, still horizontal, they had entered two small side-by-side tunnels. They could see light at the end of their respective tunnels and could hear a faint buzzing or burring sound. They were moving towards the light. The man they had been with previously was waiting for them as they emerged from the tunnels. They could smell disinfectant, like in hospitals. Throughout all this they were wearing their summer clothes. In response to questions about the nature of the tunnels, we were told the machine didn't need a motor. They had to go through the tunnels so that they could "go on". The process had been carried out "for them (meaning the perpetrators) to use us". *They were unable to influence humans, who had to come to a realisation first. Lots of other people had gone through these experiences and they would know when the time was right.* Then, they emerged from their tunnels and were out, lying on two tables. They were given sweet, medicinal, liquid to drink ("about half a cupful"), "to counteract the effect of the machine". Rosemary observed that there were now three people with them, but, when asked who they were, she said she would have to confer with Valerie before answering. She then placed her right hand on Valerie's left arm and became silent. Soon after, Rosemary became very agitated and the hypnotherapist decided to close the No.2 session.

On being released from their trance, Rosemary and Valerie looked at each other and then burst out laughing in a rather embarrassed way. They were finding it hard to believe what they could remember describing and saying during the hypnosis.

Where had all that come from? After they had had time to recover from the ordeal, a full discussion with all present in the room developed, with Rosemary participating very actively. At some point, they were asked if they felt able to draw that map they had been shown. They turned to each other, as if hoping for inspiration. Rosemary became very exercised by the question. It was like a window in a wall with blackness beyond. In the blackness were white things, but they seemed to be represented three-dimensionally. It had been difficult to determine which of these collections of items had been in the distance and which ones were closest, because the scale had seemed to be variable. Valerie was making noises of assent during Rosemary's attempts to describe a very complicated scenario. (All this was beginning to arouse great interest in the rest of us, especially me). "Was it like a hologram?", someone asked. Yes, it was something like that. *Then, after a pause, Rosemary 'made my day'. She said, leaning back, that the nearest thing she had seen, that was like it, was what you see in a planetarium.*

Of course, we were all disappointed that the session had had to be wound up before it had revealed the sequence of events linking the emergence from those tunnels, being shown the map and being returned to the car. It was agreed we would have another session with the same hypnotherapist at a mutually convenient time in the near future. This turned out to be one month later, in April, 1991.

**No. 3**

We assembled again at the Harris home on the afternoon of April 24th, 1991. As before, the observers present, besides myself, were Harry Harris, Linda Taylor, Elaine (Rosemary's sister) and the hypnotherapist. This session had been given the objective of trying to fill in the still unknown elements of the CE4. Each of the previous hypnosis sessions had lasted for just over an hour, after which time the two subjects had been deemed by the hypnotherapists to be too stressed to proceed further. For that reason, it was agreed that the questioning on this occasion should commence by asking Rosemary and Valerie to tell us of their arrival in those strange rooms and then proceed, as quickly as possible, to the time they were first aware of each other's presence. It was our collective hope that they would recall passing through those 'tunnels' and then be able to move into the next stages of their forgotten experiences.

Elaine, Linda and Harry had all asked the hypnotist to open up the entranced women's minds to receive questions from them. I was content, at outset, to continue in my role as an observer, but I had come equipped with an A4 drawing pad and a sharpened pencil, just in case an opportunity arose to request drawings of that intriguing map.

No problem was experienced in re-establishing the women's minds within those strange rooms, remembering the awareness of their first meeting within one of them and bringing them to tell us about those tunnels. But, thereafter, with several questioners beginning to participate, things went very wrong. Both Valerie and Rosemary became confused and stopped answering the questions. In fact, they flatly refused to divulge any more information. As I had my transcript of the No.2 session on my knee, against which I was checking out their answers on this occasion, I knew that they were now withholding information they had given freely before. When all other attempts to break into the wall of silence they had created seemed to have failed, I asked the hypnotist if he would introduce me to them, as a questioner. This he was happy to do and, when I spoke to Valerie and Rosemary, they seemed to accept me. I began by pointing out that I had made a transcript of all they had said in the previous session and I was puzzled to know why they were now unprepared to tell us things they had told us previously. Rosemary's quiet response was curt and to the point --- "You haven't asked the right questions". Realising that perhaps the eagerness of the previous questioners to move to scenes beyond those tunnels had created their confusion by, effectively, putting their memory processes into closed loops, I decided to break them out of those loops by asking a different type of question. Did they remember telling us about a map they'd been shown? "Yes", said Rosemary. I had brought with me a drawing pad and pencil. Would they try to draw what they could still remember of that map for me? Another affirmative answer was received from Rosemary. The hypnotist stepped forward and handed the pad and the pencil to Rosemary, instructing her to open her eyes and to draw what she could remember of that map.

We all held our breath in silence as, for the next three or four minutes, Rosemary struggled to draw her diagram on a pad which seemed too flexible, without a firm table beneath it, for any accurate drawing to be made on it. However, I dared not interrupt the precious process and allowed her to continue. After the hypnotist had been told she had finished, he thanked Rosemary and passed the pad and pencil back to me. I decided not to try to analyse it there and then and, in fact, hoped for an opportunity, later, to obtain a drawing from Valerie.

The questions from others became directed mainly towards Valerie, who had sat there unmoved for some time. It was all to no avail. We could not get Valerie out of that 'loop'. Eventually, I asked if she thought she could draw me a picture of that map. She responded, in a quiet voice, with, "I'll try". Once again, the hypnotist introduced the pad and pencil to her, told her to open her eyes and to draw what she'd remembered most about that map. A suitable cardboard box was hurriedly inserted beneath the pad as Valerie began to draw. Once again, we held our breaths in silence as the entranced Valerie slowly and deliberately applied pencil to paper for several minutes. At one point, near the end of this demanding process, she stopped, saying she couldn't remember, but, when the hypnotist encouraged her to try harder to remember, she added several additional markings. Then, it was finished. The pad and pencil were again handed to me. I was excited and amazed by what I saw, but I knew the drawing would need careful scrutiny and analysis, as would Rosemary's very different impression.

Well, the session went on to fill the usual hour, but no further progress was able to be made. Neither woman was able to 'break out' of the memory loop centred on those tunnels. This seemed to be triggered by a memory of seeing three people in the room as they lay on their tables after leaving the tunnels. When they were asked, in different ways, "Who are these people? Have you seen them before", they just 'clammed up' into silence. Further questioning about 'people' usually led back to the 'man' seen before and after travelling through the tunnels.

Clearly, the No.3 session had failed to realise its overall objective, but for me, it was a triumph. Now, at last, I had material to analyse. It was agreed to have a fourth session, but, as far as I know, that was never arranged.

Before going on to deal with my analysis of those maps, I want to first of all to state that the Shropshire CE4 date and time combinations turned out to have fitted the global AT patterns very well, even though the start and end times had to be necessarily approximated from the journey times given by the participants.

The women had estimated they had lost about half-an-hour during their journey from Shrewsbury to Telford. This was based on past experience of similar early morning trips, when traffic was light. According to Rosemary, she had looked at her watch and checked the time with the clock in the cloakroom before they had left the club. It was then 1:10 am. BST. The area of the encounter is situated about 4 miles to the south-east of Shrewsbury, at Atcham.

Numerical AT analysis of possible encounter times at that location had shown that there were two candidates, one being an SAC arrival on site at 1:43 am BST and the other, a possible departure time, of 2:14 am BST. The gap between these two consecutive possibilities (which might indicate the minimum mission time) was 31 minutes. The arrival timing was produced by a sunrise-orientated track over the area and the departure, by an escape route which was sunset-orientated. Once again, another major CE event seemed to have been modelled very well by the AT.

## Drawings Analysis.

Figs.38 and 39 show the drawings produced by the two women whist still in hypnotic trance. It seemed that they had no idea of the nature of the 3-D hologram they were being shown during their time with their abductors, other than it had been intended to show where the creatures had come from. The 'hole-in-a-wall' picture had baffled them because it had had many features that seemed to be scaled differently and, perhaps, variably. It seems that Rosemary had tried to capture the overall scene during her entranced drawing session, whereas, in contrast, Valerie had concentrated on trying to represent some outstanding features of the display.

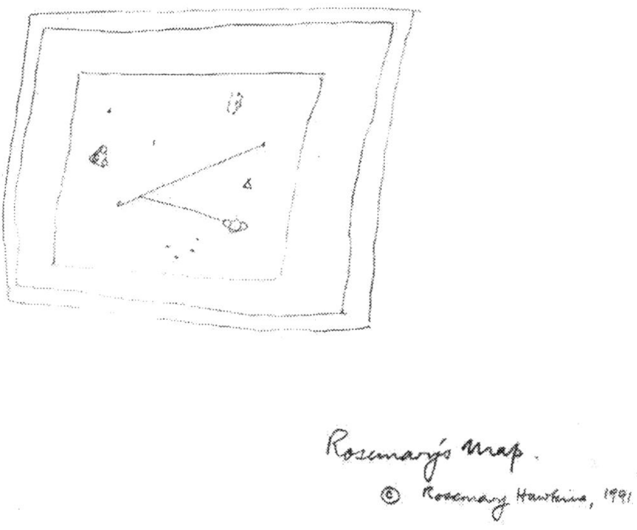


**Fig. 38**

Rosemary's drawing (Fig.38), though perfectly consistent with her description of the presentation, did not clearly identify the features being presented. Several possible interpretations were able to be attempted, but I was not impressed by any of them. The lines linking features might have represented travel routes linking planets in various regions of space, but I have been unable to make any sense of them, to date.

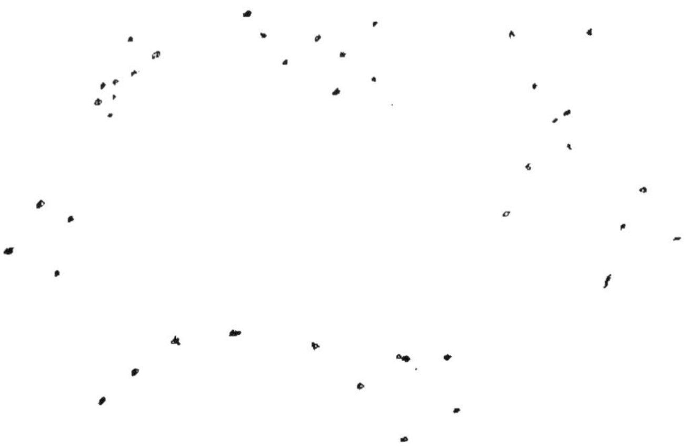

**Fig. 39**

In contrast, Valerie's drawing (Fig.39) seemed to offer more of a chance of interpretation, because she had produced a remarkable set of clustered points which looked like star constellations. Working along with that as a possible answer, I began to search the star map(s) for those shapes, but, try as I might, I could not find any matches. It was rather like searching for needles in a haystack! So, the problem was set aside awaiting further inspiration. Several months later that inspiration came in a surprising manner.

Our old home in north-east Cheshire backed, eastwards, onto open countryside which rose towards the Derbyshire Peak District. It provided a wonderful spot for star-gazing, without light pollution. One clear night I stood at our large patio window taking in the glory of the star-filled sky when I glimpsed in the corner of my left eye a cluster of faint stars just emerging over the treetops. It was the Pleiades cluster, looking brighter than I had ever seen it before. I rushed away from the window for my 20x8 binoculars and then returned to focus them on the Pleiades. ***Suddenly it***

***occurred to me that I was gazing at one of Valerie's drawings!*** Through those binoculars the brightest stars formed a tilted figure '2', just as she had drawn them. I went for the drawing and examined it at the window (presumably with a torch) as I alternatively viewed the cluster in the sky. Even the number of stars was the same and they were spaced more or less correctly. This had to be another 'Eureka' moment!

I went back to the analysis of that drawing with new eyes, reasoning if that one feature there represented the Pleiades, were there now other star arrangements in the sky, adjacent to that cluster, that might also have been depicted. The problem was one of scale. If the other collections of dots were drawn to the same scale as the Pleiades cluster, then a magnifying glass would be required. Using the magnifier, I searched the sky all round the cluster, but nothing presented itself as being remotely like Valerie's depictions. So, I decided to search that region of the sky without the magnifier. Fig.40 shows the outcome, labelled on Valerie's drawing. The first pattern to the right of the Pleiades turned out to be a good representation of The Hyades, also in the constellation, Taurus. Working further to the right, the next two arrangements of stars were found in the constellation of Gemini, Propus, Tejat, Castor and Pollux being among those represented. The trapezium of four stars to the left of the Pleiades was found in Aries, the stars Sheratan and Hamal featuring. The remaining feature of the drawing was eventually identified as a collection of stars spanning three constellations, Pegasus, Andromeda and Cassiopeia. All this is shown by Fig. 40, Valerie's drawing having been rotated through 180°.

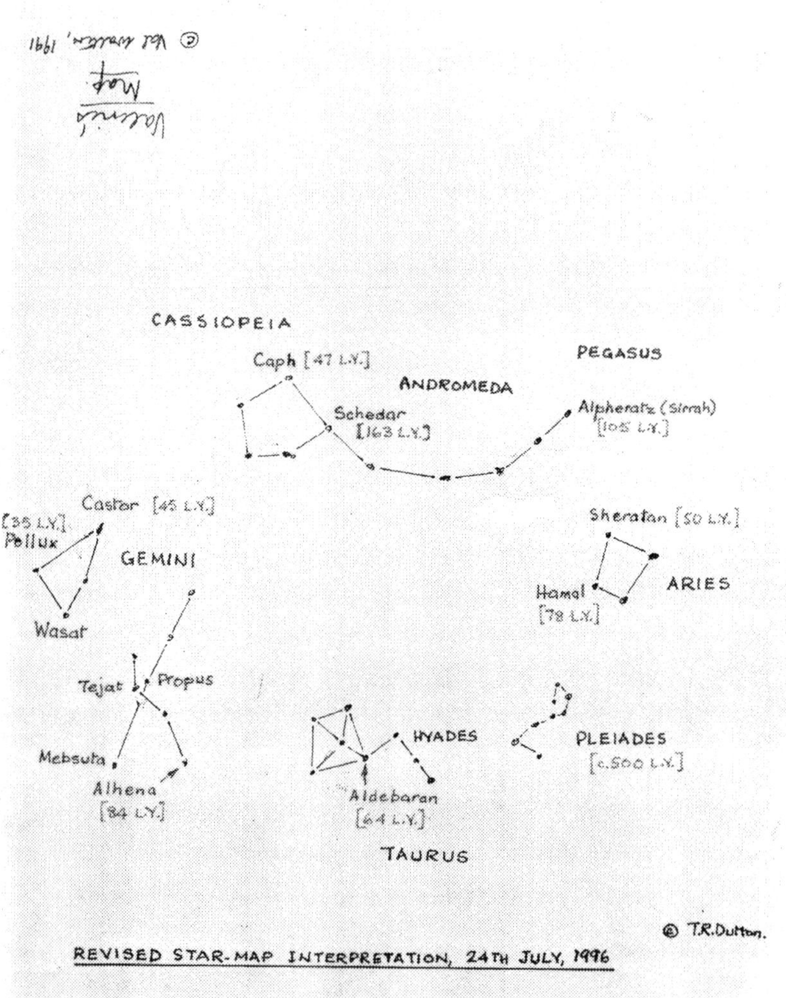


**Fig. 40**

To satisfy the sceptics (usually astronomers), who would no doubt be thinking that, in an exercise involving joining of dots, all sorts of similar patterns could be found in the sky, I produced Fig.41. Here, the star patterns are shown as they appear on a star chart. Can any reasonable person fail to recognise how well Valerie's drawing depicted these?

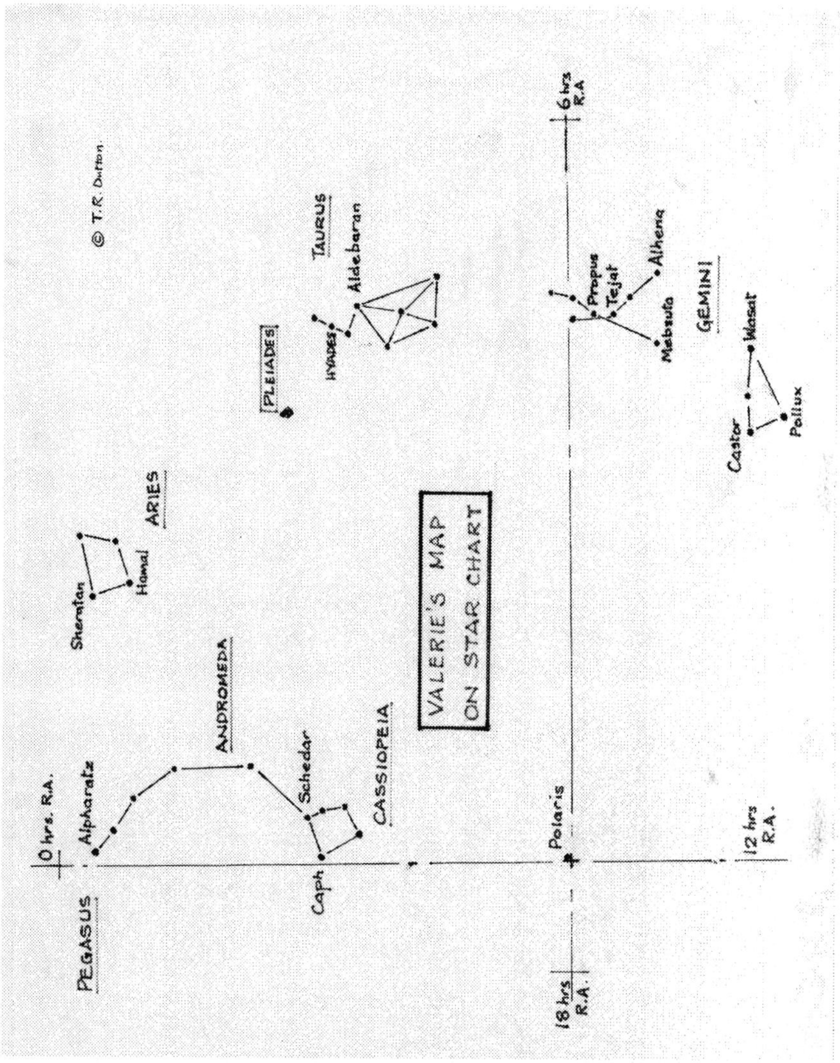


**Fig. 41**

In my view, this is the best breakthrough we have been given in our quest for understanding of the CE4 phenomenon. We may have also been given the clearest indication yet of the area of the sky from which these particular visitors had originated.

Now, it so happens to be true that several contactees (as opposed to 'abductees') have claimed that the they had received information that their contacts had come from the Pleiades area. It seemed that Valerie and

Rosemary had no knowledge of that. In fact, they claimed to have had no interest in UFO stories, astrology or astronomy before their CE4.

## The Barney and Betty Hill CE4

It seems appropriate to end this Phase 5 by considering the first and most famous alien abduction story ever to become widely known. American husband and wife, Barney and Betty Hill, lost their memories of a prolonged interruption of their journey home from a holiday stop in Niagara, after they had encountered an SAC and its alien occupants. What's more, like Valerie and Rosemary, under hypnosis Betty Hill had produced a drawing to represent something she had been shown by her abductors. She had been told it showed where they had travelled from in the universe. It was a series of small circles and dots with lines drawn between some of them. This drawing was later analysed by an amateur astronomer, who constructed a three-dimensional model of it, by means of small balls hung from the ceiling of a spare room, at different levels and distances apart. She had concluded from this construction that the ETs' home was within the constellation of Reticulum.

Although the story was extremely challenging, I had originally resisted trying to analyse the circumstances of the Hills event, because it had occurred during a long car journey from the Canadian border to their home in Portsmouth, New Hampshire. On reconsideration, I decided to have a go. Using the account given in the book, **'The Interrupted Journey' [16],** published in 1980, and a road map of that area of the USA, I was able to reconstruct the Hills journey, by marking the named stopping points on the map. Although they had been followed by an unidentified light in the sky for some time before, the actual encounter occurred in the lonely White Mountains area of New Hampshire. Using the times when each location had been reached by the couple, the AT programs enabled me to compute the data and to determine whether any of the information given by the book was in agreement with the AT's predictions for each location.

To my amazement and satisfaction, I found very good correlations between the actual events and the predictions. The following graph (Fig. 42) shows the details of the analysis and will be followed with a detailed account of its features.

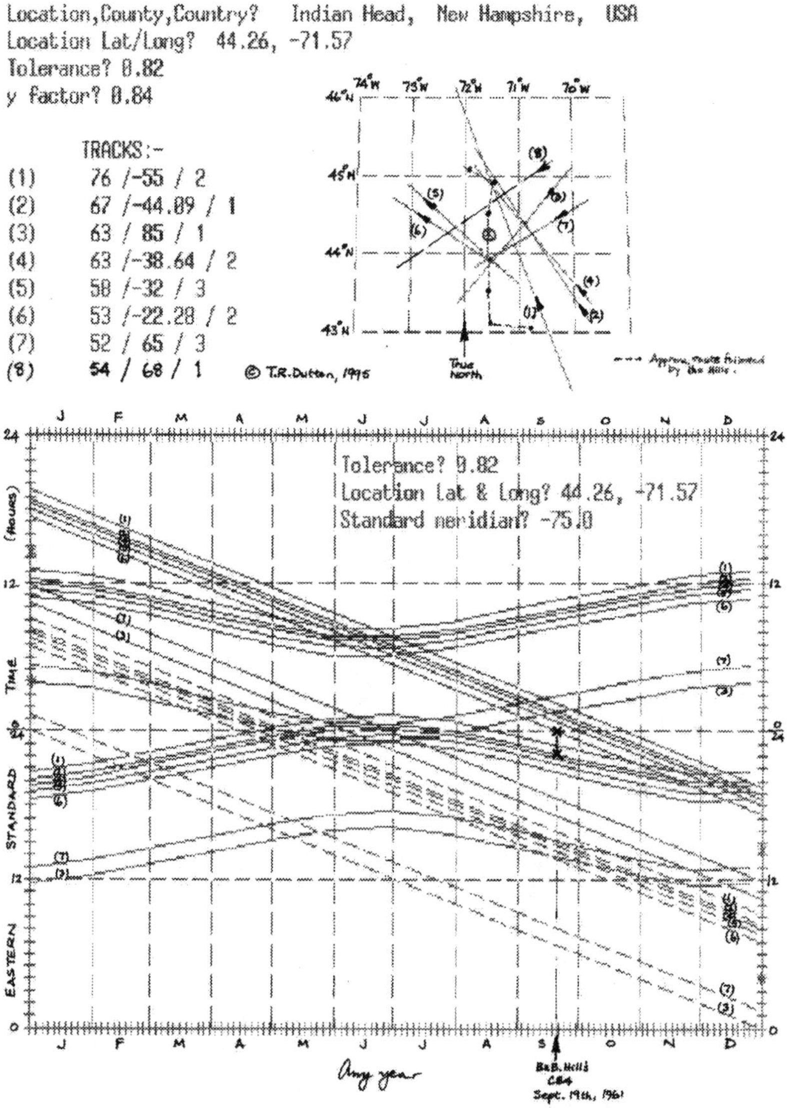


**Fig. 42**

The Hills home was in Portsmouth, New Jersey and this coastal location is marked on the latitude/longitude map. All the other key places are also marked on that little grid map. Please refer to that map whilst reading the following account of their journey.

During September, 1961, after a touring holiday, which they terminated at Niagara Falls, they began their return home via Montreal, Canada. On September 19th they turned southwards to get onto US Route 3. They left the Canadian border post at about 9pm, expecting to arrive home about 3am next morning. They then stopped at Colebrook for refreshments and left there at 10:05pm. Their route home took them through the lonely Connecticut River Valley and then into the rugged Pilot Range of mountains. The moon lit up the scenery as they progressed and a bright planet was noticed near to the Moon. This, I judged, would have been Jupiter, accompanied in the sky by another planet, Saturn. Then, they noticed another bright light in the sky and, when they stopped to walk their dog, they perceived it was moving and gradually getting larger. It seems that this unidentified light, which later became recognisable as an SAC, arrived on their scene shortly before 11pm.

They stopped again at 11pm to view the object through binoculars. It had flashing lights, but it was definitely not an aircraft. As they drove on to Indian Head, a huge craft with windows suddenly swung low across the road ahead.

Through binoculars, occupants were clearly visible behind the lighted windows. The SAC hung over the field to one side of the car. The Hills remembered driving off in terror, hearing beeping noises inside the car and then feeling drowsy. The beeps returned and the car was still in motion. A signpost told them they were then 35 miles further along the road. Only when they entered US Route 93 did they become fully aware of their situation. They had no recollection of that 35 miles from Indian Head southwards. Subsequently, they arrived home two hours later than expected.

It's time now to consider the analysis of these details.

The SAC (UFO) was first sighted at about 11pm Daylight Saving Time (DST), that is, 10pm EST as plotted on the timings graph. The lower black cross represents that event. It is lying on the Sunset Track No.2 timing line.

If we now assume that the Close Encounter occurred soon afterwards and that it was responsible for a subsequent two-hour memory loss, the CE4 would have been terminated at about midnight EST, as shown by the upper black cross. That cross is lying on Star-related Track No. 6 and this is suggesting that the SAC was retrieved by a retrieval spacecraft moving East-West in space above that location.

Looking again at the latitude/longitude grid map, the No.2 track line runs close to the Canadian border post, so that could have been where the Hills car was detected and followed by the atmospheric scout craft. The No.6 track crosses the Hills route 23 miles south of Indian Head, as drawn and, therefore accounts for many of their lost journey miles, if that track had been used by the retrieval craft.

Despite slight discrepancies, I was pleased with this result, since it explains both the missing time and the unaccounted-for journey miles.

## References

[16] Fuller, J.G.'The Interrupted Journey" (paperback) CORGI Books, Transworld Publishers, Ealing, London. 1981 Souvenir Press, 1980 (book)

# PHASE 6:
# Unexpected elements

## CHAPTER 16

## THE SMOKE-ALARM EXPERIMENT

During the summer of 1993, on June 6th, just before we moved to Devon, my wife, Marion, and I were awakened from sleep in the early hours of the morning. As we lay wondering why we'd been disturbed at 2:45 am (BST) a loud bleep from the adjoining hallway answered that question. "It's that damned smoke-alarm", I said, not in the least amused. I waited a few minutes, but there were no further bleeps from it. It dawned on me that this was unusual because when the battery had needed replacing in the past, the bleeps had continued at one minute intervals. I got out of bed and tested the offender using the test button provided. The noise from it nearly deafened me. Surely, there was nothing wrong with the battery, so I decided not to disconnect it, much to Marion's disapproval. At 3:10 am. the thing went off again and, in the same way as previously, the bleeping stopped of its own accord. Marion's disapproval of my determined non-action became rather irritable, but I was intrigued by it all --- lost sleep or no lost sleep! "It's all in the cause of Science", I said. (She'd heard that excuse before!) We then rested in peace until 4:25 am and further bleeps at 5:10 am. and 5:30 am. ensured that little sleep could be had during that period. However, we were able to sleep thereafter until our normal getting-up time. As we were both now in retirement and would be moving around the bungalow all day, I said I was going to monitor the alarm at frequent intervals to check whether it was going to continue its bad behaviour during daylight hours. But, first, I checked the voltage of the battery. I knew that standard 9v batteries needed replacing when the voltage dropped to 8.2v and, with the alkaline variety, this critical voltage was 7.8v. The normal battery in our alarm gave me a voltage of 8.6v. So, there should have been no problems.

Remembering Mrs. Linda Jones' experiences and the resulting speculation that perhaps some kind of RF scanning might account for them, (see Part 3, Chapter 21), I wondered if that might also be the cause of our smoke-alarm's strange behaviour. This prompted me to check the times of our disturbed sleep against the AT's timing graph for our area. The result was quite startling. Correlation was generally within half-an-hour of predictions and usually less than that. Significantly, there were no bleeps from the alarm during the daylight hours, but they began again at 8:15 pm, 8:35 pm and 8:50pm. (These were shown to correlate well with East Didsbury time predictions, East Didsbury being only about seven miles away). Then there was silence until 10:10 pm and 10:30 pm. Another silence followed until 11:25 and 11:40 pm. All these were in good agreement with predictions. After showing this to Marion, I persuaded her to allow her sleep to be broken for yet another night and, sure enough, it was! There were four interruptions between 2:06 and 5:10 am, before silence reigned supreme until 7:25 and 8:10 am. During that morning of June 7$^{th}$, the alarm bleeped twice during a 10-minute period at 8:20 am. and three times from 10:03 and 10:30 am. **The important feature of these interrupted outbursts was that no adjustment could be made to the battery voltage when the alarm was still mounted on the wall.** Having ascertained that correlation with predictions had been very impressive throughout, as shown by Fig. 43, I decided to work on the hypothesis that RF scanning might have interfered with the delicate electrical balance within the circuits of the ionisation-chamber smoke alarm.

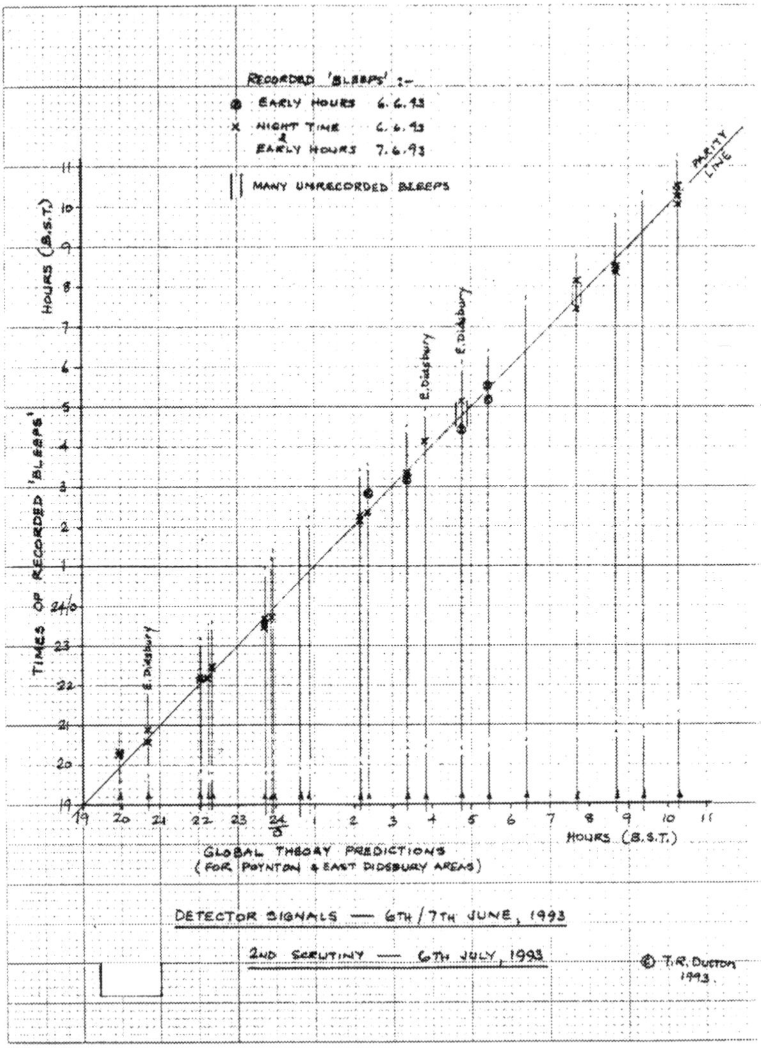


**Fig. 43**

I consulted a physicist friend, Mr. John Jones, about this possibility. He said he had no expertise on smoke alarms, except about the principles of their operation, but he remembered an incident in which an overload alarm on an overhead crane had been triggered by the operation of 'walkie-talkie' radios in the vicinity of the gantry. The functioning of that alarm would have been similar to that of smoke alarms. (John Jones later decided to set up his own monitoring station, utilising the same kind of smoke alarm, but failed to register anything unusual. We concluded that the controlled stepped-down and rectified mains supply he had used would not have simulated the peculiarities of a small battery under load.)

Having had my hypothesis supported in that way, I decided to contact the manufacturer of the smoke alarm. When I told the technical director of my suspicions of RF interference, he became very concerned. The company had tested their alarms through the complete range of radio frequencies and had found no peculiarities. He wanted me to return my alarm for investigation and to have it replaced. I thanked him but said I was keen to resolve the mystery for myself. He was not a happy man but reassured himself by saying perhaps the measured voltage of 8.6v was too close to the critical voltage to be of great concern to the company. After this, I decided to replace the alarm with another and to install the interesting one in a wooden box with a sliding lid. An on-off switch and a voltmeter were also installed, together with a potentiometer to control the voltage delivered to the alarm. As we were soon afterwards involved in the process of finding somewhere to live in Devon, the boxed alarm was never tested in Cheshire. It travelled with us to Devon and remained inactive for some months after our arrival here. A UFO report in the Torbay area caused it to be put into action during 1994.

Monitoring began on May 14th, the day after a UFO event had been reported by a person living in nearby Paignton. The spectacular UFO had apparently hung low over the adjacent tree line for over an hour and had been located somewhere in the general direction of Paignton Harbour. The event had occurred in the early hours of the morning of the 13th, which was unfortunate, because that area is visible from our hillside home and the object might have been able to be observed from here, if I hadn't been sleeping at that time. I visited the witness and afterwards switched on the boxed alarm. Marion and I then proceeded to monitor that box, day and night, for the next 12 months. We took it to bed with us and whenever we went out in the car, it travelled under the driver's seat. It was never out of hearing range if we could possibly avoid it. Well, our sleep was disturbed on numerous occasions and on several of them Marion told me that either that "thing" had to go or she would! However, I persuaded her to endure because the results we were obtaining were intriguingly good. (It was all in the cause of Science!) Throughout that 12 month period I was encouraged further by strange happenings in our area, including loud bangs from the sky, some being like explosions and others more like sonic booms. These booms occurred at times when the Air France Concorde was known not to be decelerating down the English Channel. A mysterious red flare object, a fireball event and another UFO sighting were also registered during our monitoring period.

But the results we obtained were staggering. The times when the alarm had triggered were plotted regularly onto the Torbay timings graph and were found to be usually congruent with the predicted lines. The other surprising element was the large number of times these day and night 'bleeps' had occurred. There was always the doubt that I might be deluding myself and that we were losing sleep to no avail, but several things happened to indicate to the contrary. On the night of March 29th, 1995, a loud muffled explosion was heard at 5:50 pm. BST. At 9:22 pm., a bleep from the box coincided with a violent disturbance of the TV reception. Later that night, during a routine nightly check, I discovered that the overload switch serving the motorised garage door and outside lights (which were not switched on) had been tripped. We continued monitoring during waking hours throughout the latter part of 1995 and on November 12th, during the early hours of the morning, a smoke alarm in the hall adjacent to our bedroom bleeped once, on several occasions. At breakfast time the boxed alarm was switched on and placed in the kitchen, some 12 metres away from the hall-mounted alarm. At 2:40 pm. GMT the wall-mounted alarm bleeped once, but only 12 minutes later that alarm and the boxed alarm bleeped **simultaneously**. This seemed to prove that both had probably been triggered by the same stimulus.

Returning now to the outcome of the 12 months' monitoring exercise, altogether we recorded 330 'signals' within one hour of the nearest predictions and 211 (64%) had occurred within 20 minutes. A set of 330 **random** times were then produced and allocated to the same days as the actual signals. These were then processed in the same way by the AT's numerical checking program and it was found that 62 (19%) of these had not been within even 1 hour of nearest predictions and only 176 (53%) had corresponded to the +/-20 minutes rule. Allowing for the fact that the boxed alarm almost certainly had produced some spurious signals (the voltage supply had had to be checked and adjusted quite frequently and battery life was reduced to 12-13 days), the results did seem to indicate that real signals had been received on many occasions. This indication was given further credence when, during subsequent years, the numbers of 'signals' registered during test periods were significantly reduced. It seemed reasonable to suppose that the Torbay area had been targeted for surveillance from high altitude during 1994 and 1995.

# CHAPTER 17

# Morse code revelations

During September, 1991, I had written to The Messenger, a Macclesfield based local newspaper and, in my letter, I had described the scientific investigations I had carried out into the nature of UFOs and crop formations. This had prompted an elderly lady in Wilmslow to write to me, to express her appreciation and to tell me about a remarkable happening which had occurred in her home some 20 years earlier.

She also had things to tell me about the reasons for the crop formations and to warn about the ecological crisis the earth was entering. These two phenomena were linked, she asserted. She had tried to arouse the interest of several famous scientists, but unsuccessfully.

I feel it is important that the story of that happening of August $9^{th}$, 1971, in her home, should preface the other remarkable things I had yet to learn from her.

At 6 pm. on the evening of August $8^{th}$, whilst preparing dinner in the kitchen, through the window she had seen two large UFOs, partially obscured by "their own cloud formation". She had grabbed her binoculars and had run outside to get a better look at them. The binoculars had seemed to have been knocked out of her hands. She had called her husband who arrived in time to see the objects fading away into invisibility.

It was the lady's practice to sit up late and to always say her prayers in the lounge before retiring to bed. After midnight, at 1 am., on Tuesday, August $9^{th}$, she had been engaged in that way, kneeling beside the settee with her back to the patio window, when her neck "was penetrated by a hot stinging sensation which travelled upwards to my brain and down my spine". She turned towards the window and saw a disc-like UFO hovering on the other side of the window. She estimated they would have been no more than 60 inches apart. A brilliant beam of light shot into her forehead and, as she expressed it, "I was given the heat treatment". She claimed she still had marks on the back of her neck and for two years after the event she experienced a burning sensation every Tuesday. I arranged a visit to her home to discuss her experiences in greater detail.

On arrival, I was made very welcome and ushered into the lounge. The rear garden was on view from the patio window. It was not a large garden. My attention was drawn to a corner plot some thirty feet (10 metres) away (if my memory serves me correctly), in which bushes and conifers were growing. Two conifers were quite tall and there was a gap between them. I was told there had been a third conifer between those two at the time of the happening, but the next morning the one in the middle had been found to have been bent through 90 degrees, towards the window!

Already impressed by this lady's experiences, I was eager to know more about the information she had been given and about the means by which she had received it.

With my notepad on my knee, I jotted down the information as it was given to me.

As a girl, she had lived in Manchester. When she was 12 years old, her mother had been seriously ill with pleurisy. As she awoke one morning, she had heard a gentle voice saying to her, "I'm sorry to have to tell you that, this day, your mother has passed to a higher life". Next, I heard about her spiritual conversion on June $24^{\text{th}}$, 1970, at 1:45 am. During an 'out of the body' experience she was "taken into the Light of Christ and shown the meaning of books of the Bible". She had become a devout Christian and a 'seer' after that experience.

Getting back to the topic of particular interest to me, I was handed a list of things she had been told about the UFOs she had seen. The list, which I copied, was as follows:

1. [They] Sail on the currents of Space;
6. Made in transmuted metal in one piece, like a balloon;
7. Central power column and control system;
8. Geisler tubes round outer periphery;
9. Red emission during low speed and hover;
10. Become invisible at higher frequency;
11. Fly in a vacuum with full atmospheric pressure behind;
12. 200,000 miles per second speed capability;
13. Gravity, heat, light, the vertical component of magnetism and energy that exerts pressure.

On considering this list, I asked my hostess whether she had a scientific background. This caused her great amusement. Nothing of the kind. She'd had to leave school early, when only 13 years old, and had received no further formal education whatsoever. Most of the information in her list she might have picked up from various speculative UFO books, but one item stood out from the rest. "What's a Geisler tube?", I asked. She laughed, said she didn't know and had hoped I would know. I was baffled. "How have you received this information?", was my next question. Her answer rocked me in the chair. "In Morse code". When did she become acquainted with that code? She had been a member of the Royal Observer Corps during WW2 (attached to the Alexander Park balloon barrage) and she had become a very proficient sender and receiver of messages. I was still puzzled and asked how she had received the information on her list. Once again, I was going to be left open-mouthed by her reply.

She told me that, after that event in her lounge, the telephone had developed a habit of ringing late at night. This occurred three nights per week for a period of several months. Sometimes when she'd answered it, she'd heard an indistinct mystery voice. Then the Morse code messages started. They were received in her left ear. At first, combinations of E, U and sometimes V, like call signs, were received --- and then came the messages, which she wrote down.

When I left that lady's home, promising to let her know how well her leadings were tying up with all I had collected and deduced from my other investigations, I was a very bemused man.

* * *

On my arrival back home, I could hardly wait to discover what a Geisler tube was, if such a thing ever existed. I began a search through my motley collection of science books, but after spending considerable time doing that, I was none the wiser.

So, I decided to ring our son, Paul, who has a physics degree and, during his days as an undergraduate, he had gained experience within the nuclear industry. Once again, I drew a blank. He couldn't remember ever seeing reference to such tubes. Another search of my bookshelves then bore fruit. In The Penguin Dictionary of Science (left behind by Paul) I found this:

GEISSLER TUBE. A tube for showing the luminous effects of a *discharge of electricity* through various rarefied *gases*. Consists of a sealed glass tube containing platinum electrodes. Named after H. Geissler (1814-79).

Apart from the lack of double 's' in my notepad version, there was no mistaking that this was the kind of feature of those UFOs being described in that list (Item 4). But why had such an obscure piece of $19^{th}$ century scientific equipment been listed? As will be more than amply demonstrated in Chapter 22, it seems that the ETs are playing games with us --- perhaps testing our knowledge and powers of deduction?

Another controversial item in that list is No. 8, which gives the maximum speed of the UFOs as 200,000 miles per second. The measured speed of light, 186,000 miles per second, is generally accepted as being the maximum speed attainable in the universe. This is the basic tenet on which the Special Theory of Relativity stands and most physicists would be quick to pounce on that No.8 item and to denounce it as nonsense. However it seems that most physicists are unaware that, when discussing his General Theory of Relativity, Einstein postulated that the speed of light would vary with the strength of the gravitational field the light happened to be travelling through --- and that the generally accepted light speed was a localised one, applying only to the Earth's environment. So, who knows what the speed of light is in interstellar and intergalactic space?

By telephone, I informed the lady in Wilmslow of my findings and thanked her again for all she had shared with me. But, then, she went on to tell me more. On a Tuesday or Wednesday during mid-September 1973, at 4:45 am., she had been instructed to go outside. There had been a clacking sound followed by a swishing sound. The sky was cloudy and it had been raining. The clouds above her had started rotating rapidly in a circular motion. The clacking sound had returned as she retreated indoors and she'd got the thought, "tomorrow". At the same time on the following day, the clouds above had rotated again and it had then begun to rain. But she'd received two messages (telepathically?). They were:

**"To the apples we salt we return"** --- and --- "Will not be coming back for a while".

The first of these seemed to explain why some individuals have several significant UFO encounters during their lifetimes, whilst others (like me) see only occasional unexplained lights in the sky. It seems I am not one of those salted apples!

# CHAPTER 18

# THE TUNGUSKA EXPLOSION --- 1908

During November, 2006, I received authoritative information (from Edward Ashpole) giving me the actual location of the Tunguska event and that it had occurred at 7: 14 a.m. local time. That vital new development presented the opportunity to run my AT programs. The resulting Timings graph is shown below as Fig. 44.

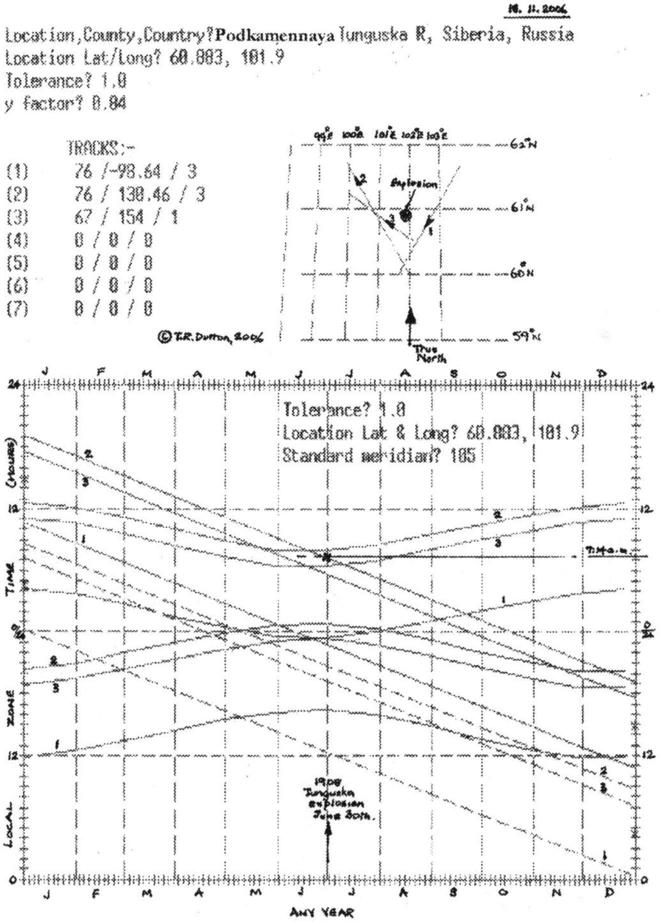


**Fig. 44**

The graph shows that the timing of the explosion closely coincided with the AT's prediction of 7:24 a.m. (numerically derived) associated with the No.2, 2130h RA (star related) track over that area. It must also be significant that the SE-NW orientation of that track conforms well with eyewitness accounts of the path of the brilliant object, prior to the explosion. Here is an extract from page 235 of Arthur C. Clarke's book, **'Mysterious World' [17].**

*"it seemed that the flying object had entered the earth's atmosphere and become visible somewhere over Lake Baykal and then travelled from southeast to northwest as it plunged downwards, though there was some suggestion that it might have changed direction. Indeed, this thought, based on eyewitness accounts, which now number more than 700, is one of the main planks of those who believe the object was a spaceship. Certainly only a controlled vehicle could have changed direction. No single witness claims to have seen the object actually manoeuvre. But there are sharp contradictions in reports of the flight path as the great 'pillar' careered across Siberia. Testimony in the more western area consistently gives a different angle of approach to that from the Baycal area."*

All that is perfectly consistent with Fig.44. From the latitude/longitude map, the qualifying track No. 2 would have passed over Lake Baykal, travelling southeast to northwest, and it is implied that the object would have veered northwards from that path prior to the explosion. This would account for a change of direction during the final approach stage.

Figs. 45 and 46 are global presentations of the situation given by the Fig.44 computations.

The three AT paths passing closest to the Tunguska site are shown extended round the globe. They are seen to pass south of the explosion. Tracks 1 and 3 were shown in green (on the original). The explosion site and qualifying track No.2 were shown red. The approach over the large Lake Baykal is clearly indicated. The Tunguska site is shown in its correct position, at 7:14 a.m., relative to the Earth's terminator on that day, nine days after the Summer Solstice.

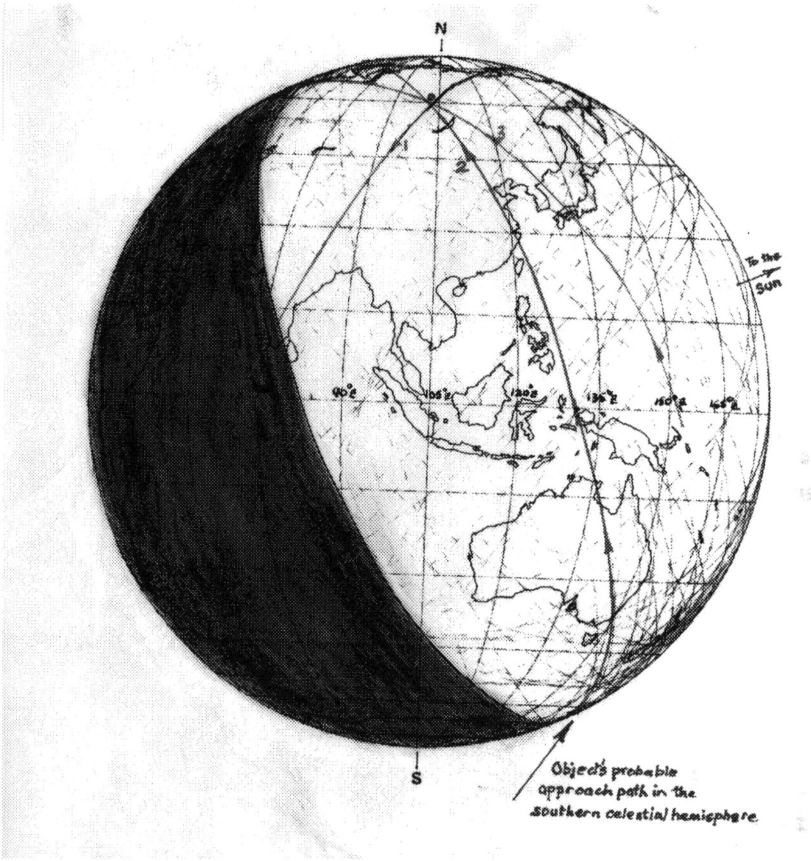


**Fig. 45**

Travelling on the computed No.2 track, the object would have probably engaged that predetermined track somewhere in the southern hemisphere and then proceeded, in a controlled manner, north-westwards.

In terms of the Theory the evidence suggests that it was almost certainly a large delivery/retrieval spacecraft, which was not intended to enter the atmosphere. It seems that pre-programmed and automated control of its progress in 'super-orbit' was lost as it passed over China and it began to descend into the atmosphere over Russia. As it's speed just before the explosion was estimated to be only about 1km./second, it would seem probable that attempts had been made to achieve a crash-landing. When this had seemed to be failing, the craft could have been deliberately destroyed.

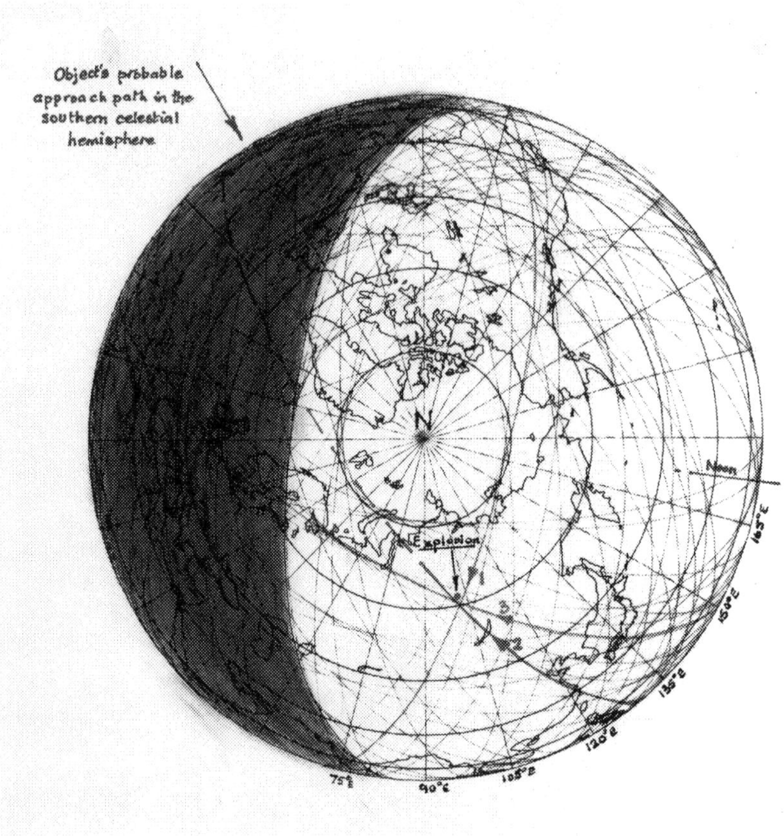


**Fig. 46**

All this can be regarded as a clear indication that the Tunguska object was almost certainly involved in the ETs' still-ongoing exploration programme.

As this book neared completion, a very informative book about this 1908 event was produced by a leading Ukrainian scientist, Dr.Vladimir Rubtsov. His book, **'The Tunguska Mystery',** published by Springer in the U.S.A., gives details of all the scientific investigations carried out in Russia since 1927. Much more is now revealed of what is known about the nature of the object and the final explosion. The AT-based analysis described in this chapter is in no way invalidated by the new evidence presented by the Rubtsov book.

## References

(17) Clarke, Arthur C. (book) "Mysterious World" Book Club Associates, © 1980 Trident International TV Enterprises, Ltd., © 1980 Arthur C. Clarke

# PHASE 7:
# Signs of growing interest

*During the forty years of this study I met with many people, some of whom have made significant contributions to its development and promotion. The chapters of Phase 7 identify most of those people and acknowledge the help they provided.*

## CHAPTER 19

## PROMOTERS AND LINK-UPS

### The London Link-ups

In Phase 5, I told of the key contribution made to my knowledge of CE4s by the late Ken Phillips of BUFORA. Ken had chaired my 1987 evening lecture in May of that year and he and his wife, Anne, had kindly provided me with overnight accommodation. He had brought me into contact with Mrs. Linda Jones and, thereafter, we became friends-at-a-distance. Ken was the BUFORA investigator for Greater London and thus received UFO reports from witnesses in that area for examination. He had established a good relationship with the Ministry of Defence through their then-collector of reports, Mr. Nick Pope. Ken had promised to send copies of the UFO reports to me for analysis, which he then proceeded to do. He also sent the same reports to Nick Pope for comment. Then I discovered he was sending my analysis reports to Nick Pope and I was receiving Pope's comments on the same UFO events. Eventually, an opportunity to contact Nick Pope, directly, opened up for me. I sent one of my published short articles to him for consideration. This had explained the nature of the AT and how it had been derived. Nick Pope acknowledged receipt of this but added he could not understand it. My (typically) direct response to this was to ask whether this meant he had 'binned' it or had passed it on to someone who might be able to understand it. He responded very politely and said he was obliged to pass on all information he received to the appropriate departments, but was unable to divulge which departments.

Communications continued via Ken Phillips for some months afterwards. Then came the announcement that a book by Nick Pope had just been published, with the title **'Open Skies, Closed Minds'**, in which he had openly declared his belief that some UFO reports were inexplicable. This conclusion had been drawn after only three years as the MoD's 'expert on UFOs'. Since then, Nick Pope has become a household name in Britain and has been, for some years, a favourite speaker at UFO conferences. ( I had the pleasure of sharing a platform with him at Kidlington, Oxfordshire, in 1996 and had long conversations with him more recently at an event to commemorate the life of the late Capt. Graham Sheppard, a well-known participant in two air-to-air UFO encounters during 1967.)

My link-up with Ken Phillips continued until his untimely death on July 18th, 1996, the eve of the 1996 BUFORA conference in Cardiff. He was sadly missed by all who knew him and for me it was great personal loss. In friendly co-operation we had processed innumerable reports over the years until his death.

Another link with BUFORA was through its President and co-founder, Lionel Beer, who, some years ago, kindly provided me with overnight accommodation after a BUFORA meeting in Kensington Library, at which I was the invited speaker. He also drove me round the sights of London on the following (Sunday) morning, before I returned home by train. After Marion and I moved to Torquay, Lionel contacted me again and it was then I discovered that he had been raised at the family home in Paignton, where his mother (now deceased) still lived. We are still in contact and visits are exchanged quite frequently.

## Edward Ashpole.

As a result of a recommendation from Ken Phillips that he should contact me, biologist, science writer and SETI author Edward Ashpole came into my life during 1988. Edward had just completed a book called **'The Search for Extraterrestrial Life'** (which was later published in1989) in which he had discussed the evolutionary probability of life existing in other parts of our galaxy, the development of extraterrestrial technology and the possibility that ETs may have developed the means of visiting this planet quite frequently. In the penultimate chapter he had ventured onto scientists' forbidden ground and speculated that perhaps some UFO reports might qualify as evidence of this.

**Plate 1**

When Edward decided to contact me by 'phone, we arranged that we should meet at my home so that he could have sight of some of my original work and we would be able to go on from there. **Plate 1** shows us together during that first meeting. Edward's publisher had asked him whether he could write a book about the UFO evidence and, when we met, he was then in the process of collecting information for that proposed book. He was clearly impressed by my methodical and objective approach to the evidence and even more so by the development of the Astronautical Theory. He asked if he could include this information into his forthcoming book and I was only too pleased to comply. At last someone respected by the scientific community was willing to publicise and recommend the work for scientific testing. Thereafter, whenever I felt I'd been wasting my time, Edward would cheer me by telling me I had produced the **only** objective theory, based on the world-wide evidence, which could be **tested** in various ways, not least by astronomers.

We kept in touch by telephone quite frequently during the new book's gestation period, spending hours discussing various difficult aspects. Having only basic astronautical and astronomical knowledge to help him, Edward found my terminology and 3-D geometry quite difficult to grasp (and he was not the first or the last!). When, in 1995, the book **'The UFO Phenomena' [18]** was finally published, Edward had devoted an entire chapter to an explanation of the AT, which was not entirely accurate, even though he had used quotations and diagrams from me to assist the readers' understanding. However, he had done his level best with very difficult material and had called for astronomical observations of the sky to look for the delivery and retrieval craft in space. I was extremely grateful for his perseverance and for the amount of space he had allocated to my work, even though he had decided to distance himself from my crop circles work and from my admission that a few claimed 'contactee' (not 'abductee') reports had been included in the database.

Throughout the years from our first contact, Edward Ashpole and I have become very good friends-at-a-distance. We are also colleagues in our quest to bring scientists (especially astronomers) to consider the evidence. Whilst some of our top radio astronomers continue to search for radio evidence of ET life out there in the cosmos, it could be that some ETs found us a long time ago and are regularly monitoring our activities, sometimes at close quarters.

## The Essay Competition.

On July 27th, 1998, I received a telephone call from an enthused Edward Ashpole. He had just read in a science magazine that a new organisation in America, calling itself **the National Institute for Discovery Science (N.I.D.S.)**, based in Las Vegas, had announced that it was running an Essay Competition for essays on our topic. The winning essays would be published on their web site and attractive prizes were being offered. He thought we should produce a joint essay and enter it. He could provide a scientific rationale to support the idea that technologically advanced ETs might have reached us in the past, and this could preface a description of my **Astronautical Theory** for SAC events and an explanation of the means by which the theory could be tested. He thought we had a good chance of winning. Still staggering from all this, I couldn't help expressing my doubts. The misgivings were further reinforced when Edward added, "The difficulty is that the closing date for entries is August 11th", but he still thought we could achieve that. I agreed to have a go, but the difficulties facing us turned out to be even greater than I could have anticipated then.

On July 30th, I received the details of the competition in the post. Neither Edward nor I were on-line via the Internet then, so all communications between us had to be by telephone, post and, occasionally, by FAX. Basically, the elements to be addressed by the essays were:

1. If ETs were to contact humans on Earth or in the solar system, what would be the probable means by which that could occur and how would we know we were being contacted?
2. Design a rigorous and innovative research project, or a set of such projects, focusing on how to detect and verify the presence of ETI.

Drawing information from previous articles and papers written by me, by August 2nd I had produced my draft contribution to the essay. In fact, I had produced a first draft by July 31st but then had to find a way of making my Wordstar output compatible with MSWord software being used by NIDS. The two software systems are incompatible. Only by converting to ASCII format was it possible to provide the script in, hopefully, acceptable form. Edward Ashpole had also been using Wordstar for some years and had found himself in the same situation. On August 3rd I received his draft contribution, on floppy disc, in the morning post. As requested, after checking through for errors and then adding my contribution to the disc, it was posted back to him on the noonday post. The draft of the complete essay, now with Edward's suggested title **'The Scientific Search for Evidence of Extraterrestrial Intelligence in the Solar System'**, was received from him on August 5th, for editing. Changes were communicated by 'phone. On August 7th Edward informed me that he had posted off the package to NIDS, in the hope it would be received before the deadline on the 11th. All we could do then was to find other things to do and to hope our late entry would be acceptable. We were relieved when we received notification from NIDS on August 22nd that our essay had been received and was being considered.

Many things then occupied me for the following two weeks. A very welcome interruption came with the receipt of a FAX from NIDS, late afternoon on September 10th, informing me that our essay had been awarded one of three First Prizes. As Edward did not have a FAX facility, I rang him immediately with that excellent piece of news. We learned later that because the judges had been unable to choose between the three essays, the (millionaire) Executive Director of NIDS, Mr. Robert T. Bigelow, had generously decided to regard them all as being worthy of the First Prize money. NIDS also asked our permission to display the essay on their web site and promised to bring all the essays to the attention of leading academics. What a result! It seemed as though at last we had broken through to academia. As things turned out, nothing could have been further from the truth. One exception to this rule was provided by two FAXes I received from Dr. Allen Tough, Professor Emeritus at University of Toronto, Canada, with queries. Dr. Tough wanted a copy of a paper I'd referenced in the Essay and, as this was not able to be readily reproduced at that time, I referred him to Jennifer Jarvis' web site, **'ORBWATCH'** (see below), where Jennifer had tried to reproduce it. (This link-up with Dr. Tough was to develop further during 2002). Otherwise, relatively little useful response was received to the essay and no one came forward offering to research my findings. Plate 2 shows the Certificate issued to each of us to commemorate our success in the competition.

Mr. Bigelow must have been very disappointed too. Later, he purchased videos from me and members of his scientific team asked for timings graphs for four specified areas in the American Far West. I discovered later their teams had not been in the right places at the right times for the short periods when they had been located in those areas. It became clear, also, that their main interest had been in finding positive links between UFO activity and animal mutilations.

## The British Interplanetary Society and the Essay.

At the end of March, 1999, 'out of the blue', Edward Ashpole received a letter (which he copied to me) from the editor of the Journal of the **British Interplanetary Society** (JBIS), informing him that our essay had been forwarded, by a learned member of the Society, for publication in the JBIS. It had been reviewed by external readers and their very critical remarks were given in the letter. Both reviewers had regarded it as a 'paper' rather than an essay. The editor then informed Edward that our 'paper' could not be published in the JBIS.

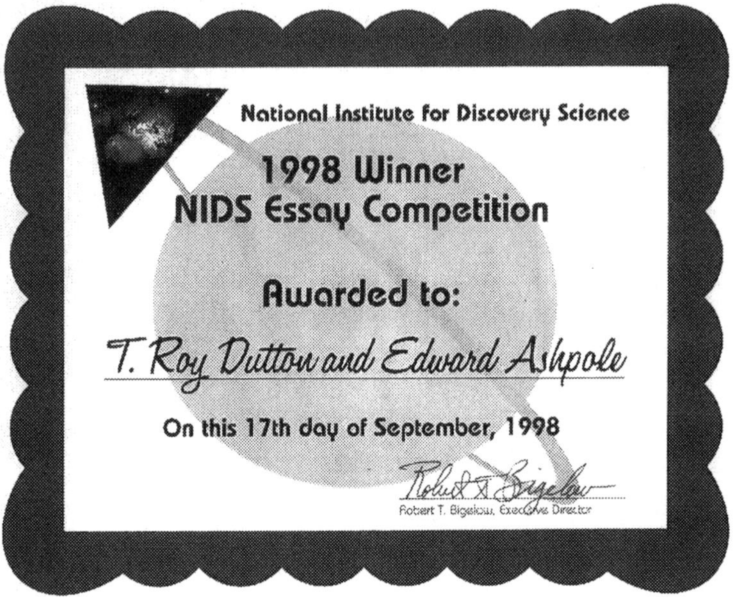

National Institute for Discovery Science
1998 Winner
NIDS Essay Competition
Awarded to:
T. Roy Dutton and Edward Ashpole
On this 17th day of September, 1998


**Plate 2**

Edward discussed this letter with me and then immediately sat down to answer it. He pointed out the essay had been regarded incorrectly and that, since he and I retained the copyright, no one should have offered the work for publication without our permission. He then went on to demolish the arguments ranged against us and, especially, against my part of the essay. He pointed out that I would have been quite capable of producing a scientific paper on the work, but that had not been the purpose for the essay. Edward received a polite but curt letter of response from the JBIS editor during April. "You never submitted your essay to JBIS. JBIS would not have accepted your essay for publication even if you had. So, no great problem after all."

I think that attitude speaks volumes for itself.

The difficulties we had experienced in communication with NIDS served to convince both Edward and me to update our obsolescent computer systems, especially, by adding MSWord software and getting onto the internet with e-mail facilitated.

## The Farmer's Contribution.

Some of the time Marion and I spent in Wiltshire and Hampshire during August 1991, we stayed at a farmhouse close to an ancient (prehistoric) pathway called the Wansdyke Path. One evening, on returning to the farmhouse after a long trek round the crop circle sites, the landlady and farmer's wife, Mrs. Bull, offered to make a warm supper drink for us, an offer we were pleased to accept. The Bull's joined us in the lounge for this period of respite and Mrs. Bull had something she needed to tell us. About midday that day she had seen, from a window, a large quantity of cut straw, lying in a field close by, suddenly swirled into the air in otherwise still conditions. The straw had been lifted high above the field and had then drifted off, like a dark cloud, towards the south-east. Neither she nor her husband had ever seen anything like it after many years of farming there. (Some time before, I had investigated a similar event in Marple, Cheshire, so I was not unfamiliar with the phenomenon.) Mr. Bull had estimated that the column had been about 12 to 15 feet (4 to 5 metres) in diameter and the straw had climbed to a height of about 1000 feet before forming a dark mass and drifting away. Mr. Bull then went on to describe something that was even more interesting. Last August, he said, he had been in another stubble field and he and his men had been burning off the stubble there, when the most memorable happening had occurred. A narrow black 'tube' had suddenly appeared only a matter of yards (metres) away from him, and it had seemed to have originated from beyond the sky haze above. I asked if , tomorrow morning, I could go out to examine the area. Mr. Bull kindly said that, after breakfast, he would take me to the location in the nearby field where the straw had been lifted.

Next morning was again a bright and sunny one, promising another hot day to follow. I met with Mr. Bull and, with the farmer standing on the area of the stubble field once occupied by the airborne stalks, I took photographs towards the farmhouse to show how close the event had been to the house. As we came together again after this, Mr. Bull very generously offered to take me to the other site, some distance along a cart track, in his pick-up truck. After we had arrived at the edge of a large field, I was guided to the approximate spot where the 'tubes' event had taken place. This was located close to the far side of the field, which was bounded on that edge by the Wansdyke Path, with an ancient burial mound just beyond it in the wood. Mr. Bull then gave me a graphic description of the phenomenon. It had occurred late afternoon on a fine, overcast and humid day in August.

The 'tube' had suddenly appeared close by without any commotion being caused. It had an estimated diameter of about 4 feet (1.2 metres) on the ground and had seemed to have the same diameter, diminished by perspective, all the way up into the sky haze. It was definitely not any kind of tornado. (He had witnessed one of those in the past.) When it had first appeared he had been some 20 yards (metres) away from it. He had become intrigued by the strange rattling noises coming from its base and had then walked to within 10 feet (3 metres) of it. There, he had been amazed to find bits of straw and dust being swirled round at high speed, within the confines of the 'tube', even though he could not see through the blackness of the tube at eye-level. As Mr. Bull stood with me reliving the event, it suddenly occurred to him that the black dust could not have accounted for the blackness of the tube, because the dust and straw hadn't seemed to be lifted from the ground. In any case the base of the tube had only moved about 2 feet (say 0.5 metre) during its lifetime of about 5 minutes, after which it had simply vanished into thin air. Within a minute or so of its disappearance, a similar one had established itself closer to the corner of the same field and had persisted for a shorter period of time. That one had disappeared leaving a thin wisp of condensation. An entry in his farm diary for August $24^{th}$, 1990, confirmed all that Mr. Bull had told to me about the time of the events and the prevailing conditions.

As these events had occurred close to an ancient trackway and site, I was keen to know more about the area. Flint and stone implements had been regularly unearthed in that field, demonstrating that a prehistoric settlement had once been established there. So, here again was another link being created in my mind between what one could correctly assume to have been ***a demonstration of advanced technology, not of this world, at an ancient site***. As for the technology being demonstrated, it seemed to me that Mr. Bull had probably witnessed the means by which crop circles were being formed, when the same equipment was used in a different mode.

On returning home to Cheshire, I began pondering how the projected beam witnessed by Mr. Bull (for I was sure that that had been the nature of the phenomenon) might be manipulated to carry out different tasks. Considering my guess, expressed in Phase 3, that a form of very high frequency gravitational radiation might be being employed for the crop circles task, how could that same beam technology be used to just swirl loose material objects on the ground, to push crops to the ground and, possibly, to uplift objects, animals and humans. The answer seemed to lie in the degree of focusing of the beam. The Bull phenomenon had been produced by a cylindrical beam; that is, it had not displayed signs of focusing. The rotating electro-mechanical energy within it seemed only to have been scanning the ground.

If, however, such a beam were to be focused in a convergent manner onto the ground, the gravitational energy within it would, conceivably, produce a gravitational force field, which might accelerate objects with mass towards the ground in an artificial way. This might account for the overpressure experienced by the crop stems. On the other hand, if the beam diverged between projector and the ground, a negative (upwards) gravity force might be experienced by material objects on the ground below. So, by using one type of projection system, it would be possible to use it to scan, to press down and to uplift whatever was on the ground immediately below the field-generating aerial craft. All this is shown diagrammatically by Fig. 47.

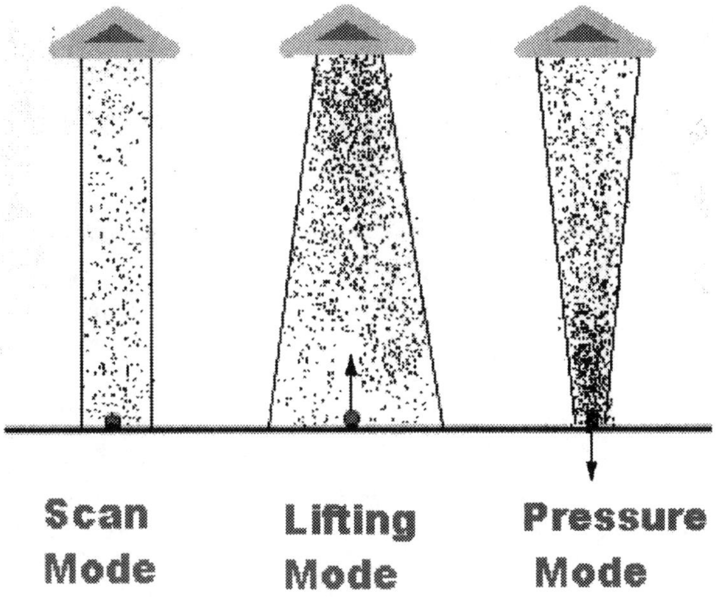


**Fig. 47**

## Jennifer Jarvis

When we were once again in Wiltshire during the summer of 1995, we booked in with the Bulls for B&B for a few nights. On arriving downstairs for breakfast the following morning, we were pleased to find we were in the company of a young, good-looking, woman. After introductions all round, it seemed that we were all on crop-circle missions. We discovered that our new contact was Jennifer Jarvis, who was taking time off from family duties in Canada to become acquainted with the strange phenomena being reported from Southern England. Having been reared in England and having parents and a brother still living in the South, she was not on altogether unfamiliar territory, even though she had been settled in Canada for quite a long time. As we were to discover, our new companion was a bundle of energy and was always ready to follow a new lead with great enthusiasm. The news about those 'tubes' witnessed by Mr. Bull in fields close to the farm a few years before was more than she could bear. She wanted to see things like that for herself and, indeed, she had already become acquainted with the then current crop-circles 'scene' by travelling round the sites. When we left and went our different ways for that day, Marion and I had three special missions in hand. We had arranged to meet up with Roy Rowlands to video the 'scene' for our forthcoming Part 2 video recording, to meet with an American tour group we had met and assisted on previous occasions and to contact the CSETI group to arrange a video interview with **Dr. Steven Greer**. By the end of that day, all that had been accomplished. The next day was devoted to the videoing task with Roy Rowlands. That was to be our last night at the farm before travelling back to Torquay, Devon, our new (retirement) home base. Next morning, Jennifer and I spent a considerable time at breakfast, comparing notes and promising to keep in touch after we had all returned home.

My lasting liaison with Jennifer Jarvis began three days after we had arrived home, with a telephone call from Jennifer in Canada. Following onto this, I supplied Jennifer with a timings graph centred on her location to the west of Toronto. As I was not available via e-mail, all correspondence between us at that time had to be mainly by Air Mail, with the inevitable lag of 5 to 7 days between postings and receipts. A card posted by Jennifer on August 18th informed me that she had decided to apply for CSETI training. There was also a request for an Alton Barnes, Wiltshire, graph, to include the Avebury area, in preparation for her next visit. Then began a stream of letters over the following few months. At first, they informed me of UFO events reported to her by Canadian associates which, invariably, had been found to lie on the lines of the graph for her Canadian location. Gradually the area of application expanded with reports from other places some appreciable distance removed from Jennifer's home area. Further graphs were produced for those areas. And there was one other graph requested for Moffat/Crestone, Colorado, USA, where Jennifer was hoping to be located during her first CSETI training course. All this within a month of our first meeting at the farm in Wiltshire. I was only too pleased to co-operate! Here, at last was someone who really wanted to test the AT in different places and in different ways. In return and to help me locate places mentioned in her reports, Jennifer supplied me with two detailed maps of South Ontario and up-state New York. The CSETI training course in Crestone took place in the summer of 1996 and Jennifer returned from it fully revitalised and wanting to get started locally. She drew together a small team of friends and associates and they began going out to selected sites for skywatches.

During the Spring of 1997, Jennifer and her small team of associates had begun to occupy, quite regularly, a site on the shore at the north-west end of Lake Ontario, which had been found to provide much aerial excitement. This location, near Oakville, looks south over the lake towards Hamilton and Stoney Creek. The famous Niagara is over there to the east-south-east. Quite frequently the group had begun to observe strange light phenomena out over the water, looking towards the south-east of their location. Jennifer brought her zoom-lens equipped camcorder to bear on these anomalous lights and began creating a video recording of them. The most outstanding and puzzling feature seemed to be a pillar of light or 'light pillar' which seemed to precede the 'light-ball' aerial activity. These light balls seemed to appear from different directions to converge over the light pillar and then to merge with it, as if diving into the lake. When Jennifer first told me of these events, I had studied the maps she had provided. The far shore looking south-east was measured to be about 24 miles away, and my first thought had been that the group may have been viewing the lights of aircraft landing at St. Catharines Airport, near Niagara-on-the-Lake, Ontario. St.Catharines was perfectly in line with the bearing I had been given for the lights. Given certain atmospheric conditions, distorted 'mirage' images of aircraft landing lights and even airport ground lighting might be produced. I suggested trying to view the same phenomena from somewhere along the lake shore to the south and west of the group's usual position. Jennifer at first rejected the idea she was spending precious time recording mirages, but she would view the scene from a different position, if only to remove that possibility from my mind. The outcome was that the lights were still seen to be over the lake from a position nearer Hamilton. From the bearing Jennifer supplied, the lights then seemed to be located midway between Oakville and St. Catharines. My next suggestion was that the group should hire a boat to explore that area of the lake and, perhaps, to view the phenomena at closer quarters; however, Jennifer was not prepared to be that brave. No way could she be persuaded to get onto that water, especially at night!

So, the saga continues. Jennifer has now amassed much video footage of those anomalous lights, which she makes available to be analysed by experts. She has also produced a web site called 'ORBWATCH' and some of her footage and stills are displayed on this. I have produced a timings graph for the area and it seems the 'boys' (as Jennifer calls them) are following the global UFO timings rules as given by the AT. All the evidence seems to imply that there is some sort of ET base under Lake Ontario, to which ET exploration craft return between missions and before being retrieved. It would be nice to think so!

## The Tactics Study.

But there is yet more to be told about Jennifer Jarvis. As I related earlier in the section on Edward Ashpole and the NIDS Essay, I did not have access to the Internet until 1999. This did not deter Jennifer from sending sightings information, sometimes garnered from the Internet, by Air Mail. One day a fat package landed on the floor of our porch. It was a printout, from Jennifer, of much sightings material extracted from the famous Filer's Files web site. She thought I might like to process some of it. On examination it seemed to be the most comprehensive set of recent reports I could have ever hoped for. The data were for the period 1996/7. I began processing immediately, because I had been presented with an opportunity to investigate how the overall strategy revealed by the AT had been used tactically during any given 24-hour period. The outcome of this prolonged study was very revealing. I discovered how, for example, different paths in space and over the ground had been used to access a given targeted location; how ground tracks with different inclinations but generated from the same equatorial generator had enabled target zones in widely separated global locations to be accessed simultaneously, and so on. One possibility I did not find evidence for was the obvious linking of several global sites by a single track during any 24-hour activity period. (As will be told later, that discovery had to wait until early 1999.)

To summarise this study, a paper was produced in March 1998 with the title,' **GLOBAL U.F.O. ACTIVITY --- A STUDY OF TACTICAL TECHNIQUES.'** A copy of this paper was posted off to Jennifer with my thanks for having made the study possible. Jennifer's excited response was a request for permission to display it on her new web site, which of course I gave very gladly. But that created a big problem for Jennifer. Because my computer equipment was so antiquated and the report was in the Wordstar format, it became necessary for Jennifer to re-type the entire 12-page report, tables and all. Such dedication! The result is that that paper can still be accessed via her ORBWATCH web site.

By early 1999 I had equipped myself with an updated computer and got myself 'on-line' for the Internet, complete with e-mail facility. A Press Release had been issued on February 24$^{th}$, 1999, by the editor of an Italian magazine, 'Notiziario UFO', providing information about a 'wave' (intense outbreak) of UFO sightings in Northern Italy the previous night. Jennifer Jarvis was again instrumental in bringing this item to my attention. The Press Release stated that sightings began at 7 pm. (European Standard Time) in two towns in the region of Venice and were followed by sightings from all over Northern Italy, including Rome. Fortunately for my purposes, details of some of these events were also given, so that they were able to be processed in detail. Sightings had ceased just before 9 pm.

Before beginning the numerical processing, I needed to satisfy myself that the details given of the sightings ruled out the possibility that a brilliant conjunction of Jupiter and Venus in the sky that night had been the cause of all the excitement. I was pleased to find that I could eliminate that solution and, so, commenced processing. The outcome was that several common tracks over the region were able to be identified as having been used during that 2-hour period. Two tracks could have been used to retrieve the exploration craft just before 9 pm. They were virtually at right angles to each other and served the Venice and Turin areas, respectively.

It so happened that I had received information from contacts in Reading, England, that UFO-related activity had been taking place between 6:30 pm and 8 pm GMT on the same night (February 23$^{rd}$). A few days later, I had been informed by contacts in the Irish Republic that UFO activity had been reported in various places between 7:30 pm and 9 pm GMT . One of those places was Boyle, Co. Roscommon, where activity had ceased at 8 pm. GMT. The processing of the Reading and Boyle events revealed that both these sites shared the same track line as that key departure option running between Venice and Valdagno in Northern Italy. What was even more interesting was that 9 pm in Italy is the equivalent of 8 pm GMT in Britain and Ireland. Here, at last, was *evidence that the same track line is sometimes used to access widely separated locations.* Fig.48 shows the single track linking Venice, Valdagno, Reading and Boyle. This has been extracted from my unpublished report, dated March 1999, with the title **'ANALYSIS OF THE U.F.O. REPORTS OF 23RD FEBRUARY, 1999.'**

My informant in Ireland had been Mr. Eamonn Ansbro, FRAS, ( see Plate 3), an amateur astronomer of some renown, who had initiated a link-up with me some years earlier.

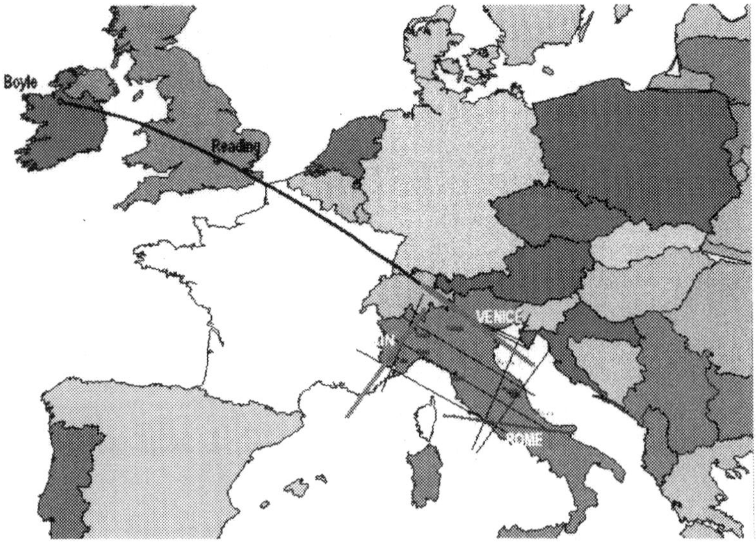


**Fig. 48**

**Plate 3**

## Eamonn and Catherine Ansbro.

Soon after the publication of Edward Ashpole's book of 1995/96, 'The UFO Phenomena', Eamonn Ansbro contacted me by letter in January, 1996. He had viewed a copy of the Part 1 video Roy Rowlands and I had produced during 1994 and had also read Edward Ashpole's account of my Theory from the book. These had impressed him and he hoped I might be able to assist him. He had seen anomalous lights in Ireland and had set up an observational group in County Kerry to observe some of those frequent visitors. They had been in touch with CSETI and had decided to adopt and adapt some of CSETI's protocols. The lights in the sky had seemed to behave in an intelligent manner and he felt sure they were the result of technology not of this Earth. As a result of this conjecture, and given his knowledge of astronomy, he had spent some time trying to fathom out where in our galaxy they might be coming from. He sent a marked up star map with his letter and asked if these paths in space tied up in any way with my own discoveries. Alas, they didn't appear to link up at all. I asked for details of his group's location and any recorded timings for the observed events and suggested we could proceed from there. I supplied him with a timings graph for that area and then correspondence ceased until about a year later. Eamonn began to report good correlation between the observations and the Theory's predictions. In fact, he had ventured to arrange skywatches based on my suggested interpretation of the graphical output and on the times that then seemed to be favoured. He'd had a lot of successes using this technique; so much so that he'd ventured to invite TV cameras to accompany his group and, apparently, the TV teams had not gone away disappointed. I marvelled at his success rate and was told that the group had been further assisted by the use of CSETI-type protocols.

Then followed requests for graphs for other parts of Ireland, especially, Dublin and Boyle, Co. Roscommon. I was able to show that Dublin was linked to Boyle by a common ground track; was linked to the Co. Kerry site by another; and that the latter was linked to an area to the east of Boyle by yet another common track. These three tracks produced an 'Irish Triangle', as shown by Fig.49. Thereafter, Eamonn began to try to communicate his experiences and the Astronautical Theory to as wide an audience as possible. Emboldened by his successes, he wanted to prepare people for the probability that a formal landing would take place in the very near future. (As far as I know, he is still waiting for that momentous event.) During this period (1997-1998) the Ansbros were spending more and more time in Dublin, where Eamonn's precision optics business and the family home were located. After his elderly mother's death, the decision was made to move out to Boyle, with a view to setting up an advanced observatory there.

## Personal Contact.

During July 29th to 1st August, 1999, Eamonn and his American wife, Catherine, were our guests at our home in Torquay. They had requested the visit because they really wanted to get a full understanding of the rudiments of the AT. My memory of that short period is one of providing an intense 'crash-course', during which I attempted to explain the 3-D geometry involved by all means possible.

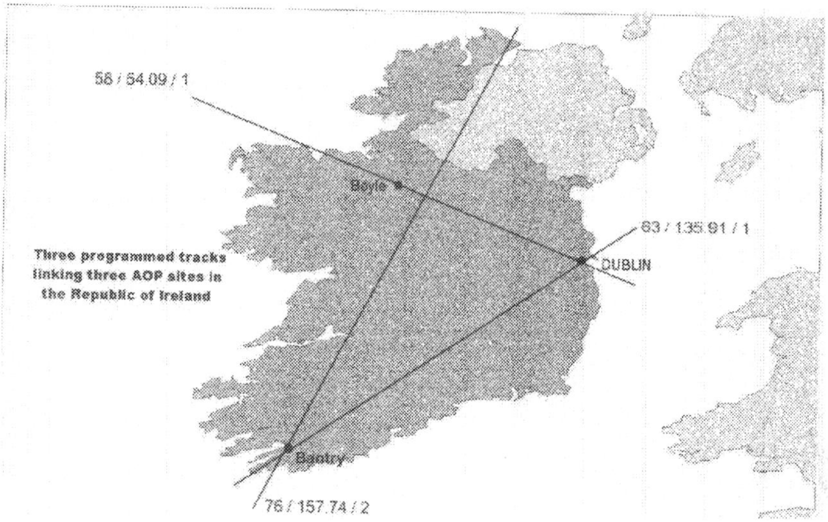


**Fig. 49**

My other impression was that Catherine was even more interested in the details than Eamonn appeared to be, as many of the searching questions came from her. (I was later to discover that her father was a professor of theoretical physics in an American University and that she was also highly qualified, which explained her level of interest.)

Following onto that visit, Catherine began urging me to produce a comprehensive scientific paper to explain how the theory developed and how the solutions 'dropped out'. I recoiled from that task because, as I tried to explain, much of that voluntary work had been done manually, with lists of data and early graphs existing only in pencilled form. It would be a major task to reproduce more than 20 years of early work to the standard now required for scientific papers.

## The OSETI III Conference, 2001

Further urgency was then given to that request when, early in January 2001, I was informed by Catherine that Eamonn had made contact with OSETI (Optical Search for Extraterrestrial Intelligence), an organisation formed by a group of American optical astronomers who wanted to expand SETI's radio-frequency activities by searching the skies for optical evidence of the presence of ETI activity in the solar system and beyond. A conference, OSETI III, was to be held in California later in January and a paper submitted by Eamonn had been accepted for presentation at that conference. What was more, the title he had selected was, **'A New OSETI Observatory to Test Dutton's Astronautical Theory'**. Understandably, the Ansbros had been greatly encouraged by this acceptance of a controversial paper, but, as Catherine pointed out, it would emphasise the need to have a paper on the derivation and nature of the Astronautical Theory ready to supply to any interested scientists.

In the face of this welcome coercion, I decided I would have to try to meet this anticipated need. So began two months of concentrated effort to produce that paper.

Meanwhile, the Ansbros flew off to California and to the Conference. Eamonn's paper was scheduled to be delivered on January 23$^{rd}$ and all communications between us ceased for a while. Some time after the event I learnt that Eamonn's paper had caused uproar among some of the astronomers present. They suspected he had been sent to devalue the conference by 'sending them up'. They had insisted that his paper could not be published in the Proceedings of OSETI III unless it could be re-written to change its emphasis. Not wishing to have his material completely left out of the official records, Eamonn then had to apply himself to the task of meeting their requirements. The revised paper was eventually published under the title, **'A New OSETI Observatory to Search for Interstellar Probes'** and the development and nature of Dutton's Theory was given only three small paragraphs of coverage. It provided yet another example of the manner in which my work has been deliberately marginalised over the years by the scientific establishment.

## The 'Basics' Paper.

To explain the 'basics' of the Astronautical Theory, I had to delve into archived material going back to my early work in the late 1960s. As I had anticipated, there was a lot of it and it was generally in scribbled note form, scribbled listings and pencilled graphs. After all, this study had begun as an interesting spare-time hobby and, at outset, I had no idea whether it was going to lead anywhere. But as the pieces of the 'jigsaw' seemed to be creating sections of an overall picture, I could not just abandon this activity, even though sometimes many months might pass between one inspiration and the next. So you can imagine how difficult it was going to be to bring all this together. To make things more difficult still, there were several 'blind alleys' followed during the course of the development and these would have to be sifted out.

I decided that the paper I was trying to produce would have to use photocopies of the old graphs and diagrams and some of the old listings. The task took two months to complete. The paper, including appendices, consisted of fifty A-4 pages. The main scripted report occupied 19 of those pages and there were, in addition, 23 tables and Figures. Two appendices accounted for the remaining pages. The title I chose for this paper was:

**'Puzzling Global Reports of Strange Aerial Craft (SAC): A Comprehensive Technical Assessment and a Testable Theory. 1. Early Work and the derivation of an Astronautical Theory.'**

I can remember that my original title had been modified after consultation with Edward Ashpole, who also suggested changes to the Abstract. When I sent a copy of this paper to the Ansbros, Catherine was especially overjoyed and wanted to disseminate it to people she knew, some of whom were scientists. I can't remember who those people were, but I'm fairly sure her father was one of them. She also purchased copies of Parts 1 and 2 of the video recordings, in (expensive) NTSC format, two of which were going to her parents. (The Part 3 video was about to be recorded later that year, but was not available then.) So, the liaison with the Ansbros had borne fruit for me and I was extremely grateful for all their help in the promotion of my work, in association with Eamonn's own.

Since then, others have got to know about the AT through Eamonn's activities. Unbeknown to me, UFO Magazine's Editor, the late Graham Birdsall, accompanied by a member of a group called Interseti, Michael (Mike) Murray, visited Eamonn's observatory in Boyle to interview him and to view the facility. They arrived there on July 23rd, 2003, and viewed the observatory the following day. A full report of the interview and photographs were published subsequently in the September 2003 edition of UFO Magazine. Before publication I received a surprise telephone call from Mike Murray to introduce himself and to tell me about Interseti. He told me he'd been over to Ireland to view Eamonn Ansbro's observatory and whilst there, he had seen a copy of a paper I had produced. He had been very interested by it and he wondered if I would allow him to publish any of my work on the Interseti web site. Well, I was only too pleased to oblige and sent him a copy of an article I had written called, **'Outrageous Discoveries'**. Very soon that appeared on the web site. Soon afterwards I received another request. He had a copy of a paper by me dealing with UFO Tactics. Of course, I guessed that would be the paper that Jennifer

Jarvis was already displaying on her web site, but I thought there'd be no harm done by having another web site displaying it. Mike Murray must have been given a copy of it by the Ansbros, but he would have the same problems of transfer to his web site as Jennifer had had. Some time later that paper also appeared on the Interseti web site, but not in an altogether accurate or complete form. The article I'd supplied was described as the 'easy' version and the paper as the 'difficult' version of my theory. (Actually, neither of these documents contain full explanations of the AT.)

Following onto the hostile reception his paper to OSETI III had experienced, my advice to Eamonn was that he should concentrate on obtaining incontrovertible physical evidence of the things in the sky for presentation to the scientific community. In my view, only if such observations were eventually to be accepted would it be appropriate to mention that they had been made at times suggested by the AT. Since then I have heard little from the Ansbros, so I presume Eamonn has taken that advice seriously. But there was to be a further favourable 'spin-off' from Eamonn's appearance at that conference in California.

## The SETV Interest

Through, I think, Dr. Allen Tough's presence there, together with several members of an OSETI sub-group promoting the idea of SETV (Search for ET Vehicles), I began to receive e-mails from two other members of that group. Besides Dr. Tough, one of those people was Scott Stride of the famous Jet Propulsion Laboratory (**JPL**) and the other was an astrobiologist, Dr. David Darling.

**Scott Stride** was intrigued by my suggestion that permanent bases had probably been established outside the solar system, in the distant past, by the ET perpetrators of the surveillance activity. In a series of lively e-mail exchanges we examined the practical difficulties that would have been encountered and the problems of communication with those bases from the ET's home planet. It was a very stimulating and thought-provoking exchange that finally 'ran out of steam'.

Allen Tough and David Darling were both keen to have copies of my 'basics' paper for consideration. After receipt of his copy, Allen Tough's e-mailed response was extremely enthusiastic, with many congratulations on having been able to create such a paper and wanting to see the work tested in a rigorous scientific way. He felt his contribution would be to create a protocol for those tests, since that was his special application. After several e-mailed exchanges, he began to realise that the task might not lend itself to the normal processes of checking out a theory, not least because, for observational investigations, the major unknown would be where and when the phenomenon might manifest itself. (I suspect that the NIDS researchers probably made the mistake of going to sites where UFO events had happened in the past, but were no longer manifesting. CSETI and Eamonn Ansbro had the right idea. *Go to the places where activity is currently being concentrated.*) The other means of testing the AT would be to create a new database from attested reports and then to process those reports following my methods. Dr. Tough finally conceded that it was not going to be easy to create a really rigorous protocol.

Dr. Darling was also appreciative and intrigued by the paper. After several e-mailed exchanges about it, he offered to install it on his web site. It turned out that he had an established and NASA-acclaimed web site with the latest news on space exploration displayed and an encyclopaedic facility from which could be accessed information on all manner of things, some of them extremely controversial. Before attempting to install that large 'basics' paper, he wondered if he could display an article on the topic, written by me. Unfortunately, after looking through the articles I had produced for special purposes over the years, I decided I would have to create a new one. With a change of title, this turned out to be acceptable and is still accessible under the 'SETV' heading of the Encyclopaedia. (Dr. Darling qualified in Britain and worked at Jodrell Bank before moving over to America, where he had settled until the time of our initial exchanges. Soon afterwards, he and his wife retired back to England, but his web site activities continue unabated. I am extremely grateful for the support he continues to give to this life-time's 'labour of love'.)

## Dr. Massimo Teodorani

It would be greatly remiss of me not to mention the interest shown in my work by Italian astrophysicist, Dr. Massimo Teodorani. This link-up originated after the Ansbros had attended a conference held in Norway to consider the anomalous optical phenomena (AOP) that had been monitored at a remote place called Hessdalen, since the 1980s, by Norwegian scientists. Some years before that conference, I had analysed the timings of the events monitored in 1984 and had communicated my findings to the leader of the project, Asst. Professor Erling P. Strand. Information about the AOP in Norway had caused Massimo Teodorani to lead a team to investigate the activity in 2000 and then to return to Hessdalen with a better-equipped team during 2001. Dr. Teodorani produced a paper reporting the findings of his EMBLA 2001 mission and, without warning, I found myself presented with a copy of that very fascinating report. The lights in the sky (AOP) had been photographed, videoed, and spectroscopically analysed and, from the information gained, the nature of those mysterious lights had been determined. They were, generally, very strange ball-plasmas, which were observed to be inexplicably complex in form and in behaviour. One specimen seemed to be, unmistakably, a structured disc, for which no natural explanation of any kind could be found. By reporting these facts truthfully, Dr. Teodorani subsequently found himself vehemently under attack from another scientist for concluding in that way.

All that was the background to the link-up, by e-mail, with Massimo Teodorani that followed my receipt of his paper. He became conversant with my Hessdalen findings and, during the course of our exchanges, told me of approximately 30 AOP sites scattered in various parts of the world. I asked for those sites to be located for me and then ran the AT programs for each one. I was able to report that at least 10 of those sites had been found to be linked by common AT ground tracks. This impressed Dr. Teodorani and, subsequently, he told me he had referred to my theory in one of his recent papers. Then followed a prolonged gap in communications, when we had no cause to write to each other again; but recently I have received, via the Internet, a copy of a paper he has produced of an analysis of UFO events in the Hudson River Valley **area** of the USA. This work had been facilitated by data provided by Jennifer Jarvis.

## Michael Hesemann.

My wife, Marion, and I met Michael Hesemann for the first time, in August, 1991, at an informal meeting of crop circle researchers assembled at the Waggon and Horses Inn, Beckhampton, near Avebury, Wiltshire. As the meeting drew to a close, we were approached by a fairly tall man, with a rounded and cheerful face and well- trimmed dark beard, who introduced himself as Michael Hesemann. He and his cameraman, Peter Heppa, had come over from Germany to collect information about the crop circles. He had been interested by my claimed discovery that they were being produced by an advanced airborne projector, which was beyond human technology and which we would be unable to replicate. He was even more interested when he discovered, before my retirement, I had been a senior projects engineer in the Future Projects Department of British Aerospace, Manchester. He wanted to know everything about the bold claim I was making concerning the creation of genuine crop circles. After a prolonged explanation from me, Hesemann wanted a videoed interview with me there and then. Unfortunately, the time was by then racing on towards midnight and the weary landlord was telling us all to leave. We left the premises, as asked, and wandered across the road to the overspill carpark (where we had parked), still talking with Michael Hesemann. He suggested he could interview me out there in the unlit car park if we used our car headlights to illuminate the scene. He was going to pretend we were in a crop-circle at night. This caused me great amusement because any slight shuffling on the gravel seemed to give the game away. There was also the little problem that it was difficult not to get a small post-box in shot. Just as we had sorted ourselves out at about midnight, drizzle rain began to fall. "No problem", I said. I had an umbrella in our car. This was taken out and Michael and I stood under the umbrella for the interview. Then Peter, the cameraman, complained that the video recorder was getting wet! On Michael's suggestion, we two could stand in the rain and Marion could hold the umbrella over Peter and the camera. After trying this out, Michael and I were getting rather wet as the drizzle became more intense and Peter wasn't at all happy with the pictures he was getting. So, Michael at this point in the proceedings decided he would have to change his plans for the following morning, because he was determined to get my story on tape.

The next morning dawned sunny and bright and Hesemann rang us at our B&B to arrange for us to meet him in Marlborough. He would then drive ahead of us to guide us to a recent pictogram at **Clench Common**, near Marlborough. We all arrived on site at soon after 11 am. for that interview, this time, to be genuinely videoed in a real crop formation. Marion and I remember that interview with some enjoyment and amusement. The sun was by then high in the sky and the shadows on my face were very dark and deep. Just when it seemed we were all lined up for the interview, Peter protested to Michael, "He looks like a skeleton!" So we all waltzed round again until the shot pleased our discerning cameraman. The results of that effort were later incorporated into the Hesemann video, **' Das Mysterium der Kornkreise'** (Crop Circle Mysteries) which, on a later occasion, Michael presented to me. (It was an English version with a German title sleeve.) The encounter with Hesemann was also described in his book with the title, '**Botschaft aus dem Kosmos**', a copy of which he presented to me with a glowing inscription when we met again in July 1993. Unfortunately, this was the complete German edition, so I had to content myself by just looking at the photographs and picking up the meaning of the odd word here and there.

On a later occasion, Marion and I sat at a table with Hesemann and his cameraman outside a deserted winery, while I tried to explain the nature of the Astronautical Theory. It was not easy, but I thought every attempt to break through into public consciousness with such potentially important discoveries was to be considered worthwhile. I discovered some years later that Hesemann had tried to summarise all I had told him in one paragraph of his book, **'UFOs --- The Secret History.**' Unfortunately, he'd got the story wrong and had implied that the words were mine by using inverted commas. When I became aware of this, I registered a complaint with him by telephone, but I don't think he ever changed anything in subsequent editions.

I think the last time we met up with Michael Hesemann was at the 1993 crop circle conference in Bath, Somerset. He introduced me in such flattering terms to his German companion that I thought he must be joking! (He had a mischievous sense of humour.) However, he assured me later that he had been very serious. Later still, he requested timing graphs for several places in Germany. These I supplied and, after that, I heard nothing more from him. However, I have received two visits from Peter Heppa to provide him with two interviews, on an 'old friends' basis, for his own video productions. Peter told us that Michael Hesemann has now returned to his original archaeological studies.

## Joyce Murphy.

We first encountered Texan Joyce Murphy and her husband, Pike, during our travels in Wiltshire in the early 1990s. Joyce had formed a tour company for the purpose of taking small groups of Americans to global locations featuring UFO and paranormal activity. With Pike acting as her co-ordinator and driver of the hired mini-bus, on that occasion they had brought a group to England to view the crop formations, meet the people involved in research and to watch the skies over the fields. Their base was the 'The Merlin' in Marlborough, a small licensed hotel which had come to be regarded as a kind of 'Mecca' for overseas visitors. (Michael Hesemann and Peter had often stayed there).

Our involvement in Joyce's activities probably began in mid-July 1994. We met with her and husband Pike at 'The Merlin'. We guided the group on a tour to Avebury and its surroundings and then proceeded to 'The Barge Inn' for an evening meal. The following afternoon we had arranged to meet again at the Merlin and were introduced to members of the American UFO research group, MUFON, and to others then present in the tree-shaded area behind the premises. It was arranged we should be there again on the following afternoon to meet up with George Knapp, a well-known American broadcaster and UFO researcher. (I think Joyce had suggested that he should interview me.) After a late arrival, Knapp and his crew duly set about recording all I had to tell them. The session was a long one and probably took up a couple of hours. I can't remember the topics we covered and I don't think I ever heard a broadcast of the material. But Knapp was the first of two well-known American broadcasters who were to become familiar with my work as a result of Joyce Murphy's efforts. The other was to be Jeff Rense, who had a regular and popular American, coast-to-coast, radio broadcast on UFOs and strange happenings, each week. More about that later.

Before we parted company with the Murphys in Wiltshire, we invited them to join us in Torquay for a short stay before they returned home. After ensuring that all members of their tour party had been safely installed onto their correct and various aircraft at Heathrow, Joyce and Pike managed to do that for a couple of very enjoyable days. On our little tours of the area, Pike marvelled at the sight of cows standing upright on the steep hillside pastures of Devon. The cattle he was accustomed to seeing were usually grazing on acres and acres of Texas flat land. I suggested that, perhaps, Devon cows must have developed adjustable legs. After hearing of the UFO activity and occasional crop-circle manifestations in our part of Devon, Joyce resolved to spend some time on her next English tour introducing her party to the area and to informal presentations of my research.

Soon after they had returned to Texas, I received a gift parcel from Joyce and Pike. This supplied me with detailed time zone information for all the states of America, and for the rest of the world, with the exception of Mexico and Canada. Also, contained were items of floppy disc software giving details of all American States. They had remembered the difficulties I'd had (before the coming of 'Google Earth') in tracing the locations of smaller American towns and determining the Time Zone each place was in; also, the periods of Daylight Saving Time applying to each location. This was a gift indeed! --- and facilitated all my processing of events reported before the year 2000. I will be forever grateful to the Murphys for that generous act.

We corresponded over the following years and one year Marion and I were asked to guide a minibus full of mostly dowsers to ancient sites in South Devon. Unfortunately, on the chosen day, a frontal system and a deep depression had settled over the South West, creating low cloud, thick mist and outbursts of heavy rain. We were at a loss to know where to go with this group because most of the ancient settlement sites were located on Dartmoor, then buried in mist and rain. However, we managed to find two sites in our coastal area to satisfy their needs. En route between these sites, the elderly Californian couple sitting behind me leant forward and asked if the rain would soon stop. I had to answer negatively and then to go on to explain how long-lasting atmospheric depressions proceeded quite frequently to empty some of the Atlantic Ocean over us. Their horrified response was, "But this is your Summer!" On that occasion, the minibus was driven by Ruben Uriarte, who turned out to be MUFON Director for Northern California. We found him to be an amicable man and very reserved. He was good company and appeared to let nothing upset him. He was to become a replacement for Pike on all subsequent trips.

Many of the link-ups with Joyce Murphy had to be done by Fax or telephone before I went on-line and Joyce usually initiated them. One such initiative resulted from Joyce's having informed a famous radio host, Jeff Rense, based somewhere on the West Coast of the U.S.A., that he would find my material irresistible. Joyce told me it would mean that I would be contacted by the show's producer on the morning of July 23rd, 1998, at 5 am., and then put 'On Air' to talk with Rense. Would I be able to manage that? Of course, I would do my best to comply, even though it would mean setting my alarm clock for 4:45 am. and I would have to have collected my wits together for that 5 am. call.

Sure enough, all went to plan (I think). I was asked very sensible questions and I hoped I was giving sensible answers! I recall there was a 'natural break, after the first half-hour and the interview then continued for the remaining half-hour until 6 am. BST. Rense seemed to be pleased to talk with me and there may have been questions from listeners for me to answer. It's all rather a vague memory now.

During the next visit of Joyce Murphy's group to Devon, the following year (1999), I had something quite momentous to report to her. I told her about Fred Lewis-Goodwin and his mysterious box on the hillside behind the Torbay Holiday Motel (see PHASE 4). Predictably, she wanted to meet him, so I arranged this with Fred, who had, in turn, got the permission of the owners/managers of the Motel to hold the interview in its main dining room. (The Booth family owners turned out to be very helpful and co-operative on several occasions.) Joyce interviewed Fred and immediately afterwards sent off an e-mail to Jeff Rense to recommend that he should have Fred L-G on his next programme. Subsequently, we were told this would take place at 5 am. on the morning of August 12th, 1999, and that I would also be contacted. Sure enough, that 5 am. call came, as scheduled. I was told by the producer that they hadn't yet been able to contact Fred and that I would be called upon to outline the happening and to answer Rense's questions until Fred could be brought on-line. Well, the first half-hour went by and they still hadn't managed to contact Fred. The concerned producer spoke to me during the natural break. Could I give them Fred's telephone number, in case they had been calling the wrong number? Very quickly I was able to oblige and, apparently, they **had** been calling the wrong number! When the programme resumed, Fred was there to answer Jeff Rense's questions for himself and I think I became more of a listener from then on. What a pity Fred was not featuring for the entire hour, as I had done previously. I was given to understand that the Rense programmes are archived and can be replayed from the station's web site.

Joyce Murphy promoted interest in her tours by issuing a magazine with the same title as her business, **'Beyond Boundaries.'** For some time I received complimentary copies of this and was even featured in one of them. Joyce also came to regard me as a voluntary correspondent for England. Suddenly, everything ceased. I was unable to get in touch by e-mail and letters and cards sent by post were not answered. Quite recently, contact was again made and I discovered that the Murphys had retired from globe-trotting and were enjoying a peaceful existence on their Texas farm though, regrettably, it seems that Joyce is now confined to a wheelchair.

## Ivan Zemanek.

In October, 1998, some months after the Jeff Rense radio interview, I received a long letter from a man living in Washington State who gave his name as Ivan Zemanek. It was a letter of almost unbelievable generosity. Ivan had listened through the Rense interview and had been inspired by it.

Having heard mention of Edward Ashpole's book, **'The UFO Phenomena' [18]** during the programme, he had gone out of his way to obtain a copy and then found it to be compulsive reading. He'd noted that I was appealing to astronomers, both professional and amateur, to test the Theory and went on, "--- I would like to help you in my own modest way if I can".

***"If your theory is proven to be correct it will change the world forever".***

He felt that officialdom had withheld that kind of information from the public for a very long time and left the burden of communicating it on the shoulders of private individuals. He had registered my suggestion that an inter-linked global network of observatories, working on an observational project, might provide one way of validating the AT. He had done some preliminary research and discovered that there were about 50 universities in the United States offering graduate degrees in astronomy. He imagined that there were probably as many, if not more, in Europe. Ivan then went on to offer to mail information about my work to all the graduate schools in the USA and to include contact details for anyone who wanted to know more. He added, ***"If you are interested in knowing why I am doing this, I can tell you that I do not expect any reward for my work --- monetary or otherwise".*** He added that his father had been a struggling inventor and his reward would be that he had helped in his own way to help me to succeed. He just needed my permission to proceed with this activity.

To say least, I was overwhelmed by this man's generosity and self-effacing manner. But how could I accept his offer without telling him about failed attempts in the past to interest academia? Without telling him about Edward Ashpole's continuing efforts to break through; about the NIDS Essay and the general lack of constructive response to it? I decided that I had to thank Ivan, profusely, for his willingness to commit himself to such a task but, also, to warn him with those caveats.

He was undeterred by my pessimism and asked for further details about myself and the project, to include in the packages he was hoping to send out. What could I do further to deter him? I decided to accept his offer graciously and to send further information about my work and background to him, as requested. So began a frequent exchange of Faxes between us (Ivan had no e-mail facility) which lasted until 2003, when I felt compelled to tell him not to waste any more of his time and money on attempts to open closed minds. By then, he'd told me he had sent out packages to over 200 academic establishments in the USA and Europe.

During that period of our co-operation, Ivan had given me the names of important contacts he felt would be contacting me, but they never did. Also, being an avid reader himself, he mailed several expensive books to me, because he felt they would provide useful information and, perhaps, help me to find more contacts. I was so overwhelmed that I had to ask him not to continue with that practice, because much of my time was being spent on further development of the AT and, simply, I did not have the time to consider the books in depth.

As a result of Ivan's activity, I imagine few academic institutions in the Western World should have been left in ignorance of the Astronautical Theory, but did they just 'bin' the information as unsolicited mail? We'll probably never know. Ivan and I are still in touch. Each year we exchange Christmas cards and, occasionally, I'm able to assist him and his brother, Paul, with studies of their own. Being so modestly self-effacing, Ivan will probably be embarrassed by this exposure of his unselfish activities, but I could not possibly leave him out of this book.

## Kazuo Ueno.

As told in Phase 3, I first came to know Kazuo Ueno during the 'Blackbird' crop-watch in 1990. He had been passed over to me by Colin Andrews, in the hope that I would be able to answer his questions. Standing on the edge of the hill fort's ditch, I had tried to explain the nature of my UFO research and used the hand-drawn timings graph of the area to demonstrate how the times of new events might be related to the lines on the graph. The outcome of my analysis of the spiralled lay of the crops in genuine circles was also of interest to him. He wanted to write an article about all this for a leading 'glossy' Japanese magazine (AZ) and, to enable him to get the diagrams and graphs on record, we arranged for him to visit us in Cheshire later. This visit took place as planned. Kazuo and his photographer, Emma Popik, from Poland, arrived at Stockport Railway Station at 2:20 pm on Saturday, June 22nd, 1991. As we got to know Emma better we discovered that she was also a writer and a publisher in Poland and was especially interested in the UFO work for those reasons. After a session photographing my material at home and an overnight stay with us, I drove the visitors to the Jodrell Bank Telescope's Visitors Centre for a morning visit.

Emma was particularly pleased and surprised to find a bust of Copernicus decorating the front lawn there. "But he's Polish!", she exclaimed and wanted to have her photograph taken next to the great man's image --- which, of course, I was pleased to do. That made her day. She and Kazuo were also pleased to know that several crop circles had been found in that rural area of Cheshire. We arrived home very late for a cooked lunch, which the (unwell) Marion had had ready for us at least two hours earlier. However, as usual, she understood and forgave us. Soon I had to deliver our visitors back to the Station. Emma returned to Poland and wrote requesting material about my studies that she might publish over there. This request was met but there were problems of translation of scientific terms into Polish. She has kept in touch spasmodically and has sent two of her published books to us. Unfortunately, we have no knowledge of Polish. Kazuo went away to write the article in the form of an interview with me and in November 1991, I received in the post a copy of the **AZ magazine**, sent to me directly from the publisher, with a 'thank you' cheque enclosed.

Having first to realise that Japanese magazines are read from back cover to front cover (right to left) and that articles were arranged in the same way, and that Japanese script is printed in vertical columns, from top to bottom, it became clear that Kazuo had produced a very presentable article, complete with the photographs and diagrams he had obtained during his visit. Of course, I had no way of knowing whether the script was accurate, but this seemed to be very pleasing way for the work to be presented to Japanese readers.

However, prior to that, on June 27$^{th}$, Kazuo had been in touch by telephone to request another visit, this time, with the editor of another Japanese magazine. This man would be particularly pleased to meet some of the local people who had had Close Encounters with periods of amnesia. Marion and I decided we'd be able to provide overnight accommodation again and a date was agreed for the visit after I had been in touch with the people concerned. When I met the visitors at Stockport Station after lunch on July 3$^{rd}$., Kazuo introduced his companion as Mitsu. Mitsu was tall and slim and had an air of languid self-assurance about him. It also became evident that he half-smoked long and expensive-looking gold-tipped cigarettes. He gave the impression of being a member of the Japanese aristocracy, an observation later confirmed by Kazuo. The first stop after leaving the Station was the river bank at East Didsbury, the site of the Jones family's encounter of 1979. As we talked and walked along the footpath beside the River Mersey, by sheer chance we met Linda Jones and her big dogs out for exercise. I had not pre-arranged this because a meeting had been arranged with Linda and others associated with Harry Harris' hypnotic regression activities later on. However, Linda was keen to talk to the visitors, so we walked with her back towards her home. One of the Alsatians kept making it clear to us that he didn't like us by growling in a threatening manner. What's more, he kept slipping his collar as we ambled along, and despite assurances from Linda, this was rather disconcerting. Anyway, our walk ended without incident and we said our farewells before returning to my home in Cheshire.

That evening we drove to Linda Taylor's home in Timperley, near Altrincham, Cheshire, to meet up with others in the Taylor's garden, which had been kindly made available to us for that gathering.

Plate 4 shows the happily gathered company, myself excluded, because I took the photograph. Mitsu, Kazuo, Harry Harris, Linda Taylor, Linda Jones, Linda Jones' husband and son are shown in the company of the late Mr. Arthur Tomlinson (standing) who, as Chairman of the Manchester-based D.I.G.A.P., had been very helpful to me over the years from the late 1960s. Kazuo and Mitsu conducted their informal interviews and took photographs. As I drove the visitors back from that meeting, I ventured to ask Mitsu, for the first time, which magazine he edited. I was quite taken aback when he told me he was the new owner of Japanese 'Playboy' and wanted to extend the appeal of the magazine by publishing interesting factual stories in it. For the sake of the people he'd just met and interviewed I wanted his assurance that the article he produced would not interface directly with sexually-explicit photographs. He gave me that assurance, but I felt I had to inform all the participants of this to see if they had no wish to be published in that way. Although I was very surprised, no one seemed to object. A photocopy of the article was eventually sent out to me by Kazuo, but there was no covering letter of thanks from Mitsu that I can recall.

**Plate 4**

Towards the end of this period of intense Japanese interest in my work, I received word from Kazuo that he had been talking to the science correspondent of ASAHI, a leading Japanese newspaper, about my analytical work on the crop circles and had aroused some interest. Following onto that I had a call from that person, Mr. Keiji Takeuchi, suggesting a meeting at my home, late afternoon, on July 24$^{th}$. I was extremely pleased that he was willing to travel from London to North Cheshire on the strength of Kazuo's recommendation and, of course, said I would be glad to meet him. What's more he would not require a lift from the Station to our home. He would hire a taxi. Well, the chosen date was a rather unfortunate one, because Marion had an important prior engagement to keep that afternoon and that meant I would also have to act as host to my visitor. And I am the worst host imaginable! When Mr. Takeuchi arrived on my doorstep I found him to be a be-suited, quietly spoken, cultured, gentleman who spoke perfect English. After welcoming him warmly and settling him at a table on which I had laid out the presentation material, rather foolishly, as it turned out, I offered to prepare refreshments for him. His response was that he would appreciate just a cup of Earl Grey tea. I knew we had that kind of tea, but not liking it much myself, I had never tried to make it. The result was that I presented my distinguished guest with what was probably the worst cup of Earl Grey tea he had ever tasted! (I apologise, Mr. Takeuchi. I am embarrassed even to this day.) While I was out in the kitchen, he had been poring over my material and he then began asking questions. The mathematical analysis part of the presentation was quite straight forward, but then we came to the question of how these patterns were being created. I knew we were about to enter forbidden territory. Fortunately, just about then, an avionics expert, ex-colleague and friend from British Aerospace, Mr. David Baldwinson, arrived to lend moral support. (David and I had talked about the kind of radiation that could possibly be responsible for laying down crops so precisely, without damaging the cell structure of growing plants. Like me, he had been unable to identify such radiation and had listened patiently when I talked of high frequency gravitational radiation. He had agreed to try to join us at that meeting.) Well, I did my best to break my ideas gently to Mr. Takeuchi, but I could tell that as soon as I talked of ET technology, not of this world, he began to lose interest, even when David tried to put the argument that there didn't seem to be any obvious 'other' solution. So ended a very important interview. I felt sorry that Mr. Takeuchi had probably been very disappointed with the outcome after he'd gone to the trouble and the expense of paying me a 'flying' visit. I heard nothing further from him after that and Kazuo told me he'd not seen anything published.

Probably the final attempt made by Kazuo to make my work known to a wide audience in Japan was to contact leading TV producers over there with a view to producing a programme centred upon me. One day he told me he had persuaded a major TV company to invite me. They had suggested a programme on location in the fields below Mount Fuji, from where UFO activity had been reported, in the hope that UFOs might turn up again at times predicted by my timetable. Not having the confidence in my ability to be in the right place at the right time (unlike Eamonn Ansbro) I told Kajuo that particular proposal was probably not a good idea. But, overriding all that was the sad fact that, on my early retirement, I had been medically advised not to fly long distances at high altitude and that the stress of presentations would be an added hazard. So, I missed out on what could have been a very fruitful and exciting venture. (Subsequently, I had to turn down a number of invitations to go on lecture tours in the USA ---eg. in Florida, Texas and California --- for the same reasons.) That's life, I suppose!

Kazuo has continued to communicate with us over the years since then and has been our guest in Torquay on several occasions, but, in recent years, he has returned to his archaeological studies. He is particularly interested in comparing the development of mathematics in Japan and in the West, from ancient times.

## References

[18] Ashpole, E. 'The UFO Phenomena' (book) Headline Book Publishing, London. 1995, 1996.

# CHAPTER 20

# Other supporters and promoters

## At the Beginning

Throughout the long years of my affliction with this study I have gathered many good friends and supporters of my task. Some of those encountered years ago are still keeping in touch, if only by Christmas cards or e-mails. Others have passed out of this world and yet others have simply passed on to do other things. One of my first benefactors was the late Mrs. Joan Nelstrop of Bramhall, Cheshire. She was the honorary secretary of **DIGAP** (Direct Investigations Group for Aerial Phenomena) when I became a relatively short-term member of it. As mentioned in earlier PHASEs, I met, Chairman, Arthur Tomlinson there and created a friendly link-up which lasted until Arthur's fairly recent death. Joan Nelstrop and I investigated a number of UFO reports in our area and travelled regularly together to the DIGAP meetings in central Manchester. As told in PHASE 1, I also encountered Peter Rogerson at those meetings and thereby was given access to his international catalogue of selected good quality UFO reports, INTCAT. Selected cases from this collection started the database for my global study. Joan Nelstrop and DIGAP facilitated expansion of this database by allowing me to copy the contents of another catalogue they had obtained from a Staffordshire group. Without the co-operation of these people and the efforts of the cataloguers my global study could not have got started.

Another key player in my early research period was Mr. Anthony Pace, an amateur astronomer, who became BUFORA's Research Director for a number of years. He and another amateur astronomer, Roger Stanway, had produced a detailed report of the Staffordshire sightings of 1967, **'UFOs --- Unidentified, Undeniable' [1]**, which I purchased a copy of from their Newchapel Observatory, Stoke-on-Trent. This report helped me a lot in those analyses described in PHASE 1. Tony created a new kind of magazine for BUFORA to encourage people to write scientific papers on aspects of the UFO topic. Alas, when Tony went on to do other things, the magazine ceased publication and BUFORA reverted mostly to organising talks, annual conferences and issuing a 'club' magazine. However, I was grateful to the organisation for allowing me to give a first lecture on my findings at the 1976 conference in Birmingham. Instrumental in creating that opportunity was Ms. Jenny Randles, who was responsible for organising it that year. For some years after that I kept in touch with Jenny and, when she became a well-known author on the UFO topic, I questioned why she had never mentioned my work in her books. She replied she thought highly of the work, but couldn't understand it --- even after I had spent an afternoon at her home going through it with her! BUFORA provided me with a few other lecture platforms between 1976 and that important one of May 1987 but, unfortunately, I cannot recall having had any positive follow-up from any of those earlier lectures.

## As Time Went By.

During the 1980s I became well-acquainted with a professional colleague, Mr. David Cayton, who was head of the Non-Destructive Testing (NDT) Department at the same Hawker Siddely Aviation Ltd. site. Although we were scheduled usually to have lunch at different times, David's personal schedule had to be more flexible and this meant sometimes we found that our lunchtimes coincided. Very soon he became aware of my scientific interest in the UFO enigma and, from then on, many of our conversations were about the latest developments. This arrangement continued after the formation of British Aerospace, later to become BAe, PLC. In fact, it continued until my early retirement in 1991 (just before the formation of BAE Systems). Soon after my retirement and move to Devon, I heard from David that he, too, had opted for an early retirement and had decided to offer his services to Quest International (then, the producers of the, British, UFO Magazine) as their voluntary UFO investigator for the North-West Region. So we were now working on the same problems and were often in touch. David and I still inform each other of new developments, including his investigations into strange animal deaths and mutilations. Another area of investigation receiving his attention in recent years has been the crop circles scene in Wiltshire and Hampshire.

In the latter application David has been ably accompanied by another of my Cheshire contacts of the 1980s, Mr. Robert Hulse. Marion and I first met Robert after he contacted us following an article about me published in a local Macclesfield newspaper. Robert confided in us that he had experienced strange things personally and, from one of his customers he had been told of a recent UFO event near Macclesfield. Also, being a resident of Macclesfield, he had met with others with UFO stories to tell. As a result of this information, I visited one of the witnesses and tried to understand the circumstances of the encounter. Robert's recent investigations of crop-circles with David Cayton have been recorded, by him, on video with commentaries. I have been the fortunate recipient of some of these recordings, sent to me for consideration. I am pleased by the continued interest and willingness to share information with me.

My hopeful encounters with an experienced South West journalist, Mr. Arthur Blood, began in early 1997. I had been told by Edward Ashpole that the editor of the Plymouth Evening Herald had asked Mr. Blood to review his book, **'The UFO Phenomena' [18] a**nd Edward suggested that it might be helpful if I were to contact the reviewer personally. He was able to give a telephone number for that purpose. On February 19th, I made that telephone call and discussed the chapter on my work for some time. Arthur Blood seemed to be enthused by it all and suggested that Dr. Percy Seymour, Professor of Astronomy**,** University of Plymouth, might be willing to discuss it with me. He had recently written up an interview with the Professor and had found him to be helpful. Equipped with the Professor's telephone number, I followed that suggestion immediately, on the same day, referencing Arthur Blood and requesting a meeting for a short presentation. I was more than pleased when Dr. Seymour suggested we might be able to meet during **Science Week** in **March, 1997**. Subsequently, we arranged to meet at the planetarium during the lunch break, between 12 noon and 1 pm., on Thursday, March 20th. As requested, I informed Arthur Blood of this as I knew he wanted to arrange to be present.

A loose-leaf presentation pack of diagrams was prepared for the interview and Marion and I travelled to the university during the morning. We were met at the door of the planetarium by a welcoming Professor and asked to sit down on one of the curved benches in the fully-lit display room. As I began the on-the-knee presentation, we were joined by Arthur Blood, who then listened intently throughout. Dr. Seymour seemed to nod assent as I presented each astronomical step taken towards the final outcome. I can't remember any interruptions. We had some discussion at the end and then I asked if he could help me to arrange for this material to be presented to a scientific forum. The response was that he was sorry he would not be able to do that, because he had been unable to arrange one for his own work on interplanetary magnetic fields. By then, the allocated hour had flown by and we had to leave. I expressed my thanks to Professor Seymour for sacrificing his lunchtime break. As we left the planetarium, Arthur Blood wanted to find somewhere to talk immediately. The refectory seemed to suggest itself to me as being an appropriate place, so all three of us went there.

Once seated in the refectory, with Marion and I stuffing our faces with much-needed food, which Arthur apparently had no interest in, he proceeded to ask further questions, scribbling my answers into his notebook. He had been clearly impressed by all he'd heard and seen. He thought the Professor had been, as well. He believed he had heard of a discovery that was even bigger than Galileo's, if it could be confirmed. Retiring soon, from a lifetime of journalism, he thought this story would bring his career to a satisfying conclusion. I hoped so, sincerely. Arthur intended his story to go directly to the national press. He'd had long-standing links with members of it and felt certain they'd share his enthusiasm. Following onto this, Arthur paid us a visit in Torquay and viewed the rest of my material. He left us even more enthused than before.

Sometime later I received his first draft for my approval. He had written a very readable and interesting article and few changes were necessary to the technical details. The article was then sent off to his favoured newspapers in London. In view of his past successes, he was surprised by the lack of response he received. He resorted eventually to telephoning one of his contacts with The Daily Telegraph and asking why he hadn't had any interest shown in the article. The answer he'd received was, "It's just too incredible!" Arthur hadn't been able to believe his ears! He believed the article to be one of the best he had ever written. What a disappointing way to end a career. But he did not give up entirely. He wrote a version of the article and sent it to another old acquaintance, the editor of The Western Morning News. One day soon afterwards, I received a 'phone call from someone at the 'News' office telling me that they had received an article from Arthur Blood about my UFO studies and needed to just check up on a few details. Within days, an article was published with someone else's name under it. Understandably, Arthur was incensed by what seemed to be blatant plagiarism! Thereafter, his efforts to communicate my work ceased and he took up his life-long hobby of fine art painting, in retirement. We still keep in touch occasionally and exchange Christmas cards. Arthur Blood still has hopes of a UFO breakthrough in his lifetime. Fairly recently he and his wife witnessed an unidentifiable object moving slowly over the River Tamar. Arthur was even more enthused when I told him the timing of this sighting concurred with the nearest prediction on the Plymouth timings graph.

There have been many other, more casual, supporters over the years. It would be impossible to list them all, but I'm extremely grateful for all they have done. Notable among them has been Mr. Brian James, who was once Chairman of Contact International, an Oxfordshire-based research group until it was disbanded a few years ago. I supplied Brian with graphs for his area and he used to send me copies of the group's magazine. He also kept me informed of any happenings reported to him. My presentation at **the Kidlington UFO Conference in 1996**, where I shared a platform with Nick Pope and others, was on the invitation of Brian James. We were also allowed to sell a few of our videos at that conference. In addition, I met two people from Reading there and so began an amazing association that continues to this day. Their name is **Glanville** and you will be reading more about our momentous link-up, later, in PHASE 7.

My membership of the **Centre for Crop Circles Studies (CCCS)** and the magazine **'The Circular'** provided an outlet for some of my material, and CCCS conferences sometimes provided another. For several years I participated in the local branch, down here in Devon, and was involved with investigations of a few local crop-circle events. Local man, Mr. George Bishop, headed the group and, for a while, became Chairman of the CCCS nationally. He was also, for several years, the editor of the magazine and provided information about national and international developments. It was a useful association but I grew tired of the growth of hoaxing, which was throwing a big smokescreen over the scene, and decided to resign my membership.

People we met in the crop fields of Southern England also figured in our lives.

Americans David and Shana Roulison we met early in the 1990s in the company of Colin Andrews in the Winchester area fields. We then discovered we were booked in at the same B&B in Middle Wallop and so had further time to become acquainted. The last time they were in touch was in 1996, when they generously mailed me a copy of the late Paul Hill's book, **'Unconventional Flying Objects'**. Two other American friends, Mrs. Terry Schaefer and **Ms.** Thera Querner, who were members of Joyce Murphy's original party we guided around the Avebury area, are still keeping in touch, mainly at Christmas, but occasionally they send me items of interest.

On a later visit, Joyce Murphy brought her party to Torquay, to meet with me and to visit ancient sites in the area. One of the visitors simply called herself Valenya and, during an excursion to an ornamental glass factory after a presentation given by me on the previous evening, she drew me to one side and told me of her deep interest in my claimed discoveries. She wanted to do all she could to propagate them and to participate in testing the time predictions in her own home area, which turned out to be deep within the thickly forested slopes of Northern California. We corresponded for some years after that and I supplied her with a timings graph for that area, which was near a small town called Hoopa. I received a pleasing letter from Valenya some time later. Something creating a very strange noise had overflown her home during the night and the timing of that event had been as predicted by my graph. From then onwards, for some weeks/months, she'd frequently sat on the roof of her home to carry out night-time skywatches, but when nothing unusual had been seen or heard, this practice was discontinued.

After a gap in communications of some years, Valenya again got in touch with me as this book neared completion. She had been writing articles on various topics for a magazine called **'FATE'** and these had been published. The three magazines were posted to me and Marion and I found her writing style to be very interesting. She felt she wanted to offer an article that would make my discoveries known to a wider audience. I responded rather negatively, but this did not deter her. Soon afterwards I received a draft copy of her proposed article with a request for comments and corrections. She had managed to capture the essence of my research and discoveries in such an informal and entertaining manner that caused Marion to comment that Valenya had explained some aspects better than she would have been able to do (after all these years!). I added a few corrections and returned the script by e-mail. That article was published subsequently in the March 2007 edition of **FATE** magazine, with the title, **'The discoveries of T.R.Dutton'**. Since then, Valenya has had published an article in the **American 'UFO Magazine'** (Iss 150, Vol.23, No.9). It was in the form of an interview with me, with the title **'UFO Revelations'**, for which I am very grateful.

In recent years, I have been often assisted by interested people here, in the Torquay area; notably, Mrs. Stephanie Wilson and her husband, John and a Close Encounter witness, Mr. Andrew (Andy) Bell, who lives near Ashburton. Andrew has a blog telling of his encounter(s), promoting my efforts and including a lot of UFO news.

Local newspapers have also played important parts over the years, eg. **The** Manchester Evening News, The Messenger (Macclesfield), The Sunday Independent, The Mid-Devon Advertiser, and so on. Over such a long period of time it is difficult to remember them all, but their interest has always been welcomed.

# PART 3

## EXTRA-EXTRAORDINARY EVENTS & CONCLUSION

# PHASE 8:
# Confrontations with paranormal elements

*My studies have led me into areas of investigation that were sometimes far removed from the original aerospace scenario. I proceeded to probe those challenging offshoots, because they seemed to have relevance within my overall objective. Some of them led me into the realms of the so-called 'paranormal', which the scientific establishment shies away from. Even Edward Ashpole wishes to distance himself from these elements, though he accepts that there is much we have yet to learn about physical reality. That stated, I have not been deterred from exploring them.*

## CHAPTER 21

## THE CE4 FOLLOW-ONS

Two of the people who had experienced CE4 time-losses and with whom I had become well-acquainted, Mrs. Linda Taylor and Mrs. Linda Jones, both told me of strange experiences they had had after their SAC encounters.

Before addressing those, **I want to put on record that both these women had suffered severe physical after-effects and neither of them had dared to tell their medical specialists of their suspicions about the probable links with the CE4s.** They told me of the nature of those physical and metabolic changes, in confidence. It will have to suffice here for me to emphasise that those pathological changes were indeed severe and, being aware of all the known and revealed features of their encounters, I was able to deduce that there could have been causal links.

I can remember only one of the strange post-CE4 happenings that Mrs. Taylor told me about. She had been out in the garden on a fine day, working near the rear boundary. As she turned to look towards the house, she saw, very clearly, a tall man dressed mostly in white standing beside the garage. As she gazed towards him it seemed he had no face. Suddenly he disappeared and the bewildered Linda reasoned that he must have leapt out of sight behind the garage. Extremely puzzled and curious, she had walked slowly to the closed area behind the garage. On arrival there she could find no trace of him. He seemed to have simply disappeared into thin air.

My knowledge of Mrs. Jones' happenings is more extensive because I was consulted on several occasions. On Saturday, December 12th, 1987, I received a telephone call from a very concerned Linda Jones. Something completely unheard of had occurred on the afternoon of December 9th. She explained that, as an auxiliary helper at a local school, she had been issued with a plastic lapel badge bearing her name. After she had walked home from school on that afternoon, she had taken off her winter coat and had been taken aback to discover that her name had disappeared from her badge, which she had been wearing beneath her outdoor coat. The next day, she had requested a new badge and displayed the old one to her colleagues. They, too, thought it all to be very weird, because some had worn theirs for years and one badge had even been through a washing machine and emerged in perfect condition. Did I think it was some kind of omen? I thought not --- and then went on to ask questions about the nature of the badge. It consisted of a hard white plastic base providing a small safety pin for attachment to a garment. The name tape was 'Dymo Tape'®, a thin white plastic tape with a coloured top-side, into which alphanumeric characters could be impressed by means of a special tool. The embossed characters on the tool were pushed into the white side of the tape so that the characters showed up white on the coloured side. I asked Mrs. Jones whether the rubbing of her outer coat could have been responsible, but I was told that it had never had that effect previously and she'd worn that coat frequently. I promised to look into the matter.

At 6:15 pm. on the following evening, I received another excited call from Mrs. Jones. The name on the Dymo Tape was disappearing even as she talked to me!

I asked her to remove the name plate from her lapel and to place it on the telephone table as we talked. We discussed the issues until 6:30 pm. Then I asked if the name had disappeared. No, it was just as it had been when she'd removed it from her clothing.

It seemed, therefore, that there was something about its position on Linda responsible for the effect. I asked Linda if she would be able to supply me a piece of Dymo Tape from the same reel and whether I could borrow the tape from her badge. I wanted to experiment with the untouched tape to see if I could reproduce the effect. On the following Friday I travelled to East Didsbury to pick up the specimens. There were then three specimens, because a third had been added to the collection the previous night. If I remember correctly, after returning from school Linda had applied herself to writing Christmas cards. During this, her son had come into the room and noticed that her name was again disappearing from the badge. Remembering the previous episode, Linda had unfastened the attachment pin and in so doing had found the Dymo Tape to be very hot, in fact, too hot to handle as she'd laid the badge down onto the table. So, it seemed that heat was in some way involved. Had she been sitting close to, say, a radiant heater, I asked. The answer was "No", nor could she in any way account for the hotness of the tape. I couldn't wait to start experimenting, but it would be several days before I could devise an experiment and find the time to carry it out. This is how I then tested the tapes.

Having created several impressed specimens from the unused tape and having Linda's actual examples to compare with, I poured a little tap water into a small earthenware pot and placed, submerged, one of my test specimens into the pot. Then, boiling water from the kettle was added slowly to gradually increase the temperature of the water, a jam-making thermometer being used to keep the water stirred and, simultaneously, to indicate the water temperature. At exactly 75°C the markings in my red specimen tape suddenly disappeared. When I repeated the experiment with a black specimen tape, disappearance of the markings occurred at nearly 80°C. No wonder Linda's tape had been too hot to handle! But how could she possibly have generated so much heat from her body? I wondered whether there might be another explanation, so I decided to consult a scientist friend of mine.

On the evening of December 29th, 1987, I telephoned my friend, the late Dr. John Warwicker. Before his retirement Dr. Warwicker had been Head of Crystallography at the Shirley Textile Research Institute in Manchester. After his retirement he and I had had a series of informal evenings at his home, sharing our thoughts on new scientific discoveries, paranormal happenings and a wide range of topics. So, when I told John, over the telephone, about Linda's experiences and about my test specimens and the results I'd obtained, he thought that the heat explanation was probably the most likely one, even though, like me, he was baffled by the circumstances. He thought the critical temperatures I'd measured were probably significant for that kind of plastic tape. In fact, he ventured to suggest that at those temperatures that kind of plastic could be persuaded to absorb dye --- but, not being an expert in that field he would consult with one of his ex-colleagues who had such expertise. On his invitation, I visited John on the evening of January 5th, 1988. His colleague had confirmed the significance of the temperatures I had measured and had gone on to make the suggestion that some plastics could be heated using RF (radio frequency) radiation.

I checked out that idea with the H.S.A.'s chief chemist at Woodford airfield. He told me that the technique had been used on site, in the past, using radar frequencies. Curious-er and curious-er!

When I shared all this with Mrs. Jones she was both pleased and baffled. She hadn't been close to an active radar dish when her name disappeared from her name tapes, so how could that explanation possibly apply? A solution suggested itself to me some days later on hearing about another of Linda's happenings.

On the dark and stormy night of February 9th, 1988, after returning home with husband Trevor and her son, Linda had walked out into the garden, towards the shed. Suddenly she had found herself standing in a pool of blue light and, given the extreme weather conditions, she'd thought she was about to be struck by lightning. Then she saw that the light was being beamed down from a source in the base of a low cloud some distance away. At that, she turned in panic and ran back into the house, damaging an arm and a leg in the process. Had she been monitored by those she had encountered in the meadow? The thought had occurred to her at the time and it also came to me when she told me the story. Could it be that on other occasions her movements had been monitored by beamed high frequency radiation? That might account for the heating of her name badge and, in fact, it seemed to be the only explanation suggesting itself.

Other strange events followed. I received a telephone call from Linda during a subsequent summer. She'd been out in the garden sunning herself and had dropped off to sleep. When she came round her husband, Trevor, came into the garden and noticed a strange reddened pattern on one of her exposed arms. It was a series of concentric rings of red dots about half-an-inch in diameter overall, looking like some kind of vaccination mark. They were both at a loss to know how she came by it.

But Linda also told me that strange things had happened to her when she was much younger and prior to the CE4 event of 1979. As a little girl of about 4 years old, whilst visiting a relative, she and an uncle (who was only a few years her senior) had gone out to play in a nearby meadow. After a while, her companion had returned home without her, claiming that she had just disappeared. He'd searched about, but there'd been no sign of her. Of course this had created something of a hue-and-cry and further unsuccessful searches were made for her. Several hours later, Linda had arrived at her relative's home 'under her own steam'. She'd had no idea how long she'd been missing. She'd simply looked around to find her companion was no longer there and had wandered back. That mystery was never solved.

Another strange and inexplicable happening had occurred one day when she'd been out for a drive with husband-to-be, Trevor, as the passenger in his E-Type Jaguar sports car. They were travelling on the A556 road, which linked Manchester to Chester. As they approached traffic lights at the cross-roads at Mere, slowing for the lights, *the car suddenly span round through 360°, twice, on a perfectly dry road, without any fuss or squealing of tyres, and then carried on to stop at the lights.* They had been, to say least, surprised! What had happened? How could it have happened? (It seemed likely to me that, as in so many UFO related car events, their car had been gently lifted from the road and placed gently back down again). Afterwards, Linda had realised that *the event had occurred only a short distance from her birthplace*, situated on one of the intersecting roads.

*All these things seem to provide the circumstantial evidence for the often expressed suspicion that some individuals are selected from childhood as human specimens for some sort of monitoring activity. The monitoring seems to cease when female subjects are past childbearing age. There are also few reported CE4s involving men of over, say, fifty.*

# CHAPTER 22

## Strange inputs from Reading

In Phase 6, I mentioned that, after my lecture at a conference held at Kidlington near Oxford during 1996, I had been approached by a man and a woman who had introduced themselves to me as Mr. David and Mrs. Thelma Glanville. They had wanted to know whether a particular date and time combination had any significance for me. I had to say that it didn't and then to enquire whether it applied to a particular place. Yes, it applied to Reading, Berkshire, where they lived. They had been given information that they would receive further information from ETs on that date and at that time. They couldn't tell me the whole story at that first meeting, but I asked for their telephone number and said I would check out the information against the AT's predictions for Reading. On returning home I did just that and found good agreement with one of the timing predictions for that date. When I contacted them with this news, they were extremely pleased and I was keen to know more about their claimed ET connections.

### Time Loss in Florida

They told me that the source of their information was generally their 25 years' old son, **Jeff**. Then an amazing story was told to me. During 1995, Jeff, two of his friends, his mother, Thelma and Thelma's sister, Dilys, had flown to Florida for a holiday break. On arrival at Orlando Airport, they'd hired a car for the drive to St. Augustine, about 100 miles to the north. Jeff had volunteered to drive. As they had set out in the early evening, they'd estimated that they'd arrive at their destination at about 10 pm. and would be able to buy meals before retiring to bed. The weather had been fine until they'd reached near the end of a long bridge over the St. John's River. Suddenly the heavens had opened and the rain had fallen so heavily that all the traffic in front was braking hard and Jeff had to slow down further to look for the windscreen wiper switch. As he'd been doing this he'd noticed, through the rear view mirror, two bright lights approaching rapidly through the rain. Then there'd been a bumping sensation felt by everyone in the car. Jeff, who had just found the windscreen wipers' switch, had suspected a possible hijack and had accelerated rapidly away from the scene. Suddenly, the rain had ceased and they'd found themselves driving in perfectly dry conditions again. The lights previously following them had disappeared. When they'd arrived at St. Augustine they'd been puzzled to find the place in complete darkness. All the bars and restaurants were closed. They'd then discovered that they'd arrived well after midnight. They'd had no idea where they'd lost an estimated three hours on that journey.

This seems to be a good point to break into the rest of the story to explain some aspects of my, now, longstanding relationship with the Glanvilles. As I declined invitations to visit their home on the grounds that I did not want to become too involved with the high level of excitement prevailing there and wanted to remain an objective observer, all the information I have about the happenings in Reading have come to me by post or through telephone conversations. For this reason, I do not have a full knowledge of all the weird happenings that have occurred there, but I have received sufficient video evidence that such things have occurred and are still occurring there. *All these things have happened to a normal, hard-working, family who have a thriving family business.* They live on the fringe of parkland and have a large garden. Their home they have extended to include a gymnasium at the east end and an indoor swimming pool at the opposite end. I mention these extensions because they have featured a lot in the story.

## Hypnotic Regression

During a subsequent Kidlington conference I discovered that the Glanvilles had also talked with Mr. David Coggins from South Wales. Intrigued by all he'd heard about the Florida time-loss event, he'd arranged, subsequently, to visit the Glanville's home. Whilst there, he'd been told about my research and had seen some of my material. One day, towards the end of 1997, I received a video recording from him, showing interviews with members of the Glanville family and colour photographs of some of the strange happenings in the household, including one of dining room chairs stacked high and balanced in an incredible manner. The tape also showed a hypnotic regression session with Jeff. David Coggins told me he had been a stage hypnotist and had seen the time-loss incident in Florida as something of a challenge. He had been pleased to receive the co-operation of the subjects and had travelled to Reading from his home in South

Wales to carry out his investigations. When I viewed the video, I was very impressed by his softly-spoken technique and his marvellous patience.

Responding to the gentle prompting of the hypnotist, Jeff told the story of the events up to the 'bump', but David Coggins, feeling sure that the time loss had occurred between Jeff's seeing the lights behind the car and that bump, kept patiently taking Jeff back into that unknown time span, even though there was difficulty being experienced in progressing into it. Eventually, through persistent friendly persuasion, Jeff began to remember. He had been looking for the wipers switch when he'd seen three bubbles of blackness approaching the car from the front. These had then passed into the car and one had enveloped him. Another enwrapped his mother and the third went over his head into the rear seat area. Next, he was floating in a plastic bubble, looking upwards and there was bright light shining into his eyes. Then, he was lying horizontally but had no experience of being in contact with a bed. He felt so relaxed he didn't want to even turn his head to look around. A small bright light was now to his right but, otherwise, he was in complete darkness. At one point in the regression he remembered seeing lower legs and bare human feet standing beside him, but they didn't appear to be standing on anything. At one time he could vaguely see a face looking at him from the darkness. While he continued to float in that way, the small light over him seemed to grow in size and became a ring of light. Then he began to see a myriad of changing coloured lights, which he tried to identify, verbally, very rapidly. Then something hard was being passed round the back of his neck and, simultaneously, a serrated bar with little teeth seemed to be being used under his chin to keep his mouth shut. Next, he remembered sliding down a steeply inclined tube or channel into another bubble. He began to descend and could see the car below him. Then he was back in the car. David Coggins succeeded in persuading Jeff to return, in his mind, to the time when he was still lying down, in an attempt to fill in the information gap still existing. Just as things were progressing well again, the telephone rang in that room. Jeff, clearly very shocked and shaking, came suddenly out of trance and had to be persuaded by the insistence of the hypnotist to close his eyes and to be counted out of the trance state in the usual manner. Jeff was calmed by this procedure and stopped trembling, but, unfortunately, that telephone call had brought a very revealing session to an untimely end.

From information I received about another strange happening in Reading, it seems that the interrupted journey in Florida was not to be the only time when Jeff was apparently abducted. In the early hours of one morning, David and Thelma Glanville awoke to hear a loud hammering coming from their front door. On going downstairs to investigate what it was all about, they opened the front door to find Jeff, standing stark naked, in a state of some desperation. He told them he had found himself suspended horizontally over the lawn and then dropped, a foot or so, onto the cold wet grass. He'd had no idea how he'd got there and his parents said the door was still locked and bolted. It seemed possible that he had once again been a guest of the ETs.

## Poltergeists and spiritualism

From all accounts, strange happenings had begun to beset the Glanvilles soon after their 'lost time' in Florida. The first one I have been told about occurred at Easter 1996. Thelma and David had taken a short holiday in York and visited the old sites there. Jeff had gone to stay with friends. The younger son, Nick, had stayed at home with his cousin Darren. On arriving home and entering the gym they saw the heavy exercise machine balancing on top of a bar in a dangerous-looking manner. Nick and Darren denied responsibility for this. No one could account for how the feat had been achieved. Then followed a period of poltergeist-like activity, with items of heavy furniture shoved about and ornaments being moved to other locations and, sometimes, disappearing. Throughout all this annoyance Jeff had kept on with his body-building exercises in the gym and tried not to be 'phased' by these things.

At this point in the happenings, spiritualism was brought onto the scene by Dilys. Apparently she had received comfort from local spiritualists after she had lost someone in tragic circumstances some time before. The family became convinced that they needed to discover more about the entities causing the chaos in their home. The local vicar had been called in, but he'd felt unable to help them. From then on, any UFO connection became cloaked by the actions and revelations of the 'spirits' during seances. Even Jeff had been drawn in when he'd seemed to be taken over by one of these unseen influences and, thereafter, he'd often gone into trances during sessions. Throughout all this David had been an observer and at times he'd become angry about these dramatic changes to normal family life.

All these happenings had occurred before my meeting with David and Thelma at that Kidlington conference in the summer of 1996. After I had shown that the time and date of that expected happening had been close to a prediction by the AT, David began to confide in me about other strange things. I remember he rang me one night to discuss something and then said, "You would not believe the things that are happening in this house. Right now, as I'm talking to you, out there in the kitchen, there are two wooden mug stands with rounded tops --- and one is balancing vertically on the other by its rounded top!" "Jeff dismantled them yesterday and now they're balancing again". He went on to tell me how he'd seen things fly across the room and then more about events in the gymnasium.

## Back on track with Jeff's graphics.

One night, as David and Thelma had returned home, they had seen balls of light disappear into the roof of their house. They had later found the door to the gym to be jammed, but they'd been able to push it open enough to peer inside through a small gap. They'd seen the exercise machine, weights and all, balanced at an impossible angle. They'd watched with interest as a large spider ran under the door, but then came rushing out again, as if being chased. Strange things began to happen in the swimming pool area too. The filter system would suddenly burst into life, run for a while and then shut off.

This is all background for the UFO/ET elements that followed. I think the first development of interest, in the context of my studies, was when I received sheets of A4 drawings through the post, early February, 1997. These looked like symbolic sketches from a scientist's notebook and they had been drawn by Jeff in unusual circumstances. I talked with Jeff over the telephone and asked how he had produced them. What he told me was almost incredible. Those images had appeared in white outline superimposed upon his normal vision. He'd found that by closing his eyes he could remember them all and then on opening his eyes again, he'd been able to draw them.

I noticed that the earliest set of three small drawings on the first sheet (Fig.50) was dated October 11th, 1996 and the latest image had the date January 29th, 1997 beside it. The diagram at the bottom right-hand corner of the first sheet immediately caught my eye. It seemed to be a diagrammatic representation of the AT's scenario. (It is worth bearing in mind that Jeff Glanville had no knowledge of the AT whatsoever and even today has little knowledge of it). A circle with a curved band across its middle and two more, smaller, arcs, top and bottom, seemed to represent the Earth. By measurement I established that the two arcs top and bottom represented the Arctic and the Antarctic Circles to unbelievable accuracy. A small ellipse sitting just above the equator, on the right of the diagram, was attached to the right leg of a broad upward pointing arrow, which had a direction arrow, pointing to the left, attached to the base of its left leg. I considered that this directional arrow indicated retrograde (East to West) motion and that the broad arrowhead represented the northward and southward motion of an orbiter, represented by the ellipse. To the left of this Earth diagram, a stick-man alien (with oval head) was drawn, seemingly throwing a diamond-shaped something towards the Earth. As can be seen from the annotations on Fig.50, I made attempts to interpret the other pieces of obscure symbolism on that sheet and found a few common symbols that might have had physical meaning. Of particular interest were spiral symbols which, in context, seemed to symbolise 'power'.

A further sheet of drawings was received from Jeff towards the end of February 1997 (Fig.51). This was a reduced copy of the first sheet but with an appended strip at the bottom edge.

A very interesting diagram in the right-hand bottom corner caught my eye. It looked to me to be like the stylised outline of a lake or an island. It had two 'power' symbols and a broken arrow symbol within the confines of the outline. The right-hand extremity feature of the outline looked like a promontory of some kind. It was circled and an inverted triangular shape was shown linked to the circle. Jeff thought the outline might represent a star constellation, but I was not convinced by that and continued to pursue the lake or island solution by studying an atlas and other maps. After an intensive and fruitless search I gave up for a while. Then one day, soon afterwards, whilst reading a newspaper, I saw that elusive shape in an advertisement for holidays in New Zealand. The North Island had all the important outline features of Jeff's stylised drawing. Another piece of serendipity then occurred (there have been many aiding my quest over the years). During May, I met a married couple who were from New Zealand and who were preparing to return there.

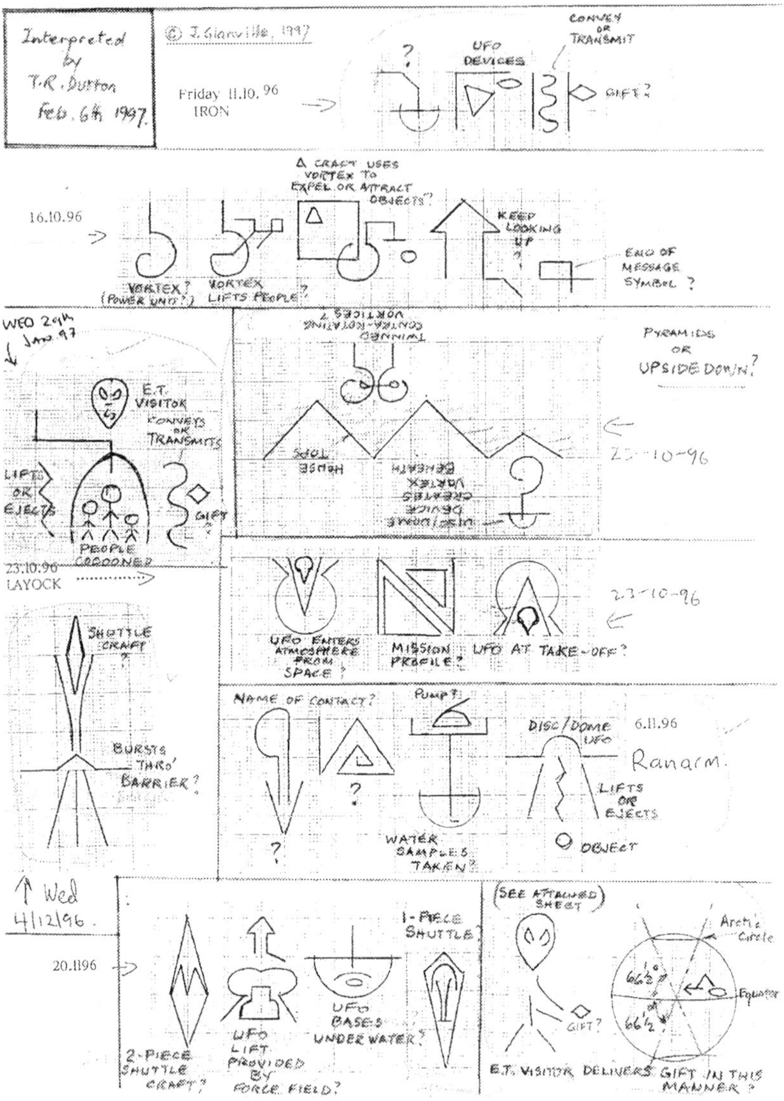

Interpreted by T.R. Dutton Feb. 6th 1997.
© J. Glanville, 1997
Friday 11.10.96
IRON
?
UFO DEVICES
CONVEY OR TRANSMIT
GIFT?
16.10.96
A CRAFT USES VORTEX TO EXPEL OR ATTRACT OBJECTS?
KEEP LOOKING UP ?
END OF MESSAGE SYMBOL ?
VORTEX? (POWER UNIT?)
VORTEX LIFTS PEOPLE?
WED 29th Jan 97
E.T. VISITOR
CONVEYS OR TRANSMITS
LIFTS OR EJECTS
GIFT
PEOPLE CROONED?
23.10.96
LAYOCK
PYRAMIDS OR UPSIDE DOWN?
23-10-96
SHUTTLE CRAFT ?
BURSTS THRO' BARRIER?
UFO ENTERS ATMOSPHERE FROM SPACE ?
MISSION PROFILE?
UFO AT TAKE-OFF?
23-10-96
NAME OF CONTACT?
PUMP?
DISC/DOME UFO
6.11.96
Ranarm.
LIFTS OR EJECTS
WATER SAMPLES TAKEN?
OBJECT
Wed 4/12/96.
20.11.96
1-PIECE SHUTTLE?
2-PIECE SHUTTLE CRAFT?
UFO LIFT PROVIDED BY FORCE FIELD?
UFO BASES UNDER WATER?
(SEE ATTACHED SHEET)
Arctic Circle
Equator
66½°
GIFT?
E.T. VISITOR DELIVERS GIFT IN THIS MANNER ?


**Fig 50**

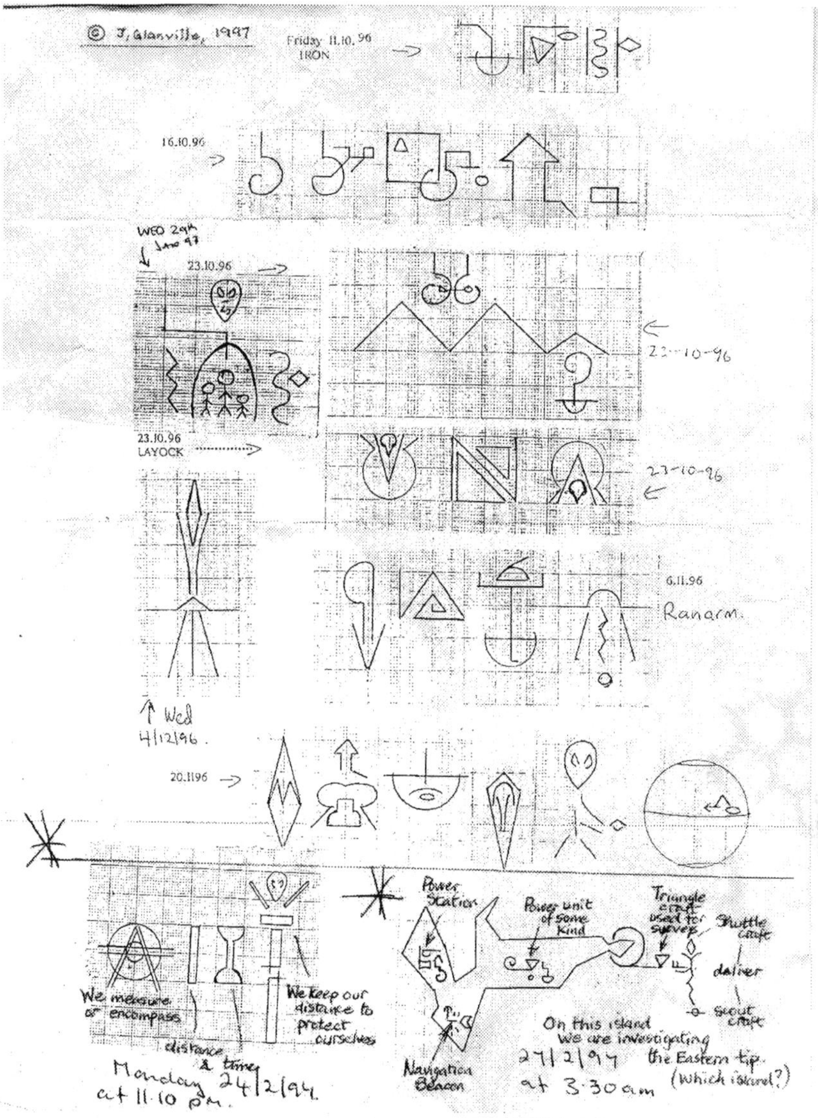


Fig 51

I showed them Jeff's drawing and asked if they knew the locations of power stations on North Island and whether the symbols shown corresponded to the positions of two of them. They were able to confirm that power stations were placed roughly in the positions shown and that the northernmost one, symbolised in a different way, was a geothermal plant. They'd no idea what the broken arrow symbolised, even when I prompted them with my suspicion that it might be a navigation beacon. However, they did know something about that ringed promontory. It was regarded as being a sacred area by the native Maori population, because they believed that their souls departed from there when they died.

(This latter revelation seemed to link strangely with some of the soul-saving activities the Glanvilles had been involved in during their seances. These are not described in this book because they appeared to be 'fringe' activities taking place in parallel with the ET's interactions.)

My next move was to enquire about that suspected aircraft beacon feature. I contacted my friend and fellow UFO investigator, the late Capt. Graham Sheppard. After retiring from British Airways, Graham had been brave enough to admit, publicly, to having experienced two UFO encounters whilst flying. After a long flying career and voluntary retirement from British Airways, he had then ventured to appear on TV programmes to support other UFO witnesses. He had become a close friend of the author, Timothy Good. Over the 'phone I asked Graham if he had any knowledge of the air traffic arrangements on North Island, New Zealand. He said he hadn't, but he would find out more about that suspected beacon. Shortly afterwards I received information in the post that a navigational beacon was located in the position indicated by Jeff's drawing. It was almost unbelievable that Jeff Glanville had somehow been given a recognisable portrayal of a location he'd never even seen!

## A remarkable Scrabble diagram

Next, I received a FAX from David Glanville on March 17$^{th}$, 1997, after I had been informed that unseen entities had taken over their dining room and messages were being left on the dining table using Scrabble pieces. A pattern of pieces had been created, which Jeff believed represented the planets of the solar system. He'd been able to guess each planet's identity from its distance from the centre of the pattern. Of special interest was a string of pieces, letters A to L, which seemed to create a trail from beyond Mars to the Earth. Jeff had tried to reproduce this pattern on paper and to sketch-in rings to represent the orbits of the supposed planets. David FAX-ed this drawing (Fig.52) to me and asked if it made any sense. (The longhand annotations are mine.)

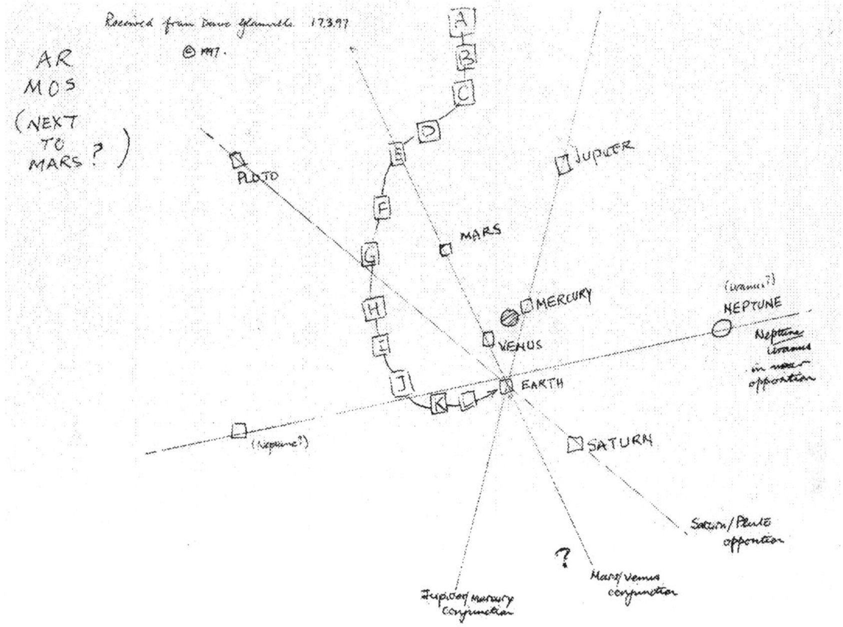


**Fig. 52**

Several features of that diagram immediately stood out. Relative to the Earth, a ruler placed on it showed conjunctions between Mercury and Jupiter and between Venus and Mars, Saturn and Pluto in opposition and Neptune and Uranus in near opposition. That was an interesting and unique set of planetary alignments. I began a long search of my astronomical software to look for a date when those conditions might have been met. Eventually, I found that **Saturn and Pluto had been in exact opposition on December 8th, 1898. Uranus and Neptune had been in opposition on July 11th, 1908, Venus and Mars had been in conjunction on June 23rd, 1908, and Mercury and Jupiter in conjunction on August 19th, 1908.** The outer planets, Pluto, Uranus, Neptune and Saturn move slowly round the sky and, therefore, they tend to remain in close alignment with each other before and after the exact conjunction or opposition date. Jeff's depictions of the alignments on the dining table were also approximate. Taking all this into account, it was quite remarkable that three of the approximate alignments had occurred during June, July and August, 1908. Uranus and Neptune will not be in opposition again until 2078-2081 and the other alignments will not coincide during that period. So, to me, it looked as if the summer of 1908 was being indicated. I wondered, ***could it***

***be just a coincidence that the Tunguska explosion in Siberia occurred on June 30th, 1908? Could it be possible that that string of Scrabble pieces represented the path of the object responsible for that enormous explosion?*** Unfortunately, on June 30th, 1908, the planets were not placed in the regions of the sky indicated by the Scrabble diagram.

That was the unsatisfactory end of the exercise until I was in the final stages of creating this book.

## The Tunguska Analysis

As recorded by Chapter 18, during November, 2006, I had received authoritative information (from Edward Ashpole) giving me the actual location of the Tunguska event and an accurate local time for the explosion. This gave me the opportunity to run my AT programs. Fig. 44 presented the resulting timings graph and the very impressive result obtained.

But there was still more to be discovered when I turned back to re-examine Jeff Glanville's Scrabble diagram.

During June 2007, whilst incorporating amendments to this book, I found myself again studying Jeff Glanville's Scrabble diagram. It occurred to me that the previous analysis of it, perhaps, had not been sufficiently thorough? So, once again, I ran the astronomical software to display the sky as it would have appeared during June 30th, 1908.

It dawned on me that much depended upon whether Jeff's diagram was showing the solar system viewed from the Northern Celestial Hemisphere or the Southern one. With Pluto and Jupiter in the same positions relative to Mars, the view had to be one viewed from the Southern Hemisphere of the sky. However, the positions of other solar system bodies did not conform at all well. The view from the Northern Hemisphere, looking down on the solar system, was much better, but this meant that the positions allocated by Jeff to Pluto and Jupiter had to be reversed and the actual position of Saturn was close to the counter marked Uranus by Jeff. (In the view from the Southern Hemisphere, nothing had filled that space).

After looking at all the aspects, it seemed that the view from the Northern Celestial Hemisphere, after swapping the Glanville positions of Pluto and Jupiter, could be regarded as a reasonable representation of the actual sky on the date of the Tunguska explosion. Fig. 53 is the revised diagram, to which other annotations have been added. The position of the

Sun provided the datum for this and some meaningful observations could then be made.

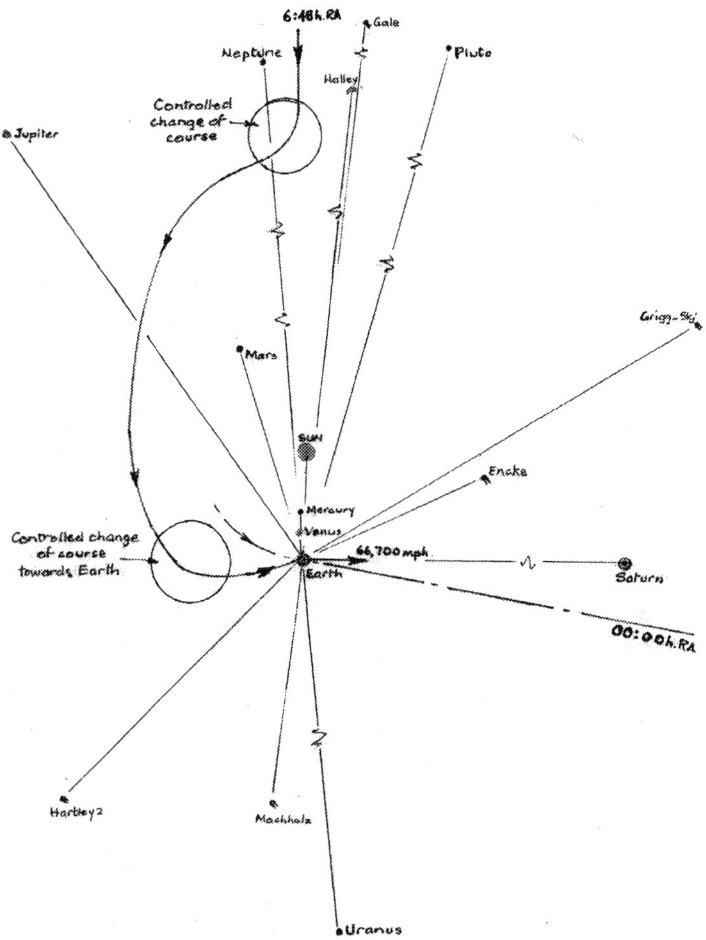


**Fig. 53**

The object is shown entering the solar system travelling towards the Sun, from the direction of Neptune. Its location in the sky would have been 6:48 h. RA. This is close to the intersection of the Galactic (Milky Way) Plane with the celestial equatorial plane. A controlled turn is then depicted, probably following the Galactic Plane, with the object now heading into the southern hemisphere of the sky to skirt round the group of planets cluttering the direct route to Earth. Whether or not the object moved in the galactic plane or entered the solar system's ecliptic plane, this controlled manoeuvre immediately tells us that the depicted path would have had to have been followed by an **artificial craft**, not a comet or an asteroid. At another point in its path facilitating this, a second controlled manoeuvre is depicted, which would enable the object to intercept the Earth at the required angle, approaching from the south. As the annotations show, the Earth would have been moving away from it at an average speed of approximately 66,700 miles per hour as the planet moved round the Sun. This means that the approaching craft would have required more than that speed to catch up with the planet (with allowance for its approach angle), plus an increment to enable it to perform its surveillance mission. Its approach speed would therefore have been in the region of 100,000 miles per hour. Measuring from the diagram, the object's approach to Earth would have been from the direction of 14:14h. RA. This path happened to be roughly aligned with the computed Track No.2 and with Comet Encke, located at 2:19 hr RA., which is in accordance with my AT observation that solar system bodies seem to be often used as navigational markers. If we now apply the other rules of the Astronautical Theory, on its northwards approach to the Earth's equator the craft would have adjusted its speed to follow the predetermined path over the Earth's surface. This would have been chosen to overfly the area targeted for the mission. Given this scenario, the track chosen could very well have been the path indicated by Figs. 44, 45 and 46 and, as suggested previously, it would then seem that control was lost in the later stages of its mission, causing it to descend uncontrollably into the denser layers of the atmosphere with reducing meteoric speed and, finally, to create an enormous (nuclear?) explosion. The integration of the amended Glanville diagram with the AT's result is something quite remarkable.

## Other dining table happenings

Many other strange things have since happened on that dining room table. Items have appeared from unknown sources and then disappeared again after lying a day or so on the table. David Glanville sent me a video recording showing some of the strange events. One such happening involved the appearance of a small telescope on a stand, together with drinking glasses in clusters. No one knew who the telescope belonged to but, presumably, it was returned to the owner after, subsequently, it disappeared from the Glanville's table. Another, startling, episode involved the appearance (not for the first time) of a large and ornate pottery chalice that was producing a flood of liquid as the camcorder recorded it. This liquid was bubbling from the chalice and creating rivulets on the highly polished table. It was then running off onto the carpet. How the liquid was being supplied to this vessel in such a large quantity was a complete mystery, but Jeff decided to get a sample on the end of his finger and then, rashly, to taste it. It was horribly salty.

More recently, whilst in conversation with Thelma Glanville on the telephone, I wondered if she would ask a question for me in Scrabble pieces on that table. She very kindly agreed to co-operate. This was the question I asked : "Why do you use comets?". The reason for that tongue-in-cheek question was to check out my observations that, during SAC/UFO activity periods, comets had been frequently found to be in alignment with the approach/departure paths determined by the AT programs. It was a leading question that I expected to be ignored, denied, or generally glossed over. A few days later Thelma had an answer for me. It was this: "Less energy required". That was an incredibly sensible answer, which seemed also to confirm my observations. In addition, it seemed to confirm that, indeed, we were dealing with space-travelling ETs.

## Sinister drawings from Jeff

The next information I received during 1997 came to me as two more sheets of diagrams from Jeff, dated March 24th and with the time given as 2:30 am. On that occasion, he had experienced them after coming round from sleep in the early hours and had just wanted to go back to sleep again, but the images persisted in his vision, and prevented that. So, he'd grabbed a cardboard box from the top of his wardrobe and spent some considerable time drawing all he could see. The amount of detail in some of those drawings was impressive. The diagrams on those two sheets were more elaborate than previous ones and seemed to me to present a rather sinister scenario. They are shown by Figs.54 & 55.

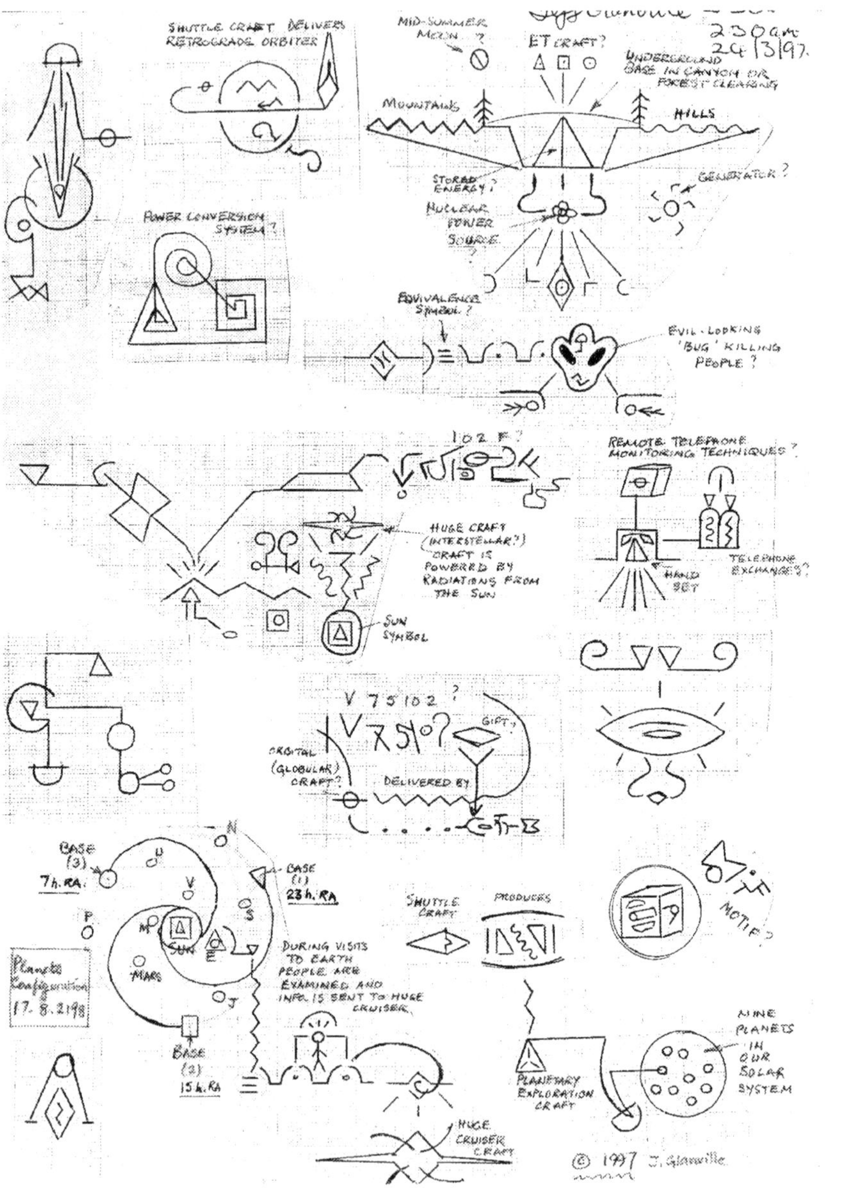


**Fig. 54**

One of the more mundane representations was a diagram in the bottom left-hand corner of Fig.54. This seemed to be another representation of the solar system. In this plan view, the planets could be recognised by their relative distances from the central sun and three equally-spaced spiralled arms (anticlockwise) connected the central sun to three symbols placed on the outer fringes. The symbols were a circle, a triangle and a square. Could they have represented bases in space? I wondered. Analysing that diagram using astronomical software, I discovered that it might represent the situation on August 17$^{th}$, 2198, about 200 years from the, then, current time. That would be the right kind of orbit period for objects orbiting the sun at the implied distances from it. In other words, those symbols might have represented the location of symbolic 'bases' in 1998. If so, two of them were located in the sky at the two of the AT's fixed star-related approach and departure path orientations. How very interesting!

But the diagram in the top right-hand corner was even more interesting. It seemed to depict a cross-sectional view of an underground base, sited in pine woods, with mountains to the left and hills to the right. The big hole in the ground contained a large pyramid shape and the hole was covered by a huge shallow dome. Above the dome were a circle, a triangle and a square with converging radial lines linking them to the dome. Underground there appeared to be something depicting a generator of some kind and, possibly, a nuclear power source. A sinister smaller diagram below this seemed to depict an evil-looking insect-like 'face' linked, on either side, to prostrated (dead?) stick people.

The second sheet (Fig.55) was an even more elaborate depiction of the **opening** of the underground 'base'.

In this very elaborate diagram the lid of the base was shown opened, the pyramid shape was shown to be half its original size, that diamond-shaped 'gift' (from the earlier globe diagram) was below it, contained within converging lines. Above the opened dome, the triangle, square and circle were linked by radial lines to the central region of the opened dome and further, upward-projecting, radial lines from them converged beneath a huge shape, which was similar to a galaxy when viewed edge-on. This shape was constructed entirely with straight lines, but from the central hump (or diamond) were generated four equally-spaced curved lines. The object looked like the depiction of a huge craft. Overall, it seemed to be in communication with the geometric objects and through them, with the underground base. To the left of the diagram, in the sky above a pine tree, a Last Quarter moon was drawn, the angle of its terminator suggesting that it was probably a winter moon in the northern hemisphere. The huge 'craft' was placed at the centre of a three-sided rectangular frame or arch. That frame was surrounded by eleven spiralled 'power' symbols, which were linked to the frame by arrow-like symbols, each made up from a triangle attached to an elongated diamond.

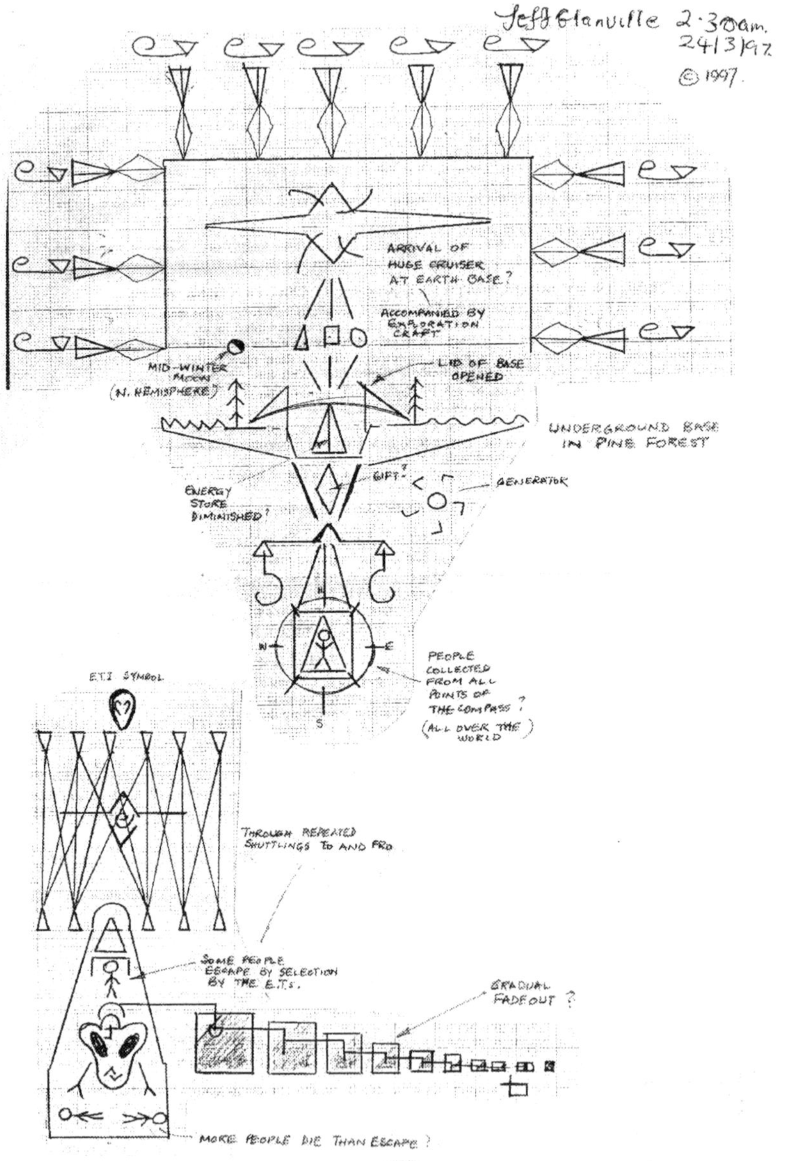


**Fig. 55**

Below that diamond, shown **beneath** the underground base, was suspended a simple balance (set of scales), with a triangle linked to a dangling power symbol at each end of the balance arm. Diverging radial lines from the centre of the 'balance' connected to a circle, the periphery of which was divided up into eight equal segments, giving the appearance of a magnetic compass. In a rectangle within this circle, a single standing stick person was shown encapsulated within a triangle. At the bottom of the same page, that insect-like face featured again, with lines on either side of it linking to prostrated (dead?) stick people. To the right of the insect head was a linked succession of shaded-in squares. These rapidly diminished in size as they progressed away from the insect face. Above this face an upright stick person was shown being apparently uplifted by a triangle. Immediately above this was a diagram giving a rough depiction of the large 'craft' at its centre, in the midst of a set of criss-crossing lines linking triangular shapes above and below it. Overlooking it all was a rough depiction of a typical alien face (an inverted pear-shape).

**This was my interpretation of the diagrams.** The first sheets were generally giving examples of the symbols to be used in future diagrams. The more complicated, follow-on, diagrams were, perhaps, intended to reveal the nature of the ETs' mission here. They seemed to suggest that humanity is going to be beset by a killer 'bug' from which no one will have immunity. The ETs' task will then become one of rescuing **selected** individuals from all points of the compass (ie. from all over the world). In other words, they will then provide the services of a space-age Noah's Ark to preserve members of the human race for the future.

## End days and the fate of human souls (?)

Having barely recovered from the amazing things delivered by Jeff's drawings just described and interpreted, I received further elaborate drawings from him during October, 1997. These had been received by him on September 18th. They are presented here as Figs.56-59. The style of the drawings was very different from that of the previous outpourings. I found the interpretation of the symbolic sketches very difficult, but I was assisted by having had telephone conversations with Jeff, getting his views about the messages he was receiving.

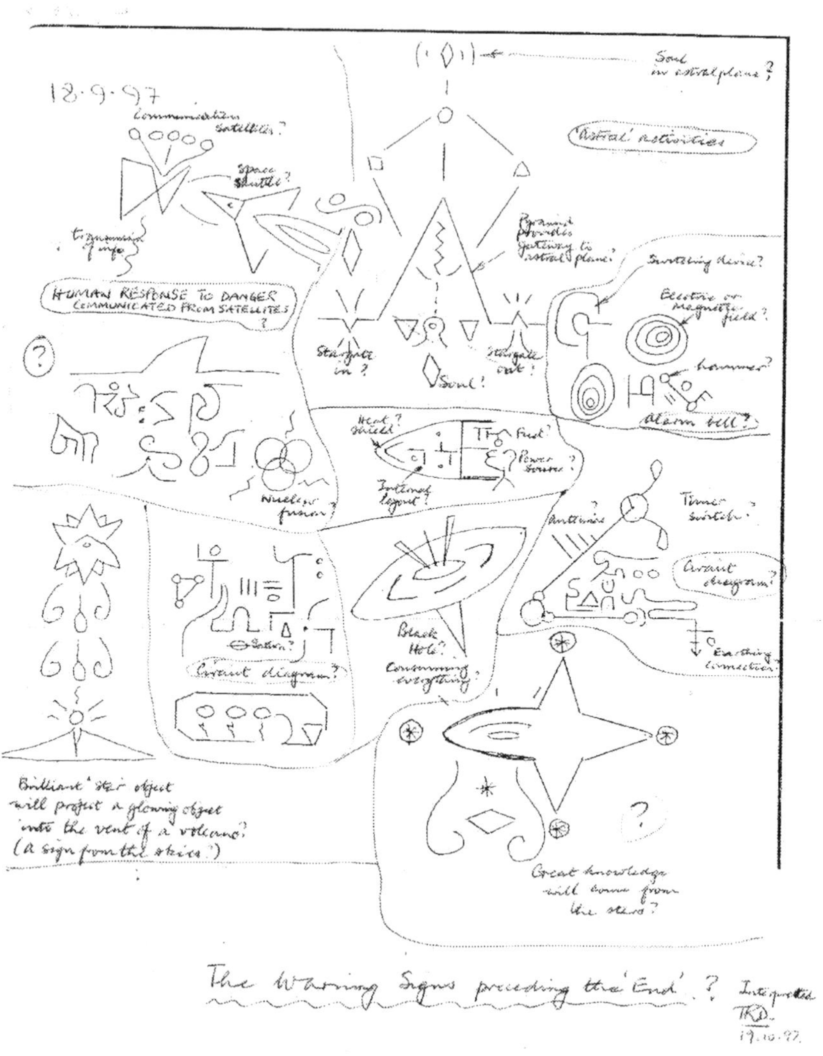


**Fig. 56**

In the middle of this first sheet there seemed to be a diagrammatic depiction of a Black Hole. To the left and right of this were drawings reminding me of electrical circuit diagrams. At extreme left, at the same level on the sheet, was what looked like a cross-section of a volcano, showing an empty central vent. Immediately above this vent was a small circle with radiating lines, which was linked to the lower of two 'tadpole' shapes. These seemed to emanate from a large ten-pointed star above them. Each of the 'tadpoles' had the familiar 'power' spiral on either side of it. The overall impression I gained was that the star object was injecting something into the vent of an extinct volcano. The upper part of the page seemed to me to be possibly portraying aspects of human space-age technology, with images resembling a space shuttle craft, nuclear fusion, and a simplified depiction of an alarm system. In the midst of these was a diagram with a large arrow-head (or pyramid). This seemed to be projecting something upwards or receiving something from above. I had no idea how to interpret that item, and this was true, also, of some squiggly symbols associated with the 'nuclear fusion' diagram.

Fig.57, the second page of this sequence of diagrams, presented some images that seemed easier to interpret.

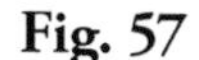

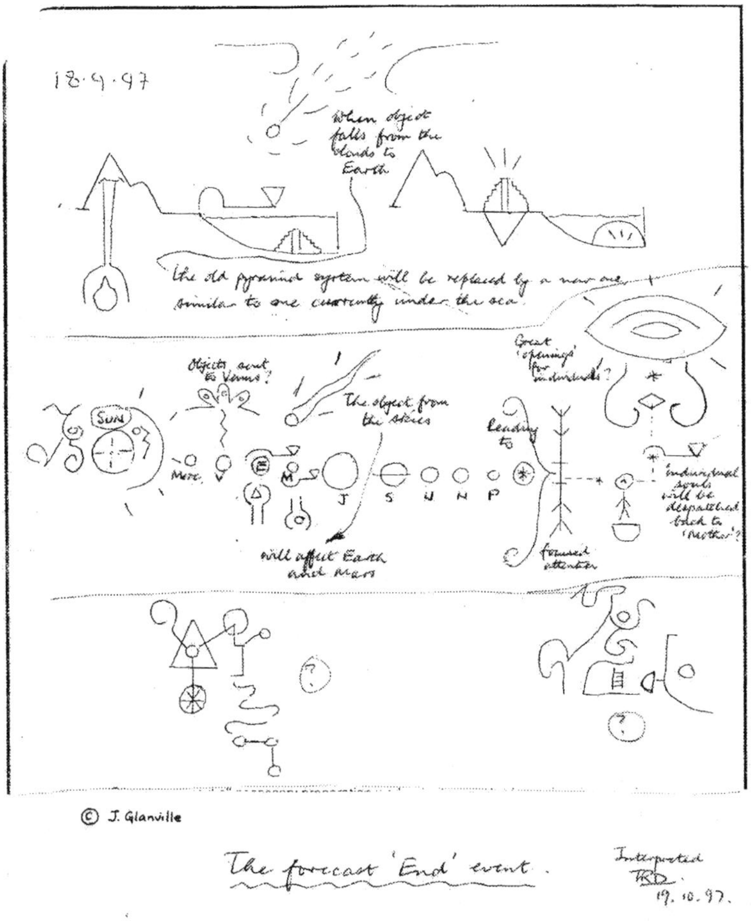


Fig. 57

There were three layers of drawings on this second page. The one in the middle seemed to provide a key to it all. A circle with associated symbols and short radiating lines was placed at the left-hand edge of the page. To the right of this, a series of nine smaller circles, of different sizes, was generated horizontally. The entire sequence suggested to me that it was a view of the sun and the nine planets of the solar system, with the planets shown distributed in the ecliptic plane. I labelled these 'planets' with appropriate symbols in accordance with their relative distances from the 'sun' (the largest circle). Above Earth and Mars, a circular object with a long and wide wavy tail was drawn directed towards Venus, at an angle to the 'ecliptic' of about 45°. Venus, Earth and Mars seemed to have been given symbolic involvement with the passage of the large meteoric object being portrayed. At the right-hand end of the planetary sequence, a further small circle was drawn and this contained an asterisk star. Surrounding this was a large 'close-parenthesis' bracket with central point pointing towards the right. To the right of this was something that suggested that it was a focusing device. A dashed horizontal line passing through the middle seemed to suggest the movement of the asterisk star through this device. To the right of this was a stick person with an asterisk in the centre of its circular head, standing on a kind of plinth. A dashed line linked this asterisk to another, via a dashed line drawn at right angles to the horizontal. This latter asterisk was depicted entering a 'door catch' symbol. Above this was drawn a large elliptical shape resembling a huge eye. A smaller ellipse was drawn near its centre, over-arched by a shallow arc. Lines radiated outwards from the outer ellipse and, below it, on either side, were two 'power' symbols. Between these, a diamond was drawn and this was linked to the 'catch' feature below it by a dashed vertical line. An asterisk had been placed between the diamond and the large eye. It looked as if the entire right-hand feature was trying to illustrate the fate of a human being during, or after, the event being depicted by the left side of the diagram. (In consultation with Jeff about that 'eye' feature, it was suggested to me that the 'eye' represented either 'great knowledge' or 'mother', the source of all souls. So, perhaps the right-hand side of the diagram tried to show the fate of an individual human soul? If so, this would link up and perhaps shed light on the complexity of the third page of drawings, still to be described).

The top layer of drawings on the second sheet seemed to be quite explicitly linking the meteoric event (shown again) to a specific change on the Earth. There were two side-by-side diagrams. They were both depicting cross-sectional views of the same geophysical features. A peaked mountain and a foothill were shown to the left of flat land and a stretch of water. In the left-hand diagram, on the bed beneath the water level, a stepped pyramid shape with a central shaft or vent had been placed. In the right-hand diagram, the pyramid had been re-located on the flat land beside the water, and radiating lines from its top seemed to suggest that it was radiating energy upwards. (With hindsight this sequence could have been linking with Jeff Glanville's later involvement with the stepped pyramids of Yucatan, Mexico, which has yet to be described.)

The meaning of the two diagrams at the bottom of this second page was too obscure and defied interpretation by me.

The third page of this collection, shown as Fig.58, caused me to consult again with Jeff Glanville and to pick up on some of his written inspirations about the fate of human souls after death.

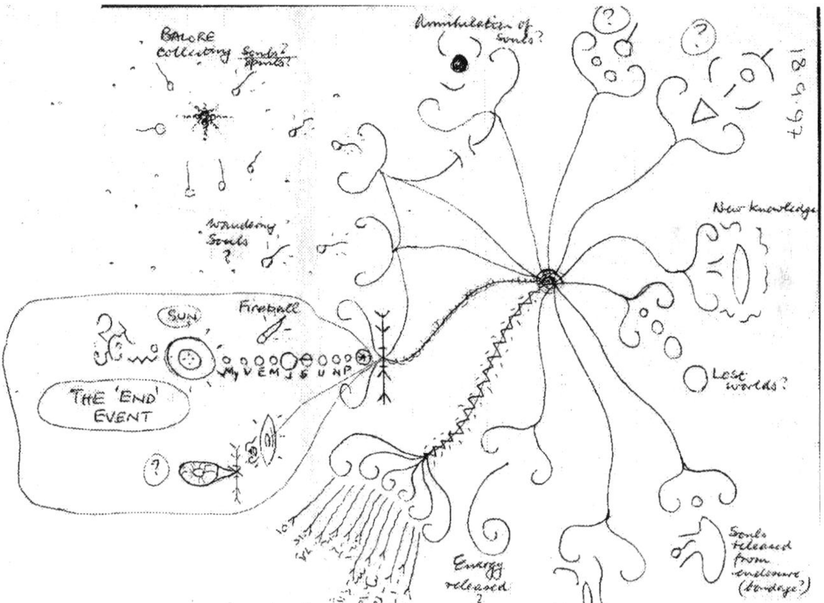


**Fig. 58**

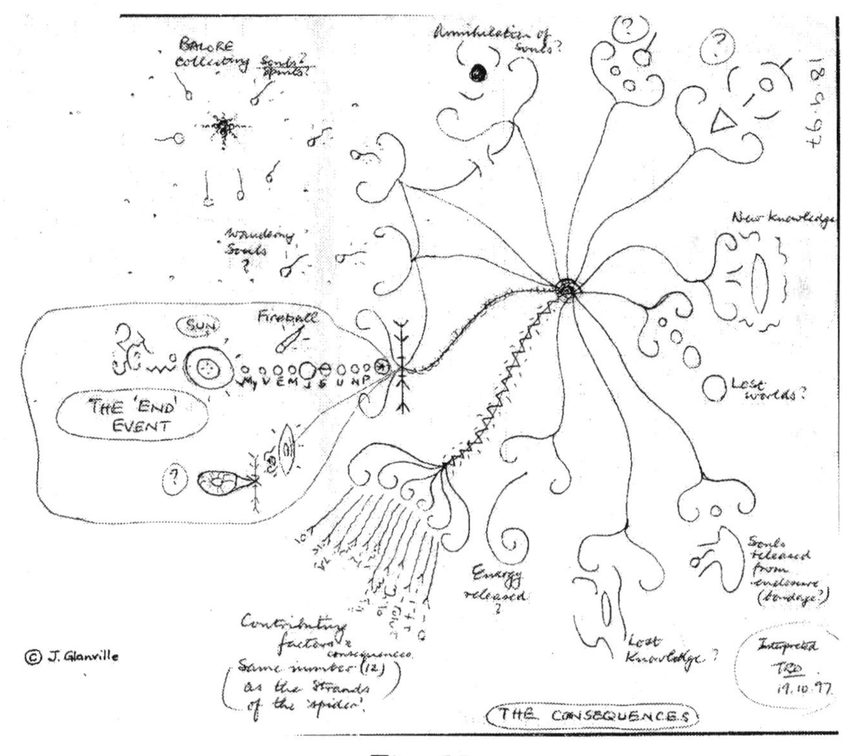

The left-hand side of this third page was a miniature reproduction of the lined-up planets from the previous page, complete with the bracket and focusing features at the right-hand end, but the right of the diagram was occupied by a huge 'spider' with long wobbly legs. One of these 'legs' connected the focusing device to the centre of the 'spider'. A highlighted leg branched out into six 'power' swirls and each of these had two snake-tongued legs and strange symbols associated with it. The remaining ten spider legs each terminated in two power swirls, forming a kind of cup. Various things were happening within these cups. Three were releasing little tadpoles and, from Jeff's writings and advice, I considered these to be representing human souls. Some were heading for a black star (called 'Balore' by Jeff). Others seemed to be annihilated, some were just heading into nowhere and others seemed to be being released from an enclosure (bondage?). If the big meteor event was intended to signal the end of the world, then this could be regarded as a warning that our souls need to be prepared for that day.

*This was all a bit off my 'beaten track', but I have an open mind and never dismiss things I have no real knowledge or experience of, even though I may feel they are highly improbable. They are being presented for the reader's consideration.*

The fourth and final page of this collection of drawings, presented as Fig.59, would have been a complete bafflement for me if I hadn't already analysed the previous page and, also, had some background knowledge of how the ancient Egyptians envisaged the progress of the human soul from this life into the afterlife.

An angular arch, mounted on two pedestals, over-spans a series of happenings enclosed within it. Outside and mounted on the horizontal top line of the arch, a large semicircle had been placed, this reminding me of the way in which the Egyptians envisaged the sun's location in the heavens. The processes depicted within the arch seemed to have something to do with the soul's progression into the astral world. To the right of and below the arch diagram were shown three horizontal lines of strange hieroglyphics, one of which was the 'tadpole' symbol. I could only guess that these hieroglyphs were intended to provide another way of explaining the progression of the soul.

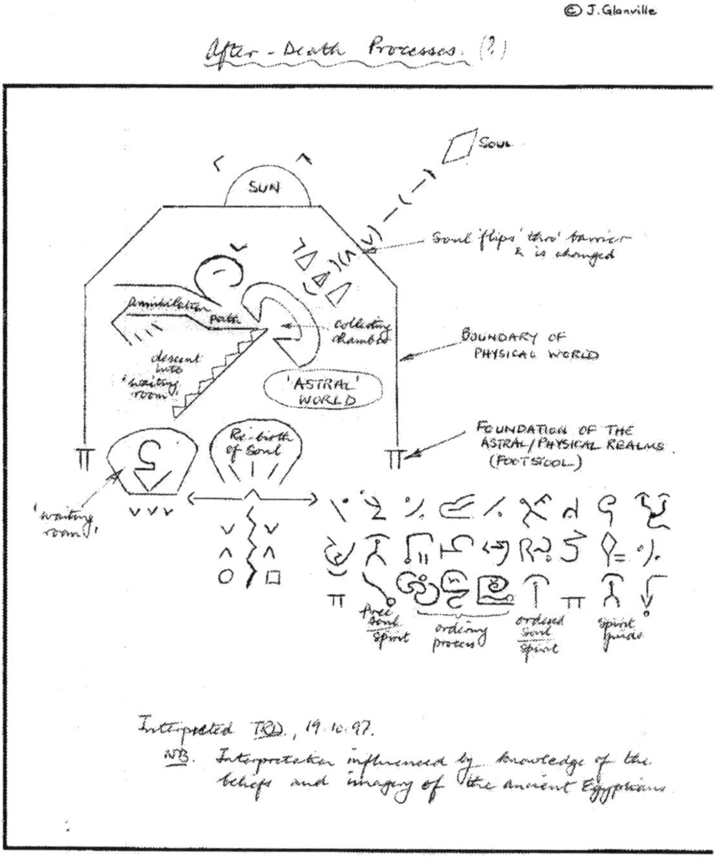


**Fig. 59**

*It is difficult to understand how a young man, who professes to have no artistic leanings, could possibly have produced such an imaginary and highly complex set of drawings as those I have described in this chapter. Even if his consciousness had been influenced by science fiction stories and films, I feel he could not have produced such drawings of his own volition.*

Were the later drawings trying to show us that the ETs are as concerned for our souls as they are to ensure that some members of our species survive when the all others have been wiped out? Could this be a link with that ringed area on the map of North Island, New Zealand, considered by the Maoris to be where their souls leave the Earth after death? I really don't know and will have to leave that judgement for readers of this book to make for themselves.

## Appointments at Rennes-le-Chateau and in South Wales

The Glanville's received messages instructing them to travel to a location in the south of France for a supposed rendezvous with the ETs. They knew the geographic co-ordinates of the place and they requested a timings graph for it. They wanted me to determine the most favourable times for a rendezvous on the given date. I complied with this request and gave them my best assessment. They then travelled to that inland area of southern France (they were already acquainted with the coastal areas) and drove to the given location on the specified day. At one of the times given by me, a brilliant flash of light had lit the sky and surrounded all present. David Glanville and another member of the group, who had both been looking in the opposite direction, did not see the flash in the sky. When I received the report from the family after this encounter, I learned that the given location had been a short distance from the now-famous village of Rennes-le-Chateau.

It was later revealed to me that this trip had taken place because they had become acquainted with Mr. Neil Hudson Newman, MSc, a civil/structural engineer and teacher. Mr. Newman claims to have deciphered an ancient code and to have determined where the legendary treasure of Rennes-le-Chateau had once been buried. After seeing some of Jeff's drawings, shown to him by David Coggins (the hypnotist), he had become intrigued and had visited the Glanvilles, travelling from his home in South Wales, to discuss his interpretation of the drawings and to share some of his work with them. About a year later (1999), Neil Newman became very much more involved with the Glanvilles. After they had left a copy of his work on their dining room table. it had then disappeared. Two messages they thereafter received related to Newman's code, very much to his amazement. This led to every new and obscure message (even though not encoded) being referred to Neil Newman and, in that way, it seems, he was drawn into the ETs' network.

That first trip to Rennes-le-Chateau arose from Newman's interpretation and synthesis of several messages, received in various ways, by the Glanvilles. From these, the location, date and time of the rendezvous had been able to be determined. It seems that I had been consulted to discover whether the

AT's predictions for that date and place confirmed the information given by the de-coded messages.

Another consequence of Neil Newman's association with the Glanvilles was that they were invited to travel to South Wales to visit a location near Caerphilly. Neil Newman had determined, from various old sources, that the site had links with the Arthurian legend. The Glanvilles anticipated that this might be to provide another rendezvous with the ETs. I was asked to provide a timings graph for an area in the vicinity of Caerphilly and to suggest the most likely rendezvous times. This was done successfully. ***During the exercise I noticed that Reading and Caerphilly shared a common ground-track line (space-lane), so there seemed to be a tangible link between the two sites.***

## Yucatan and a stepped pyramid.

During the summer of 2004, I was informed by David Glanville that Jeff had received an instruction to visit Yucatan, Mexico, during a given three-day period in July. The dates were July 23rd to 25th. Could I produce a graph for that area and advise them on the most likely times for the expected encounters? On producing the required graph I noticed a rather interesting coincidence. ***Yucatan also shared that common ground track linking Reading to Caerphilly!*** So, I would regard that track as being probably favoured for a Yucatan encounter. The global presentation in Fig.60 demonstrates how the sites are linked. Fig.61 is the computed timings graph for Yucatan.

Because Yucatan is close to the equator, the timing lines are clustered in such a way that large gaps exist between the clusters. This means that, if genuine events were to be reported at times within those gaps, the AT would be shown to be inadequate. This was to be a real test of its efficacy. I concentrated my attention on the clusters of lines containing that common ground track, which is labelled No.4 on this diagram. There were, as usual, four possibilities, with orientations of 11:00 h. RA. (dashed lines), Sunset terminator (lower curved) lines, 21:30 h. RA. (solid sloping lines) and Sunrise terminator (upper curved) lines, respectively. Each one of these possibilities was considered in turn and I decided to check whether any solar system bodies were aligned with them. The only alignment of that kind during those dates was associated with the Sunset track option, which was closely aligned with Mercury and, possibly, Mars. This I decided to regard as being a significant pointer and read off the times indicated by the lower cluster of curved lines as representing the most promising time interval for an encounter, during that period in July.

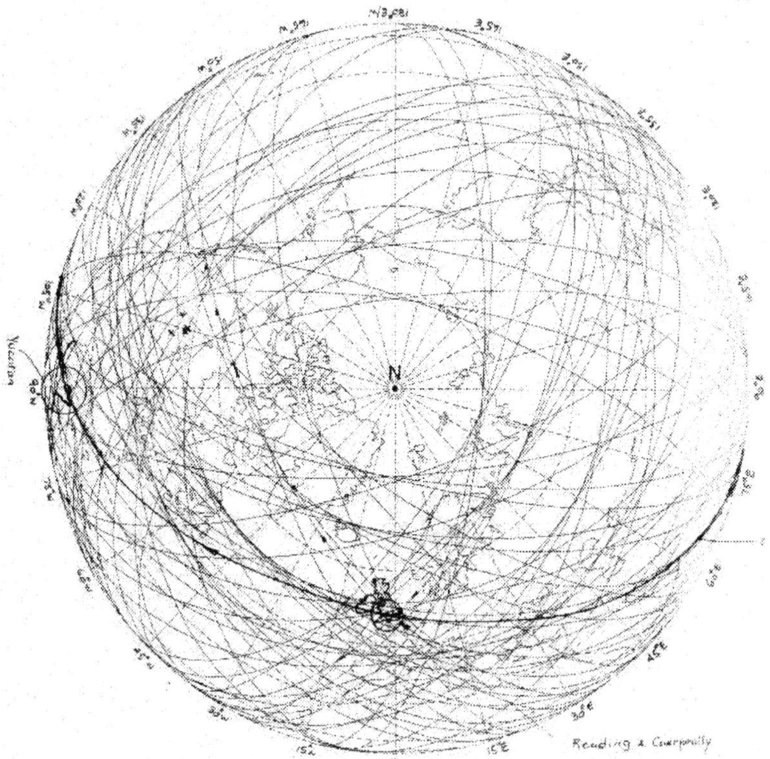


**Fig. 60**

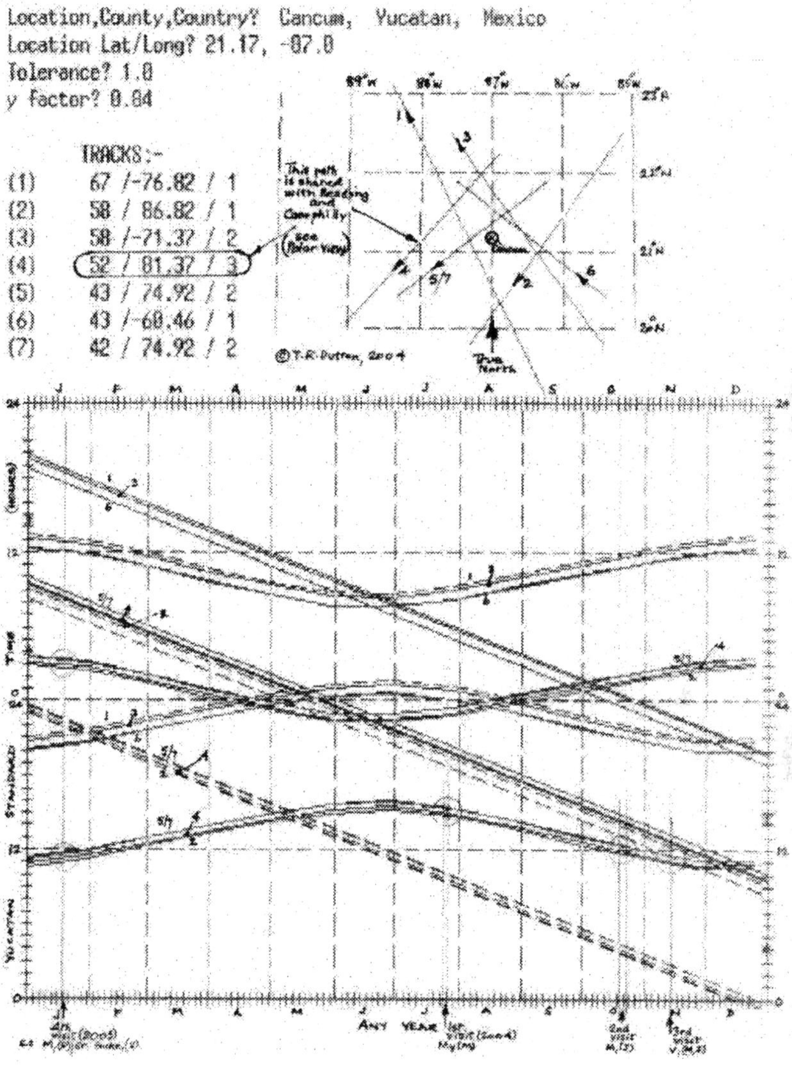


# Fig. 61

This result was given to David over the telephone. I asked how Jeff was going to arrange to fly to Yucatan during the specified period? Well, it was going to be facilitated by a strange coincidence. Jeff had recently become associated with a young woman in the travel industry, who could arrange flights to Cancum, Yucatan. She hoped to be able to arrange for Jeff to fly on flights to and from that destination during one of her respite periods, when she was entitled to have two or three days off duty. Was it all just a coincidence? David and I shared the same doubts over the 'phone. It was almost as though the ETs had arranged it to be so.

I received a report of Jeff's enforced holiday, over the 'phone, soon after he had returned. The events had occurred during the time period I had specified, between 2:30 pm. and 3:15 pm. Yucatan Standard Time. Jeff and his companion went swimming in the hotel's pool and had left their towels folded on their sun-beds. On swimming to the edge of the pool, Jeff noticed that his towel was then spread out on the recliner. He got out of the pool and folded it again, wondering who had spread it out. He returned to the water and shortly afterwards saw the towel had been spread out again. He called to his companion that it looked as if he was being required to lie on the bed. Fortunately, his companion had been forewarned that strange happenings might occur during that trip and, so, was not surprised by this. However, she would have been, to say least, surprised by what happened after Jeff lay down. According to Jeff, a large green face appeared from the sky and he became absorbed by the subsequent imagery. During this time his companion had seen him enter what can only be described as an altered state of consciousness and he began to speak in a strange, unfamiliar, language. This state had continued for something like the time period I had predicted, with the unflappable young woman taking it all in her stride. After Jeff had returned to normal, they'd been able to enjoy the remainder of their brief holiday in the sun. However, they'd discovered that a souvenir had been left on a small shelf when they returned to their room after the event. It was a little crystal skull. That skull was to feature a lot after it had been placed on the Glanville's dining table.

After all that, Jeff received three other calls to return to Yucatan. They were for the periods October 18$^{th}$ to 22$^{nd}$, November 13$^{th}$ and 14th, 2004, and January 19$^{th}$, 2005. I was asked for time predictions for each of these periods and they turned out to be correct. The October and November predictions were again Sunset ones and I was guided by the solar system alignments at those dates. Alignments with Mars and possibly Jupiter occurred in October and with Venus and possibly Mars and Jupiter, in November. The date in January, 2005, presented me with more difficulties. The Sunset option was aligned with Mars and possibly Pluto, but there was also a Sunrise option which aligned with the comet Encke and possibly with Venus. Both these indicated time periods are shown ringed by Fig.61, the Sunset time period being in the hour before midday and the Sunrise one indicating times between 2:30 and 3:30 am. After the event, I was informed that both these predictions had turned out to be accurate.

Stranger and stranger things have happened progressively throughout this sequence of trips to Yucatan. On one occasion Jeff had been instructed to climb to the top of a stepped pyramid in a period close to midday and to deliver an artefact there. He'd found his way to the very top of the pyramid blocked by scaffolding and had wondered what he might be required to do in those circumstances. Standing on an outside platform, he was then inspired to throw the object upwards in the hope that it might land at the desired destination. He'd watched it slowing down and realised it had been a vain hope. Then, to his amazement, he had seen the object suddenly accelerate upwards and disappear into the sky. The Yucatan saga may be continued, but I think the whole story (if ever complete) must be left for Jeff Glanville to tell for himself.

As a footnote, I consider that these events involving Yucatan seem to have provided significant tests of the efficacy of the Astronautical Theory in predicting the ETs' activities and, also, to have validated the suspected links of those activities with favourably-aligned solar system bodies. I am very pleased that the Glanvilles have been willing to share their strange experiences with me and to allow them to be published in this book.

# PHASE 9: Concluding reflections

## CHAPTER 23

## WHAT A QUEST !

The exercise described by the previous chapters began as the result of my inquisitiveness about phenomena that had seemed to challenge my knowledge of aeronautics and astronautics. Years before, after being handed a copy by a school classmate, I had read the Leslie/Adamski book **[15]** of 1953, containing George Adamski's claim to have met a human-looking alien astronaut, but the idea that living ETs were visiting this planet had just left me cold. In fact, it was the quality of the photographs in that book that stimulated me to look for new means of flying, other than by aerodynamics, rocketry and jet-thrust systems. So, my investigations into the nature of gravity, stemming from this interest, preceded my study of the SAC reports of 1967 by some 13 –14 years. That spare-time study of gravity took me into the realms of Relativity and Quantum Mechanics. However, it soon became apparent, even with my lay-understanding of those two major areas of physics, that our physics were not yet advanced enough to even simulate, let alone emulate, an aerial craft with the ability to modify, locally, the gravitational field it was subject to on the surface of this planet. There was a need for a new theory of gravity to unite the two major physical theories and, as the years went by, I was bold enough to try to produce one. (This was eventually brought together in an unpublished 1998 paper with the title, **'Gravity --- Some Original Thoughts'**). All this provided me with a desire to know more and, eventually, the events of 1967 provided me with the required stimulus.

Those early investigations very quickly told me that I had no hope of understanding the means by which the reported craft were propelled and controlled, but they seemed to be real enough. O.K.--- they were real, but clearly not of this world. So, what were they up to? Which features on the ground below them had been of interest and was there any pattern in the supposed surveillance activity? The story of how that Phase 1 investigation led into Phase 2 has been told earlier in this book. The progression was all very logical and led to surprising results; results supplemented by my involvement in those crop-circle investigations described in Phase 3. Then followed a period when I concentrated on trying to communicate my findings to others for independent assessment. That was an up-hill struggle. Then, some years later, the breakthroughs into the consciousness of a few scientists began, the guarded acceptance of Edward Ashpole providing a stimulus for others. This opening up of the work is summarised in Phase 6. However, the testing of the Astronautical Theory (AT) for SAC/UFO Events was not taking place. I was told on several occasions by university professors that funding would not be made available for such studies.

In the face of this 'stone-walling', there seemed to be nothing further to do than to computerise the Theory myself and to test new reports in that way. Typical results of those exercises were demonstrated by the cases considered in Phases 4 and 5. The AT seemed to apply to all manner of SAC encounters, including those involving time loss and possible abduction of the witnesses. As the checking procedure progressed, it was revealing to me, and to all who would heed the results, that ***this planet is under continuing surveillance by ETs*** and that humans and human activities are being closely monitored. The technology being used we cannot reproduce. It is, by our standards, 'magical', but why shouldn't it be?

As considered in Phase 3, there are all the indications that ET technology achieves that elusive control of gravity that, doubtless, many seek behind the closed doors of laboratories funded by military establishments. But the ability to convert matter into forms of radiation, as demonstrated in Phase 4, is the stuff of a new kind of physics.

***My speculation is that the visiting objects may travel in nearly straight lines, at near light speeds, across our solar system, by the continual conversion of mass energy into kinetic energy.***

Perhaps, when they decelerate, they merely absorb energy from their surroundings to feed their minimised atomic structures and, in that way, to re-create full size solid craft? Thus, their arrivals and departures in the atmosphere may be the witnessed luminous clouds from which solid objects emerge or the brilliant flashes of light into which they are sometimes reported to disappear. The ability to traverse space at near-light-speeds would be consistent with those alignments with solar system objects I observed. These seemed to link with the arrival and departure paths identified by the AT's processing. It seems that the ETs' craft have no dependence on celestial mechanics during trips across our solar system. They just aim, shoot and arrive in the vicinity of their targeted planet! How they travel interstellar distances is another aspect of their technology we are unable to fathom.

Phase 8 demonstrates that SAC phenomena seem to be capable of stimulating psychic phenomena. From this evidence I infer that the ETs know more about the human mind than we know, currently, ourselves. They certainly seem to know how to manipulate human perceptions, how to produce magical effects, how to bring people into altered states of consciousness and how to persuade them to follow their bidding. The Glanville's experiences are, in my view, obedience tests. They have been persuaded to co-operate in all sorts of ways ---- for example, they have allowed their dining room table to be taken over for the ETs purposes and they do not care to interfere with the objects appearing on it. The intruders' influences in the Glanville home seemed to release 'spirits' from the past. The Glanvilles were being persuaded that some of those restless spirits were of people who perished in a fire within an old inn that had once occupied that site. Their task would be to release those entities from their bondage to the site. All this led into Jeff's 'inspired' drawings that mixed technological objects, 'End Days' imagery and, perhaps, profound information about the fate of human souls. As I commented in Phase 8, all this took me 'out of my depth'. I do not know how real these things are, but I do know that they have taken over the lives of a lively and, otherwise, normal family.

On reflection, the doomsday idea might very well be reinforced by the escalation of the number of SAC encounters and other UFO reports during the latter half of the twentieth century. A survey of my database produced Fig. 62. The rapid growth in sightings is clearly shown.

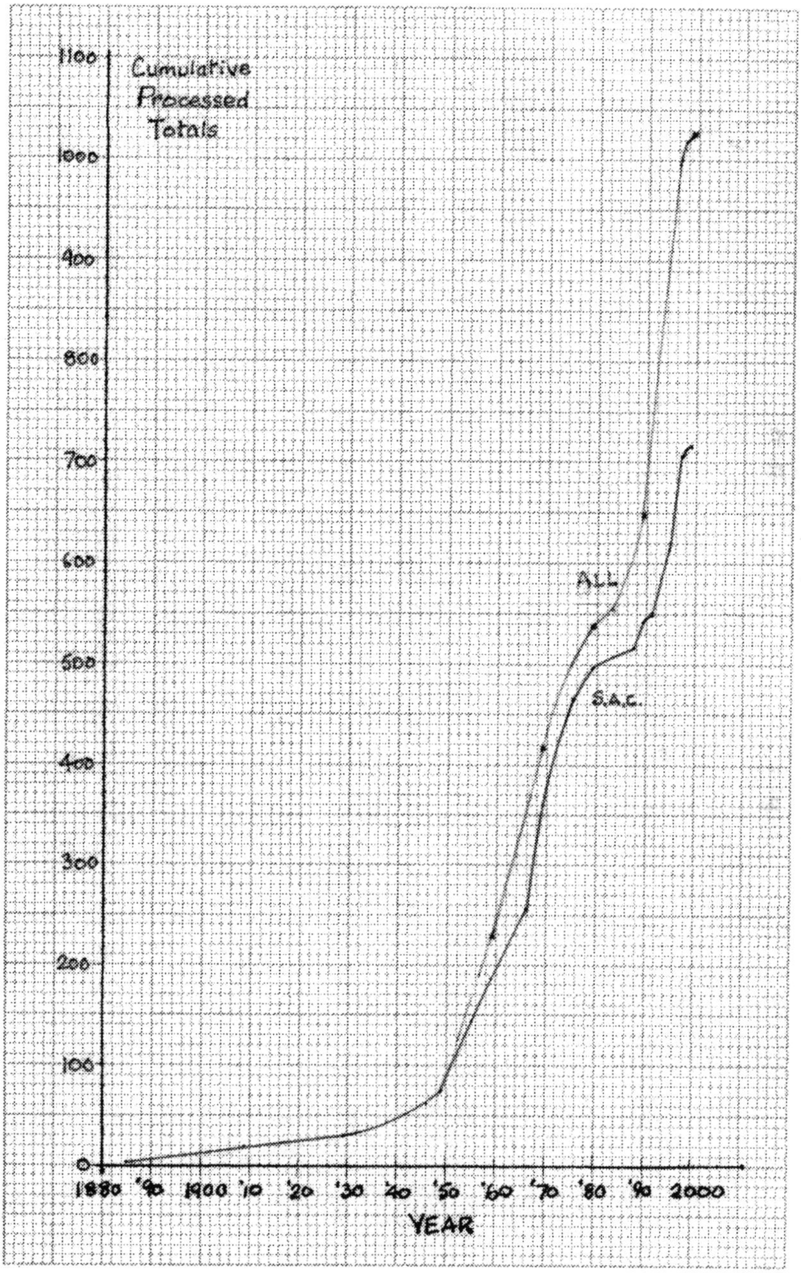
Cumulative
Processed
Totals
1100
1000
900
800
700
600
500
400
300
200
100
0
ALL
S.A.C.
1880 '90 1900 '10 '20 '30 '40 '50 '60 '70 '80 '90 2000
YEAR

Fig. 62

This feature may be partially caused by the great advances made in communications technology during the past sixty years, but, even so, the numbers of AT-validated events now in evidence are staggeringly large. Perhaps the ETs are developing expertise for their doomsday roles?

The NIDS Essay Competition of 1998, from which Edward Ashpole and I gained a first prize (see Phase 7), posed the question, **"If contact were to be made between humans and extraterrestrial intelligence (ETI) on Earth or in the solar system, what is the most probable means by which that would occur and how would we know that the interaction had taken place?** I'm sure that whoever framed that question would not have been able to anticipate the answer now being presented by the Glanvilles. What's more, Edward Ashpole and I would not have been able to put forward such an answer. But I think that the revelations given in this book and, particularly, by Phase 8, are indicating that the ETs are communicating with us in bizarre and unexpected ways, including by means of genuine crop formations (Phase 3). I think it would be extremely presumptuous for anyone to think that such advanced creatures could possibly meet with us on equal terms. Their intellects and knowledge seem to be literally light-years beyond our own.

My lifelong quest for understanding of the SAC/UFO phenomena began as an innocent investigation of reported aerial technology and has progressed to become a profound life-changing experience. I want to conclude by introducing a very different element, which seems, possibly, to relate to all the other elements of this amazing story.

# Epilogue

Throughout history, people have been awed by strange things seen in the sky and by the messages of angels (ethereal messengers). Leading religions have been founded on the information given by such messengers. On consideration of those ancient encounters, which, in the most admirable cases, seem to have been attempts to ***tame*** emotionally-volatile humankind, I discovered that the most profound and meaningful set of revelations, ***forecasting the complete failure of all such experiments***, was to be found in the Bible, presented in the book of **Ezra 4** (Second Book of Esdras in the King James' Apocrypha).

During the 6th century BC, the Hebrew high priest Ezra (Esdras) asks in earnest prayer for information about his captive people's future. He receives a visit from an angel called Uriel (Light of God), who opens the dialogue by saying Ezra's pleas had been heard by 'the Most High' (*NB.* not 'God') and that he (Uriel) had been sent to address them. He begins by presenting Ezra with a series of tests obviously intended to explore Ezra's knowledge base. Most of those things we would be able to demonstrate today; for example, ***"Measure me a blast of wind", "Show me the image of a voice".*** Of course, Ezra has no such knowledge. Uriel comments that if he lacks such basic earthly knowledge, how could he expect to understand things not of the Earth? But Ezra persists and then is shown more than he'd been able to anticipate or to understand. At one point in the narrative, he is asked to stand beside Uriel and not to be afraid of the noise he was going to hear, because it was in the nature of the thing he was standing on. [A piece of technology, perhaps?] He was then shown coloured imagery that could have represented either the beginning or the end of the cosmos. After this profound experience, Ezra continued to pray for further enlightenment. This resulted in several further visits from Uriel. During these, Ezra is informed that the troubles he sees around him, in Babylonian captivity, will be greatly exceeded in times to come; for example, bodies will be piled as high as the camel's hough [hump?]. Then there will be a time when people will be visited [by the 'Most High'?] and subjected to inquisition. Ezra is shown moving scenes depicting things yet to come in the 'last days'. *He is told that (premature) babies of three months will be born and will survive; that there will be widespread killing and looting involving ordinary people; that there will be a general state of violent anarchy existing within the nations of the world and that no nation will be exempt from this;*

*that there will be wars waged against Babylon [Iraq?] by a coalition of military forces and that while victuals will be plentiful and cheap for many, there will be widespread famine prevailing for others.* As he receives this information, Ezra is presented with animated images of 'flying swords' [finned missiles, like the Skud used by Iraq?] and 'horrible stars' [nuclear weapons?] being flung that can flatten cities and mountains. There is a very telling piece of information that Asia will not be exempt from all this violence and a chilling poetic forecast is made: *"And the glory of your power shall be dried up as a flower, when the heat shall arise that is sent over you". [The atomic bombs dropped over Japan were exploded in the air, high above Hiroshima and Nagasaki.]*

In my view, this 2500 years' old revelation is a forecast that can only be for our time in history of the world. The book of Ezra 4 was relegated to the Apocrypha as **Esdras 2** by King James' scholars, because it was regarded as being a corrupted book, written in Greek. Corrupted it seems to have been, perhaps by over-enthusiastic early Gnostic Christians, but I am convinced the main contents of the book are truly authentic. These questions then arise: *"How was Uriel able to present to Ezra such detailed information and images of our times? Is everything in the history of the world predetermined or, like the fictional Dr. Who, was Uriel a time traveller?"*

It seems to me that the entities interacting with the Glanvilles of Reading, England, and with that lady in Wilmslow could be of a similar kind, but the information being given today is for the humans of today. The CE4 evidence suggests that human bodies are being investigated, human minds interrogated and, sometimes, extended to take onboard concepts never before able to be communicated. Several beautifully-executed crop formations of the early 1990s seemed to be warnings of impending doomsday. One very memorable one appeared to represent the planets and orbits of the inner solar system, with the orbit currently occupied by Earth shown to be unoccupied. Another finely executed pictogram seemed to represent an analogue clock with its final indicator close to midnight. Could it be that the programmed scrutiny of the Earth and humankind, described previously, has been established to facilitate doomsday operations? In such an emergency, all 1010 (perhaps more?) paths over the Earth could be utilised, virtually simultaneously, to maximise the number of our species able to be rescued. That's quite a thought.

With these 'signs of our times', readers must now be left to ponder these things for themselves. Having arrived at conclusions beyond anything dreamt of at outset, I feel it is now time for me to sign off.

# FURTHER ACKNOWLEDGEMENTS

Although the main contributors and promoters of the work described by this book have already been named, there have been many others who have, over the years, played their parts --- by providing information or by testing the Astronautical Theory by observation during amateur skywatches. Their invaluable interest and support is now acknowledged with my thanks. Deserving of special thanks are the important contributions made, during 1980, (Phase 2, Chapter 7) by Messrs. George Pickford, Denzil Hallam, Frank Page, and Henry Williams (graduate trainee), then members of the Computer Services Department of Hawker Siddeley Aviation, Ltd. at Woodford Airfield, Cheshire.

# ADDENDUM

## The Astronautical Theory for UFO Encounters

Astro-navigational links with correlated tracks --- A Pilot Study
by T.R.Dutton, *Oct. 1999 & Nov. 2000 (Appendix)*


### *Abstract.*

The Astronautical Theory for UFO Encounters has been derived from a prolonged and objective scientific study of **the most inexplicable UFO encounters** reported worldwide. The theory demonstrates that a highly-programmed astronautical activity can be regarded as underlying virtually all such events --- but **especially those involving unidentifiable aerial craft**, which, by implication, appear to be fully automated. The postulated astro-navigational nature of the programming involves reference to the stars, the Sun and, possibly, other solar system bodies. The pilot study described in this paper investigates navigational links with the latter group of objects which, hitherto, could only be suspected.

The initial pilot study considered 174 good cases selected from processed data for the years 1957, 1977, 1990-1993 and 1995. (The reason for this selection is explained.). These cases had been previously processed (successfully) through PC programs created to simulate the programmed activity identified by the Theory. The inferred spacecraft approach and departure tracks (supposedly, two per event) which had been found (by computation) to be linked to the encounters in the atmosphere, were then examined to identify whether any solar system bodies might have been referenced to aid navigation during those operations. If two tracks had been identified for each event, and each track had referenced a single solar system body, then 348 referenced bodies would have been involved. In fact, 333 such links were discovered, but this number had been enlarged by conjunctions and oppositions on some occasions. A particularly surprising outcome from this exercise was that six comets had featured 90 times during the periods considered. On no less than 60 occasions those comets had been very closely referenced. A summary table of results is presented.

The positive results of the initial exercise stimulated a further, more specialised study, which is described **in the Appendix**. After 27 cases involving alleged encounters **with unidentifiable craft and their alien occupants** had been processed successfully through the PC programs, another solar system alignments exercise seemed to be a pertinent extension of the previous study. Given the smaller size of the data sample, it became possible to examine the outcome for each case in some detail. (In view of difficulties experienced by readers of the initial pilot study, it was decided to preface these results by diagrams explaining the astro-navigational nature of the Astronautical Theory. These diagrams are now referenced in the main part of the report.) As an afterthought, after the successful initial PC processing of the data provided for that case, it was decided to include the highly-controversial Roswell crash event at the end of this investigation.

Some of the results of this extra study are truly remarkable and seem to validate the suggestion that personal involvement of the ETI perpetrators, of the postulated automated global surveillance activity, occurs from time to time.

## Introduction to the Theory.

Thirty years' objective analysis and global synthesis of reported UNIDENTIFIABLE aerial objects, collected from a period of more than a century, resulted in the formulation of an Astronautical Theory for UFO Encounters. The logical process by which this was achieved and the subsequent proving activity are summarised below:-

- Technical analysis of selected British UFO reports during the period 1967 to 1973 had emphasised the **physical reality** of **the craft** being described by witnesses. Furthermore, an overview of the affected sites indicated that specific areas had been selected for close scrutiny during each period of activity.
- The objects had often been seen to descend from the sky and to return into it, with performance capabilities far beyond anything achievable by man-made craft. These observations

implied that, since they had not been humanly-contrived, **the artefacts had originated from an extra-terrestrial source**.

- A database of selected cases, reported world-wide during the period 1885 to 1971, was then used to try to identify **modes of operation in space** which could have been linked with the observed aerial activity. This exercise was begun in 1973 and, by 1980, there were indications that a programmed surveillance and exploration activity had been carried on throughout that historical period.
- Further development during the 1980s resulted in definition of the programmed activity in the form of an **Astronautical Theory for UFO Close Encounters.** (Reports defined by Dr. J Allen Hynek **[Ref 7.1]** as Close Encounters (CEs) were preferred data, because the location of the observer could be regarded as being virtually the same as that of the object being described.) **Well-established approach and departure tracks in space were identified** --- and those tracks were found to **be orientated celestially in any of four precisely defined ways.**

  *Key navigational **reference points on the Earth's equator** were also identified, as were the **pre-set angular speed of transit** and the **East-to-West direction** of the approaching (albeit hypothetical) delivery and retrieval spacecraft. These features defined a predetermined set of **'ground-track' lines,** which were projected onto the Earth's surface directly below the tracks followed in space.
- The four celestial orientations, plus the other characteristics just described, opened up the possibility that the **timings of future events at any location beneath the favoured tracks, might be predictable**. If this were to be tested by data not already in the database, then a positive result would, also, prove the validity of the global theory. A PC database was then created and special programs written to facilitate this proving work. **Almost 900 test cases later, the Theory has been very effectively proven.** In addition, direct observation groups, using the timing predictions, are reporting very high success rates.
- 

## 2. Further observations and development.

The four celestial orientations, mentioned above, consisted of two sets of spacecraft approach, or departure, tracks fixed relative to the stars; and two other sets connected with the Earth's terminator, either at Sunrise or at Sunset. These latter sets moved round the sky, following the Sun, during any year. [See **Appendix Figs. 1 to 3**]

Close scrutiny of ten Close Encounter cases (which had also involved reported encounters with alien creatures) revealed that, during the days of those occurrences (1952 – 1988), major planets were positioned in the sky such that they were aligned with at least one set of star-related tracks. This led to the speculation that perhaps the planets (and, possibly, the Moon) were being used as navigational markers to facilitate the trips to and from the Earth. Clearly, much more processing would be required to check whether this feature was present on all other significant occasions. Furthermore, it raised the question of whether similar alignments would be found for the Sunrise and Sunset tracks.

The need to know where the sunrise and sunset orientated tracks intersected the Ecliptic Plane (the plane in space in which the Sun appears to move during a year, and which is approximately shared with the major planets) created more work. It was realised that the Sun-related tracks model, derived from the processing of the original data, linked those tracks directly with the Sun only at the Spring and Autumn Equinoxes. In fact, the defined tracks were keyed to the celestial meridian passing through the intersection of the terminator and the northern-most point on a 53° inclined track, on any day of the year. [See **Appendix Fig. 2**] A program then had to be created to calculate where the 53° sunrise and sunset tracks intersected the Ecliptic, before work could commence on the planetary alignments exercise. Tracks with other inclinations to the equator would then intersect the Ecliptic to the left or right of the nominal 53° intersection position in the sky, within a range of displacements of approximately +/- 1.0 hr of Right Ascension (R.A.).

## 3. The Pilot Study.

When the 'stand-alone' program had been written and tested, it was incorporated into the Correlated Tracks Database programs for each decade from 1950 onwards. Trial runs were then carried out for the years 1957, 1977, 1990-93 and 1995.

This choice was made to explore the effects of noticeable movements of the outer planets over the overall timescale and the year-to-year effects of changes in the positions of the inner planets. The search for aligned planets was carried out in the following way:-

Firstly, the program was run so that every track serving a UFO event (ie. that had been found to correlate with the timing of that event) was processed again. If that track had been identified as being linked, say, to the sunset terminator, then the azimuth intersection (R.A.) of a sunset-related track with the Ecliptic was printed out. The Opposition intersection position in the sky was also output. This procedure was followed for tracks with any of the other three orientations.

Next, using the output data from this program, commercial astronomical software (Expert Astronomer) had to be run and searched, visually, for planets and other solar system bodies lying in the intersection zones on the day and year of each UFO event, for each of the years given above. This turned out to be a long, tedious and testing exercise.

The numbers of processed and correlated UFO events for the chosen years, and the number of associated astro-links, were as tabulated below:-

**TABLE 1**

| | 1957 | 1977 | 1990 | 1991 | 1992 | 1993 | 1995 | Totals |
|---|---|---|---|---|---|---|---|---|
| Total no. of cases | 33 | 28 | 10 | 13 | 6 | 66 | 18 | 174 |
| Cases without astro-links | 3 | 0 | 2 | 0 | 0 | 2 | 0 | 7 |
| Qualifying activity days | 26 | 25 | 8 | 12 | 5 | 46 | 15 | 137 |
| No. of track orientations | 58 | 49 | 20 | 20 | 9 | 93 | 25 | 274 |
| No. of astro-links | 48 | 72 | 28 | 25 | 13 | 107 | 40 | 333 |

In this table, the activity days apply only to those cases which were found to have astro-links associated with the tracks identified by the previous processing. **'Cases without astro-links'** gives the number of cases excluded and not affecting the figures given in the rest of the table.

The number of track orientations always exceeds the number of qualifying processed cases, because the program tries to identify **two tracks per event** --- a closely-correlating track, which could be a delivery or retrieval track, and another track possibility within one hour of the reported time, which could have played a complementary role. The number of possible links with solar system bodies (astro-links) generally exceeds the number of track orientations because, on a considerable number of occasions, Conjunctions and Oppositions were indicated. This influence on the totals was, to some extent, offset whenever several events had occurred on the same day and had shared common astro-links.

After having processed all 274 track orientations and identified 333 alignments with solar system bodies, the next problem was to decide how to interpret the results. A table, listing all the correlated bodies and giving the number of times a given body had been aligned with each of the four track orientations, seemed to offer the most easily-understood presentation. It is presented here as **Table 2.**

Solar System links with identified track orientations — Processed data samples. **TABLE 1**

No. of Correlations listed for Track Orientation and Year

| Aligned Solar System Body | 11:00h RA [illegible] | [illegible] | [illegible] | [illegible] | [illegible] | [illegible] | [illegible] | Close | Total |
|---|---|---|---|---|---|---|---|---|---|
| Moon | 1 | 3 | – | – | – | – | 2 | 4 | 5 |
| Sun | 1 | – | – | – | – | – | – | 1 | 1 |
| Mercury | – | 1 | – | – | – | 1 | – | 2 | 2 |
| Venus | 2 | – | – | – | – | – | 2 | 2 | 4 |
| Mars | 1 | – | – | – | – | – | – | 1 | 1 |
| Jupiter | – | 1 | 2 | – | – | – | – | 3 | 3 |
| Saturn | – | – | – | – | – | – | – | 0 | 0 |
| Uranus | 1 | – | 5 | 3 | – | – | – | 6 | 9 |
| Neptune | – | – | – | – | 2 | – | – | 0 | 2 |
| Pluto | – | – | – | – | – | – | – | 0 | 0 |
| Ceres | – | – | – | – | – | – | – | 0 | 0 |
| Pallas | – | – | 1 | – | – | – | – | 0 | 1 |
| Vesta | – | – | – | 1 | – | – | 6 | 7 | 7 |
| Juno | – | – | – | 1 | – | 1 | 3 | 5 | 5 |
| Hartley 2 | – | – | – | – | – | – | – | 0 | 0 |
| Grigg-Sk | – | – | – | – | – | – | – | 0 | 0 |
| Encke | – | – | – | – | – | – | – | 0 | 0 |
| Gale | – | 6 | – | – | – | – | – | 6 | 6 |
| Machholz | – | 2 | – | – | 2 | 2 | – | 3 | 6 |
| Halley | – | – | – | – | – | – | – | 0 | 0 |
| Total Correlated | 6 | 13 | 9 | 5 | 4 | 4 | 13 | 30 | 62 |

| Aligned Solar System Body | 21:30h RA [illegible] | [illegible] | [illegible] | [illegible] | [illegible] | [illegible] | [illegible] | Close | Total |
|---|---|---|---|---|---|---|---|---|---|
| Moon | 2 | – | – | – | 1 | 3 | – | 4 | 6 |
| Sun | 2 | 1 | – | 1 | – | 2 | 2 | 8 | 8 |
| Mercury | 1 | 6 | – | – | – | 2 | – | 2 | 9 |
| Venus | – | – | 2 | – | – | 3 | – | 4 | 5 |
| Mars | 2 | – | 2 | 1 | – | 2 | – | 6 | 7 |
| Jupiter | – | 3 | – | – | – | – | – | 2 | 3 |
| Saturn | – | – | – | – | – | – | – | 0 | 0 |
| Uranus | – | 11 | – | – | – | – | – | 10 | 11 |
| Neptune | 1 | – | – | – | – | – | – | 1 | 1 |
| Pluto | – | – | 3 | 2 | – | 5 | – | 0 | 10 |
| Ceres | – | 1 | – | – | – | 5 | – | 5 | 6 |
| Pallas | 1 | – | – | – | – | – | – | 1 | 1 |
| Vesta | – | – | – | – | – | – | – | 0 | 0 |
| Juno | – | 1 | 3 | – | – | – | – | 3 | 4 |
| Hartley 2 | 7 | 1 | – | 2 | – | – | 2 | 6 | 12 |
| Grigg-Sk | – | – | – | – | – | – | 3 | 0 | 3 |
| Encke | – | – | – | – | – | – | – | 0 | 0 |
| Gale | – | – | – | – | 1 | 19 | – | 19 | 20 |
| Machholz | – | – | – | – | – | – | – | 0 | 0 |
| Halley | – | – | – | – | – | – | – | 0 | 0 |
| Total Correlated | 16 | 24 | 10 | 6 | 2 | 4 | 7 | 71 | 106 |

| Aligned Solar System Body | Sunrise [illegible] | [illegible] | [illegible] | [illegible] | [illegible] | [illegible] | [illegible] | Close | Total |
|---|---|---|---|---|---|---|---|---|---|
| Moon | – | 2 | – | – | – | 1 | 2 | 5 | 5 |
| Sun | – | – | – | – | – | – | – | 0 | 0 |
| Mercury | – | – | – | – | – | 1 | – | 1 | 1 |
| Venus | 1 | 2 | 2 | – | – | 2 | – | 6 | 7 |
| Mars | – | 1 | – | 1 | – | – | – | 0 | 2 |
| Jupiter | – | – | – | – | – | – | – | 0 | 0 |
| Saturn | – | – | – | – | – | – | – | 0 | 0 |
| Uranus | – | – | – | 1 | – | – | – | 0 | 1 |
| Neptune | – | – | – | – | – | 1 | – | 0 | 1 |
| Pluto | 1 | 1 | 1 | 2 | – | 1 | – | 0 | 6 |
| Ceres | – | 3 | – | – | – | 1 | 3 | 4 | 7 |
| Pallas | 1 | – | – | – | – | – | – | 1 | 1 |
| Vesta | – | – | – | – | – | – | – | 0 | 0 |
| Juno | 1 | – | 1 | – | – | – | – | 1 | 2 |
| Hartley 2 | – | – | – | – | – | – | – | 0 | 0 |
| Grigg-Sk | – | – | – | – | – | – | – | 0 | 0 |
| Encke | 1 | 8 | – | – | – | 1 | 3 | 11 | 13 |
| Gale | – | – | – | – | – | – | – | 0 | 0 |
| Machholz | – | – | – | – | – | 2 | – | 1 | 2 |
| Halley | – | – | – | – | – | – | 1 | 1 | 1 |
| Total Correlated | 5 | 17 | 4 | 4 | 0 | 10 | 9 | 29 | 49 |

| Aligned Solar System Body | Sunset [illegible] | [illegible] | [illegible] | [illegible] | [illegible] | [illegible] | [illegible] | Close | Total | Overall Close | Overall Total |
|---|---|---|---|---|---|---|---|---|---|---|---|
| Moon | 2 | – | – | – | 2 | 2 | – | 4 | 6 | 15 | 22 |
| Sun | – | – | – | 1 | – | 2 | – | 3 | 3 | 12 | 12 |
| Mercury | 3 | 3 | 6 | 1 | 1 | 1 | 2 | 14 | 17 | 17 | 29 |
| Venus | 1 | – | – | 1 | – | 6 | – | 8 | 8 | 20 | 24 |
| Mars | 3 | – | – | 1 | – | – | 1 | 1 | 5 | 10 | 15 |
| Jupiter | 2 | 2 | – | 1 | – | 6 | – | 7 | 11 | 12 | 17 |
| Saturn | – | 1 | – | 1 | 1 | 7 | – | 4 | 10 | 4 | 10 |
| Uranus | 3 | 1 | – | 1 | – | 1 | – | 4 | 6 | 19 | 27 |
| Neptune | 1 | – | – | – | – | 1 | – | 2 | 2 | 3 | 6 |
| Pluto | 1 | – | – | – | 1 | – | – | 1 | 2 | 1 | 18 |
| Ceres | – | – | – | – | – | 4 | 1 | 2 | 5 | 11 | 18 |
| Pallas | – | – | – | – | – | – | – | 0 | 0 | 2 | 3 |
| Vesta | 1 | 5 | – | – | – | 1 | 2 | 6 | 9 | 7 | 16 |
| Juno | – | – | – | – | – | 13 | 2 | 11 | 15 | 17 | 26 |
| Hartley 2 | – | 2 | – | 3 | 1 | 4 | 1 | 6 | 10 | 12 | 22 |
| Grigg-Sk | 1 | – | – | – | – | 1 | 1 | 1 | 3 | 1 | 6 |
| Encke | – | 1 | – | 1 | – | 2 | – | 2 | 4 | 13 | 17 |
| Gale | 3 | 3 | – | – | – | 1 | – | 3 | 7 | 28 | 33 |
| Machholz | – | 1 | – | – | – | – | – | 0 | 1 | 4 | 9 |
| Halley | – | – | – | – | 1 | – | 1 | 1 | 2 | 2 | 3 |
| Total Correlated | 21 | 19 | 6 | 10 | 7 | 32 | 11 | 91 | [illegible] | 211 | 333 |

| Yr: | [illegible] | [illegible] | [illegible] | [illegible] | [illegible] | [illegible] | [illegible] | Totals overall |
|---|---|---|---|---|---|---|---|---|
| Processed Orientations | 58 | 49 | 20 | 26 | 9 | 93 | 25 | 274 |
| Total Correlated | 48 | 72 | 28 | 26 | 13 | 107 | 40 | 333 |

The Astronautical Theory for UFO Encounters. © T.R.Dutton. Oct. [illegible]

## 4. Interpretation of Table 2.

The four Track Orientations are given as headings --- 11:00h RA, 21:30h RA , Sunrise and Sunset, respectively. Each heading has below it a column for each of the seven years considered --- seven columns altogether, for each orientation. The list of solar system bodies, down the left-hand side, is a complete list of all objects found to be in, or approximately in, alignment with the tracks. To be considered as being in **Close** alignment, a body had to be within approximately 0.3h RA (4.5°, azimuth) of the target position. Bodies being situated further away from the target position were allowed to be up to 1.0h (15°, azimuth) displaced, especially, if the track inclination was other than 53°, but they were usually well within those limits.

Referring to the totals given in the table, the outstanding ones are shown underlined. They may be underlined because, either, the numbers of occurrences were large, or, the number of 'Close' alignments was a high percentage of the total for a particular body.

**The Moon** seemed to have been referenced in all track orientations, with a total number of 22 references throughout the years considered.

The **Sun** had been aligned only for 21:30h and Sunset orientations.

**Mercury** had been referenced 29 times, but mostly by 21:30h and Sunset tracks.

**Venus** showed up in all orientations, but with 11:00h RA tracks being the least often aligned.

**Mars** seems to have been referenced mostly by 21:30h RA and Sunset tracks; **Jupiter** by all but Sunrise tracks and was favoured by Sunset tracks. **Saturn** had been referenced only by Sunset tracks.

**Uranus** was least referenced by Sunrise tracks. It was linked with 21:30h tracks 11 times during 1977. (NB. That total does not only reflect the almost-stationary position of Uranus in the sky during any given year. It also reflects the number of times a 21:30h track was identified with the events of that year.) The total number of references (27) is considered to be significant.

**Neptune** features only marginally for all orientations.

**Pluto**, although having only one Close reference, nevertheless featured 18 times.

The most unexpected outcome of this exercise turned out to be the many times that minor planets and comets were found in close alignments.

The minor planets Ceres , Pallas, Vesta and Juno, were all featured, Pallas being of least significance. **Ceres** was referenced in 1977 and 1993 by all but 11:00h RA orientated tracks and, overall, featured 18 times, 11 of those references being close alignments. **Vesta** had been referenced 16 times in total during four of the selected seven years, linked only to 11:00h RA and Sunset tracks. **Juno** featured during six of the seven years, variously, in all orientations. It was referenced 26 times, 17 of those being in the 'Close' category.

During scanning with the astronomical software, a number of well-known comets seem to feature often in the target areas of the sky, **Gale** being the one most frequently signalled. It featured during the years 1957, 1977 , 1992 and 1993, and 33 alignments were noted. 28 of those were 'Close' alignments, of which 19 were accumulated during 1993 and were linked to tracks with 21:30h RA orientations. Of the remaining six comets referenced, **Hartley2** seemed to have featured most often, but only with 21:30h RA and Sunset orientations. It had featured in scans for every year, except 1990. A total count of 22 alignments, with 12 in the 'Close' category, made this comet's presence difficult to dismiss.

## 5. Assessment.

### 5.1 Important considerations.

The results of this study have to be assessed taking account of the degree of probability of track alignment with each of the solar system bodies highlighted. There are two considerations**, viz.** whether the tracks considered are fixed in space or move with the Sun during the course of any year and, also, the direction and rate of movement of the aligned body.

With the exception of Pluto, the major and minor planets move along paths close to that followed by the Sun. Mercury and Venus, being closer than the Earth to the Sun, appear to move round the sky with the Sun. The planet Mars, the Minor Planets (in the Asteroid Belt), Jupiter, Saturn, Uranus, Neptune and Pluto appear, from the Earth, to travel through only limited zones of the sky during the course of any year. Whilst Mars, being the closest of these, moves through approximately one-third of the sky, the apparent movement of the others gets progressively less with increasing distance from the Sun. The three outermost planets are so distant that they appear to be virtually fixed against the background of stars during any given year.

Comets have less well defined orbits, which are usually very elongated ellipses, and the frequency of their appearance depends on how often they sweep close enough to the Sun to produce their characteristic visible plumes of vapour and dust. Their orbits are generally inclined relative to the paths of the Sun and planets in the Ecliptic Plane. The astronomical software used in this exercise gives the predicted position of each of the long-established comets on any given day of each year, whether or not that comet is close enough to display a visible plume. It does not take account of more recently identified comets, such as Hale-Bopp. *This means that, **if the comets are genuinely linked with outbreaks of UFO activity, the picture presented here is far from being complete**.*

***Following onto the close approach of Hale-Bopp, a JPL photographic analyst discovered that a sequence of Hubble Telescope photographs of the comet seemed to show that it had had several highly-mobile small companions during that short period of time.*** As these supposed fragmentary 'satellites' of the comet did not appear on any other photographs taken, their reality has not been accepted by cometary experts (**Ref. 7.2).**

***One of the asteroids (or Minor Planets) called Ida also featured in Ref. 7.2. A photograph taken by the space probe Galileo shows the 35 miles long, potato-shaped, asteroid to be in possession of a small spherical companion.*** It has been assumed to be a satellite of Ida and has been labelled 'Dactyl'. It remains to be seen whether Dactyl is still accompanying Ida or whether the recording of their association was (possibly) an important fluke.

## 5.2 Possible Implications of the Results.

The above considerations serve to demonstrate that extreme caution must be exercised when attempting to assess the possible implications of the results of this study.

However, it is also important to remember that this Pilot Study was undertaken as a first exploratory step towards the general validation, or negation, of indicators provided by ten Close Encounter cases, which were recorded during the period 1952 – 1988. Those indicators were alignments of planets and, sometimes, the Moon with the fixed (star-orientated) tracks which had featured on the days of those 'high-strangeness' reports that had involved unidentifiable craft accompanied, sometimes, by alien creatures.

The **processed events database** used for this study is a general record of all high-strangeness UFO events gathered from the period 1950-1999, many of which do not fall into the same category as those providing the initial indicators. In fact, relatively few of the events processed involved alien creature reports or the prolonged period of amnesia which characterised those of the initial set. However, they have all qualified for inclusion in the database because they have been found to correlate well within the constraints of the Astronautical Theory. They include, for example, everything from Close Encounters with craft (whether or not alien creatures were involved), reports of aerial discs , saucers, triangles, etc., and inexplicable lights-in-the-sky reports.

Given such a wide-ranging set of data, it has to be acknowledged that some of the reports may be invalid and only correlated well with the Theory by accident --- so, some allowance must be made for this possibility. In view of the general success in finding alignments during this exercise, it might be permissible to surmise that only the seven exceptional events were invalid data --- but the validity of the other correlating events must first be assessed before that conclusion can be drawn. The **Table 2** results will now be considered in some detail.

## MOON AND PLANETS

### The Moon.

The Moon moves round the sky once every lunar month and, consequently, comes into conjunction and opposition (R.A.alignment) with each of the other planets and minor planets twice per month. It also cuts through each track line twice every 28 days. This means that the chances of lunar alignments, both with tracks and with other solar system bodies, are very high. In view of this fact, it is quite surprising to find alignments with the Moon featuring so little in **Table 2**. This might mean that we can regard lunar alignments as being of little consequence; but when 15 out of the 22 identified alignments were of the 'Close' variety, this is a difficult conclusion to draw. The best assessment seems to be that the position of the Moon may, at times, be referenced for navigational purposes.

### The Sun

The Sun appears to move once round the sky each year and, in doing so, passes through each of the star-related track sets twice during its annual journey. It aligns exactly with the sunrise and sunset tracks only at the Spring and Autumn Equinoxes and between those dates is only in approximate alignment, if at all. This state of affairs is reflected in **Table 2**. 8 of the 12 (all good) alignments were associated with the 21:30h tracks. This seems to have been a significant result since Sun alignments occurred in all but two of the seven years considered. It seems to imply that the Sun may be used as an aiming point from some distant source in the sky which is aligned with the 21:30h RA set of tracks.

**Mercury**

Mercury revolves round the Sun approximately three times during each Earth year. This means that the planet is in conjunction with the Sun approximately six times per year – leading to the same number of times per year when it might share track alignments with the Sun. This could account for the 9 approximate alignments with 21:30h RA tracks recorded. Being often displaced some short distance from the Sun, Mercury could be expected, also, to align frequently and equally with sunrise and sunset tracks. **Table 2** shows that it appears, overwhelmingly, to have been aligned with sunset-orientated tracks for all the seven years------ which must be considered to be a significant result.

**Venus**

Venus comes into conjunction with the Sun once per year and at other times is situated on either side of it, to the East or to the West. Therefore, it is ideally placed to align with sunrise and sunset tracks at dates between the equinoxes. Being linked with the Sun, it may also align with the star-related track sets four (or more) times per year. **Table 2** shows that Venus had lived up to these expectations. The overall count of 20 close alignments, out of a total of 24, indicates that this planet could feature from time-to-time in the navigational procedures adopted.

**Mars**

Being one of the outer planets, with an orbital period round the Sun of just less than two Earth years, the variations in its apparent motion among the stars, from Earth,are quite large.

The positioning from year to year is progressively different. The possibilities of alignment with the, fixed, star-related tracks and/or the sunrise and sunset-related tracks might vary from year to year. **Table 2** shows that 21:30h RA tracks had referenced it 7 times out of 15, with 6 of those being close alignments. These events occurred during years 1957, 1990, 1991 and 1993. Presumably Mars did not occupy the alignment positions for this fixed orientation during the remaining years. The fact that the planet was referenced so few times by the terminator-related tracks

(Sunrise and Sunset) seems to imply that, unlike Mercury and Venus, it has not been favoured for activities linked to those tracks.

**Jupiter**

This brilliant planet takes almost twelve Earth years to travel once round the Sun. During a typical year its position in the sky, when viewed from Earth, varies by only 2.0hr RA, approximately --- which means that only very occasionally, as years pass, it will align with

the fixed star-related tracks. However, it could align with terminator track options, once or twice per year. Given these circumstances, the alignments with Jupiter shown by **Table 2** must be regarded as being probably significant. Strangely, although it shows up very well in the Sunset columns, references to it are completely absent from those for Sunrise-tracks.

**Saturn**

Like Jupiter, Saturn is one of the big visual targets in the Solar System and, therefore, it could be expected to be used for navigational purposes whenever it aligned with any of the four track options identified. Unfortunately, Saturn's period round the Sun is almost 30 Earth years --- so this planet moves through even less of the sky per year than Jupiter.

Alignments with any of the star-related tracks during only one year could be regarded as being fortuitous, rather than significant. The **Table 2** record shows that Saturn featured only as a Sunset-related marker for five of the 7 years examined, and not very significantly.

**Uranus**

Uranus moves round the Sun once every 84 years. During any year, its motion in the sky is less than 0.5hr RA. The comments made about Saturn apply equally-well here. **Table 2** shows that it aligned with identified 11:00h RA tracks during 1957, 1990 and 1991, for a total of 9 times, 5 of those being in the 'Close' category. It was also referenced by 21:30h RA tracks 11 times during 1977 – and 10 of those were close alignments. Its links with the terminator tracks, given their greater opportunities for encountering the planet, seem to be incidental.

**Neptune**

Neptune takes almost 165 years to move round the Sun --- so, in any year it seems to be almost stationary in the sky against the fixed stars. In recent years Neptune and Uranus have been situated close to each other in the sky --- therefore, both planets might have been expected to have been referenced in, more or less, equal numbers of times throughout the period 1990-1995. This seems to have been borne out by the results. Overall, Neptune has not been referenced appreciably.

**Pluto**

This planet follows a highly eccentric orbit round the Sun. The orbit is inclined at 17.2° to the Ecliptic Plane and the path followed by Pluto cuts across the orbit of Neptune. The period of rotation round the Sun is some 248 years. Pluto is also a small target for navigational purposes, having a diameter which is only about five-eighths the diameter of the Earth. Nevertheless, it could have been referenced 18 times, even though only once, closely. Its involvement is problematical.

## THE MINOR PLANETS

**Ceres**

This is the largest of the Minor Planets in the Asteroid Belt. It has a diameter of 760 km. and orbits the Sun every 4.6 years. The orbit's inclination to the Ecliptic is 10.6°. Being so small, it is not an obvious visual beacon to aid astro-navigation, so its alignments with the celestially-defined UFO activity tracks might be only incidental. Even so, as **Table 2** shows, it was signalled 18 times, 11 of those being close alignments, for three of the four track orientations, the exception being 11:00h RA. This raises the question as to whether the smaller bodies of the Solar System are sometimes used as convenient and suitably-aligned staging points on journeys to and from Earth, an idea which seems to be substantiated by all that follows.

**Pallas**

Pallas, the joint-second largest minor planet, has a period of 4.61 years and an orbit inclination of 34.8°. Perhaps its very inclined orbit was responsible for its not being signalled significantly during this exercise.

**Vesta**

Vesta has the same diameter as Pallas, but being closer to the Sun, it has an orbital period of 3.63 years. The inclination of the orbit is only 7.1° relative to the Ecliptic Plane --- and this could have had some influence on the results obtained for it in **Table 2.** It was aligned 16 times, but only 7 of those were close alignments. However, 6 of the latter were recorded for Sunset-tracks from the total of 9 for the set with that orientation, which could indicate that Vesta was found to be a useful staging point for such operations during 1957, 1977, 1993 and 1995.

**Juno**

This is the smallest of the referenced minor planets, having a diameter of only 200 km. The inclination of the orbit is 13° and the period is 4.36 years. Juno was referenced 26 times during this study, variously, for all orientations, and there were 17 close alignments. The fact that it was found to be in alignment with Sunset tracks 13 times during 1993 and twice during 1995, seems to indicate, as with Vesta, a preference for its use as a staging point for Sunset-orientated operations.

## COMETS

Comets, whether or not displaying plumes, being very small targets, figured unexpectedly in the alignments registered. The frequent referencing of some of those considered below can be, perhaps, only regarded as further indications that they may be used as convenient staging points, as has been speculated, previously, in Para. 5.1. The long-term comets follow very elongated orbits which are inclined steeply to the Ecliptic Plane. Their points of origin seem to lie somewhere within the plane of our galaxy, the Milky Way. Given these circumstances, to be found among track-alignment bodies close to the Ecliptic, a comet has to be close to cutting the Ecliptic whilst following its inclined path round the Sun. The number of times this occurs in a given period of years will depend on the orbital period of that comet. Since the long-term comets have periods of between 3 and 100 years, they cut through the Ecliptic at intervals of years which are half their orbital periods. In such circumstances it could be expected that the chances against any comet aligning with the fixed star-related track orientations on the Ecliptic Plane would be very high. The terminator-linked tracks, which effectively sweep all round the sky in any year, might be expected to fare better, but the presence of a comet close to the Ecliptic would be the prerequisite feature. The alignments of the comets listed in **Table 2** are considered below.

**Hartley 2** *(Period: 6.28 years)*

This comet had 22 alignments, 12 'Close', registered during the seven years (thirty-eight years' time-span) considered, a surprisingly large number. It seems to be a strange coincidence that it was found to align with 21:30h RA tracks during 1957, 1977, 1991 and 1992 --- and with Sunset tracks during all years except 1957 and 1990. Some of these alignments can be attributed to the opposition of the comet to the Sun in intervening years, but the number of counts is still quite remarkable.

**Grigg-Skjellerup** *(Period: 5.1 years)*

The results for this comet cannot be considered to be significant.

**Encke** *(Period: 3.28 years)*

Encke scored highly with terminator-linked tracks but, during any of the chosen years, did not align with the star-linked ones. From **Table 2**, it showed up especially-well with the Sunrise-linked tracks and, overall, was aligned during the years 1957, 1977, 1991, 1993 and 1995.

**Gale** *(Period: 11.24 years)*

During the astronomical software scans, Gale featured frequently, especially during scans for 1993. With the 21:30h RA orientated tracks it featured 19 times. **Table 2** shows this to have been a significant percentage (46%) of the total of all 41 alignments with that orientation registered for that year. To add to the possible significance of this comet, it also scored 6 (close) 11:00h RA alignments and 3 Sunset ones during 1977. The overall score of 33 was the highest recorded for any of the qualifying bodies, and it was accumulated during the years 1957, 1977, 1992 and 1993.

**Machholz** *(Period: 5.24 years)*

Machholz logged up a score of 9 alignments overall, only 4 of which could be categorised as being 'close'. It performed best as an 11:00h RA marker during the years 1977, 1992 and 1993 --- which may be of some significance.

**Halley** *(Period: 76.3 years)*

The results for this famous comet were not remarkable, other than that 2 of its 3 terminator track alignments were of the 'close' variety.

## 6. Overall Review and Conclusion.

As the discussions in Para 5.2 have adequately demonstrated, this is a difficult exercise to assess. The overall result (**Table 2**) has demonstrated that the reported events, over the period of years considered, have definitely favoured the track orientations 21:30h RA (106 counts) and Sunset (126 counts).

The favoured solar system bodies for the former orientation are shown to have been the Sun, Venus, Mars and, occasionally, the Moon. Alignments with the outer planets can only occur when those slow-moving bodies are in the required zones of the Ecliptic, but there are some indications that they may be referenced in those circumstances. (This applies equally to 11:00h RA tracks.)

During Sunset-orientated operations, it seems that all the major bodies may have been referenced from time to time, as might be expected, and Mercury seems to have figured largely, with 12 close alignments out of 17 in total. Strangely, Mercury was referenced only once as a Sunrise-related marker, so this seems to point towards there being a significant bias.

The overall bias against the Sunrise-related option, which Table 2 clearly reveals, is the result of that option's not being signalled very frequently during the initial event-processing stage. That applies, similarly, to the smaller numbers of 11:00h RA links than those for the 21:30h RA orientation. But, even so, the evidence suggests that alignment with planets could be generally arbitrary in both Sunrise and 11:00h RA operations --- even though Venus showed up well as a Sunrise option.

The featuring of minor planets and comets in this exercise raises the question as to whether such bodies might feature regularly in the operations being investigated --- and, if so, are some of them 'preferred' bodies. Since they are not necessarily good visual markers, it has been suggested in Section 5 that they might be used as convenient and aligned staging points on journeys to and from Earth. Transient photographic evidence of small companions associated with the asteroid Ida and Comet Hale-Bopp might provide support for that idea.

In summary, ***the pilot study just described has provided ample evidence that the positioning of solar system bodies could play a key navigational role during some of the surveillance operations being carried out from extra-terrestrial sources. It has also given insights about the bodies that have been most favoured for the various***

***operations. These insights may help to indicate, during any year, which track orientations are likely to be adopted and so aid the use of the timing predictions graphs already in widespread use.*** The rules so far discovered do not help with Sunrise and 11:00h RA predictions, but should help improve 21:30h RA and Sunset predictions, by indicating which days of any year can be regarded as being favourable for direct observation work all over the world.

## References

7.1 J. Allen Hynek 'The UFO Experience --- A Scientific Inquiry' Abelard-Schuman Ltd, 1972. (Book)

7.2 ASTRONOMY (Magazine) August, 1998, Stern, S. Alan. Southwest Research Institute. 'Baby Bopps?' p.24

William K. Hartmann. 'The Great Solar System Revision' p.44

**End of Main Report**

# Appendix

**Close examination of all processed CE3 and CE4 cases in the database at November, 2000.**


## APP. 1 Introduction.

Twenty seven Close Encounter cases involving alleged encounters with alien creatures had been processed and had been included in the Correlated Tracks Database listing by the end of October, 2000. However, as this additional alignments exercise progressed, further details became available on the controversial Roswell (New Mexico, USA) crash event of 1947, which (several witnesses of that time had alleged) had involved the retrieval of alien bodies. The new information received supplied both a **date** and a **time** for the crash event, which enabled it to be processed through the programs described in the main report. As the given time was in close agreement with predictions for that date, the Roswell event qualified for tentative inclusion in the database listing, and consideration of the case was then added to the end of this CE3/CE4 exercise. As previously (in the main report), solar system bodies in close alignment with the tracks linked closely to each event, by date and time, were sought for, using the Expert Astronomer software. The outcome of this exercise is summarised by **Appendix Sheets 1 to 7**, but that summary is preceded by **Appendix Figs. 1 to 4**. These have been introduced to give better understanding of those important features of the Astronautical Theory which were described in the main report.

## APP. 2 Basics and the definition of star-linked tracks (Appendix Fig. 1)

**App. Fig 1** basically depicts the Northern and Southern aspects of the Heavens, as presented in most astronomical text books. Each circular diagram represents a flattened view of each hemisphere of the sky. The outer boundaries represent the celestial equator, which is an extension, into the sky, of the Earth's equatorial plane. Similarly, the centres of these diagrams are directly above the true North and South Poles of the Earth, respectively.

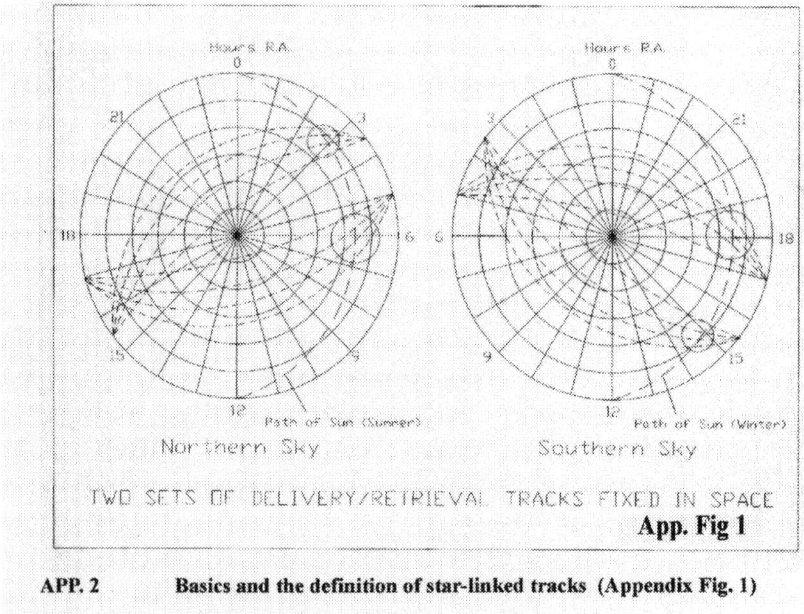


**App. Fig 1**

The sphere of the sky (Celestial Sphere) represented in this way is divided up into equivalents of latitude and longitude and they are called, respectively, 'Declination' and 'Right Ascension'. **The concentric circles shown in each diagram pass through the stars directly above those same latitudes on the Earth's surface below them**. As drawn, these circles are at 15° (latitude/declination) intervals --- and other latitudes/declinations can be found by interpolation between them. Right Ascension (R.A.) differs from longitude in that the 15° divisions shown are fixed in the sky, the zero datum being determined by the position of the Sun among the stars at noon on the Spring Equinox. During the course of a day, the Earth rotates once beneath this fixed framework, which also locates the positions of the fixed stars. Each 15° division of R.A. represents (approximately) 1 hour of rotation of the Earth beneath the stars and, therefore, **R.A. is, by tradition, progressively measured in hours (of sidereal time) to the East of the zero datum**. Unlike the planets of the solar system, the stars are so far away that they appear to be fixed --- which means they can be regarded as fixed reference points in space, relative to which the Sun and other solar system bodies are observed to move, in an orderly and predictable manner. **These features together provide useful navigational aids for mariners, aviators and astronauts. The Astronautical Theory suggests that even the ETIs use them for that purpose.**

**App. Fig.1** superimposes onto the starfield background those **fixed** identified approach and departure tracks in space, which seem to be used by ETI probes engaged in routine monitoring and exploration of Earth. When these tracks were first identified, most of the events relating to them had occurred north of the equator, so it was convenient to label them with the R.A. of the meridian (radial line) which corresponded to **the northern-most** point on each track of a set. Each set consisted of 10 well-defined 'hoops' round the world, the planes of which were found to be inclined to the equatorial plane at angles of 42, 43, 44, 52, 53, 54, 58, 63, 67 and 76 degrees. Only four of these options are drawn, and they represent the range of all ten in each set. As can be seen from the diagram of the northern sky, one set is linked to the 11:00 hr. R.A. meridian and the other, to that corresponding to 21:30 hr. R.A.

Because the Earth is tilted at a fixed angle to its path round the Sun, the Sun appears to follow a path among the stars which is similarly inclined to the Celestial Equator --- and its path across the sky during the northern summer and winter periods is shown as a dashed arc to the right of each diagram. Clearly, the planes of the star-orientated tracks (hoops) intersect the plane of the path of the Sun (the Ecliptic Plane) in well-defined areas of the sky. Since a range of angles appear to have been used, the actual intersections are spread throughout those limited zones, as shown. **However, as will be explained later, the 53° track intersections seem to be of particular significance.**

As explained in the main report, the indicators that ETI activity stemmed from locations in the solar system --- and that alignments of certain bodies with the identified star-related track lines had already been noted on several occasions previously --- meant that a search for such bodies within the indicated 'capture' zones might be revealing. A positive result would be indicated if were to be found that such alignments existed, given the orientation of a track identified, by the date and timing of an event, to be in accordance with the programming of the Theory.

## APP. 3 Definition of Terminator-linked tracks

**The terminator-linked track sets** were found to have the same number (10) optional track angles as the fixed, star-linked, tracks --- and were arranged similarly, on either side of the poles --- but they moved round the sky in a manner related to the movement of the Sun. **App. Fig. 2** illustrates this for two example dates of the year.

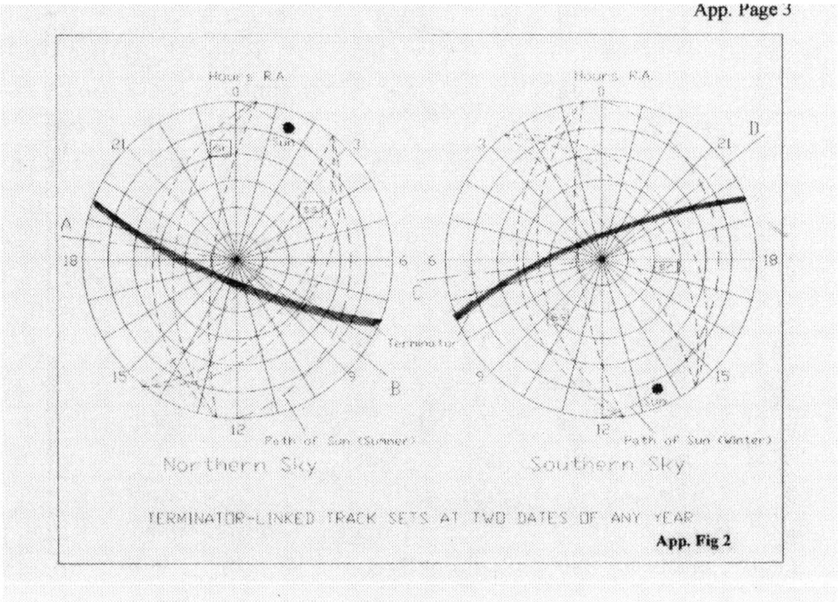


**App. Fig 2**

To understand these diagrams, it is important to be orientated correctly.

Remembering that the framework in the sky is very much linked to the latitude/longitude framework on the Earth's surface and, particularly, **that the circles of constant Declination are in the sky directly above the Earth's Latitude circles**, we can effectively project the arc of the Earth's terminator into the sky in a similar way --- as shown in these diagrams.

During the evolution of the Theory it was found that tracks with this kind of orientation were most frequently inclined with 52°, 53° and 54° angles. These terminator-linked tracks were found to be **celestially-orientated** (despite being linked to the Sun's position) and **the celestial meridian corresponding to the point on the terminator at 53° north**

**latitude seemed to fix the orientation of all other tracks in the set of ten, on every occasion.**

**App.Fig. 2** shows the Sun located at two dates of the year, one being in late Spring and the other in late Autumn. Only three tracks are drawn to represent each set of 10, the middle one of each set being inclined at 53°. This track has been located so that its most northerly point (and most southerly one) coincides with the terminator. A radial meridian has then been drawn through each of the northern intersections (A or B) and this has determined the orientation, in the sky, of the tracks with other inclinations. Each track set has been identified as being linked to either the sunrise (sr) or sunset (ss) --- and to understand why it has been identified as being such, it is necessary to imagine that one is looking upwards from the centre of a transparent Earth, towards the north, in the left-hand diagram. However, the actual track labels have been reversed in the southern hemisphere, because **the sunrise and sunset labels always refer to the northern situation**, as they did for the fixed star-linked tracks. (In other words, an inclined 'hoop' linked to northern sunrise is also linked to southern sunset --- and vice-versa --- but **its northern link always determines its label.**) This diagram indicates how the R.A. range of the respective ecliptic intersection zones can vary with the position of the Sun during its annual movement round the sky --- something which needs to be borne in mind when the results of this exercise are being considered.

## APP. 4 The 'Stargate' periods.

**By comparison of App. Figs. 1 and 2** it becomes clear that the Terminator track sets and the star-linked sets can become superimposed at four periods of each year. Small variations in the motions of the bodies of the solar system cause small annual variations in these periods to occur, which are not accounted for by the simplified model adopted for the Astronautical Theory.

The theoretical model predicts that the **11:00 hr. R.A. track set** coincides with the **sunrise set** on **15th November** each year; and with the **sunset set** on **15th May.** Similar dates for the **21:30 hr. R.A. set** are **15th June** and **15th December**, respectively. (However, it should be remembered that the actual dates when these coincidences occur could vary by several days on either side of the nominal dates.)

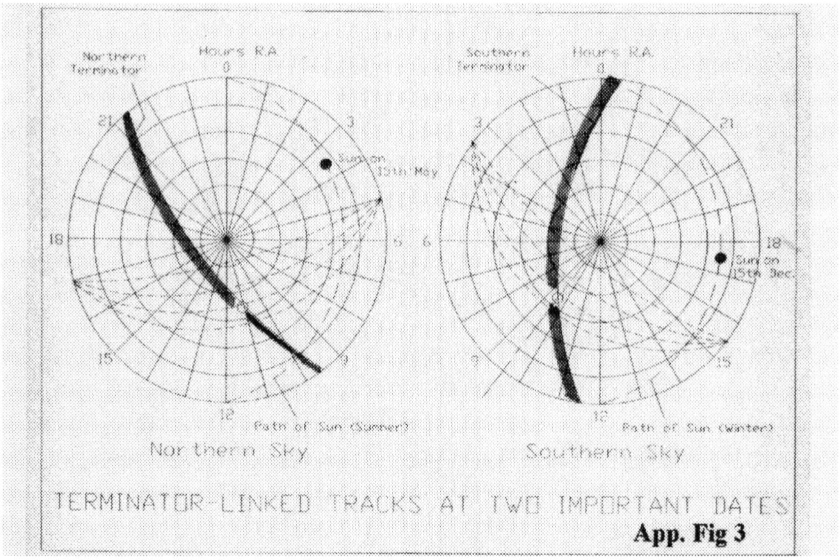


**App. Fig 3**

To demonstrate this feature of the model, **App.Fig. 3** shows the coincidence of the sunset track set and the 11:00 hr. R.A. set on 15th May, in the left-hand diagram, and a similar coincidence of the sunset track set and the 21:30 hr. R.A. set on 15th December, in the right-hand diagram. These date periods of coincidence (taking account of the expected scatter of days from the nominal dates) may have some special significance, because they have been associated with 4 of the 28 CE3/CE4 cases considered in this exercise. Further evidence for this suspicion was provided, unwittingly, by a British 'contactee'( Mr. J. Glanville), who was unaware of this feature of the model. During 1997 and 1998, he informed me, progressively, whenever he had received prior information that he would be contacted. Some of the given dates were in those special track-alignment periods. Mr. Glanville informed me that a sci-fi associate had labelled them **'stargate'** dates. All the Glanville dates and times were in accordance with the predictions of the Theory for the locations specified --- and, almost invariably, the expected activity actually did occur, as forecast, on each occasion.

## APP. 5 The CE3/CE4 Alignments Study

**It is important to stress that the CE3 and CE4 reports now included in the Correlated Tracks Database listing --- like all other cases in that**

**computed database --- were not included in the manually-processed database used to establish the Astronautical Theory.** They have been included in the current database after having been processed and found to correlate with the theoretical predictions.

The first case considered is the very controversial Adamski encounter of 20th November, 1952. Despite the rather incredible story which developed from this event, the date and timing details given by George Adamski were in perfect agreement with theoretical predictions for that place, date and time.

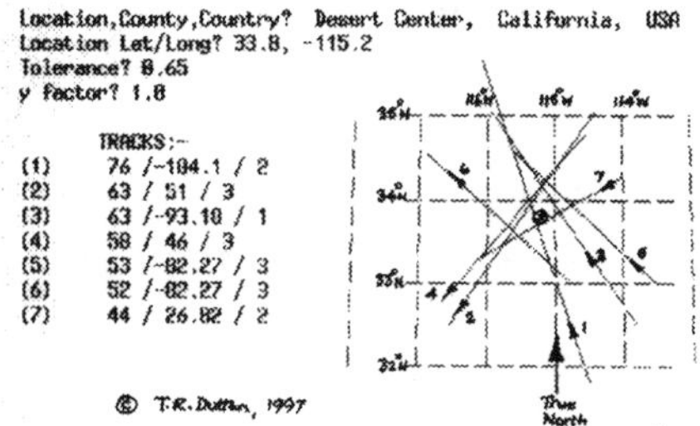


App. Fig 4

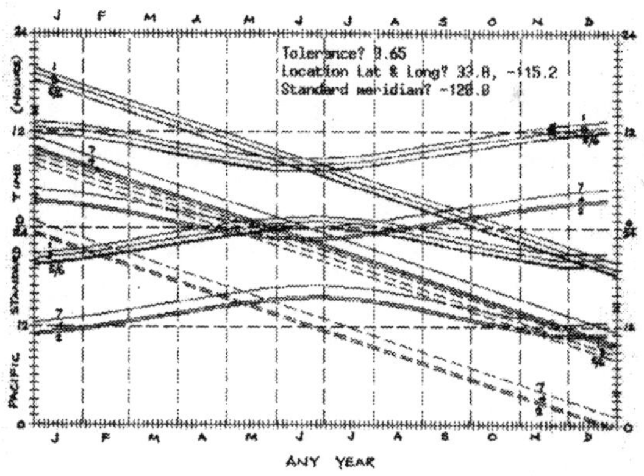


**App. Fig. 4**

This is shown, graphically, by **App. Fig 4**, which also provides a typical example of that kind of graphical presentation.

The two diagonal crosses represent the stated arrival time and the implied departure time of the objects seen by Adamski and his associates. They coincide very well with time predictions associated with **two sunrise-related tracks, Nos. 3 and 1**, respectively. A glance at the latitude/longitude grid map (above the timings graph) shows that the ground-track lines (* see

Introduction in the main report) associated with these two tracks in space run, closely, on either side of Desert Center. Referring, next, to the lower half of the timings graph, it can also be seen that the arrival time (Noon, PST) also fits an **11:00 hr R.A.timing associated with No. 1 track** --- and the inferred departure time, of approximately 1 pm. PST, is seen to have been close to the prediction for **21:30 hr R.A. orientated track No. 4**, the ground-track of which is seen, on the grid map, to pass closely to the north-west of the location. This example will serve to explain the manner in which the astro-alignments exercises have been conducted.

To proceed further ---- this evidence suggests that we should now be looking for solar system bodies in alignment **with two sunrise-linked tracks of 63° and 76° inclination; with a 21:30 hr. R.A. track of 76° inclination;** and, possibly, **with an 11:00 hr. R.A. track inclined at 63°.** In fact, this event occurred during one of those 'stargate' periods, during which the 11:00 R.A. set may coincide with a sunrise-related set. The result of that search is shown as the first item presented on **App. Sht. 1**.

Before further comment is made about the results to be presented, some further explanation is required. As commented previously, the 53° inclined track lines had seemed to determine the celestial orientation of each terminator-linked track set. During this alignments exercise, it was discovered that solar system bodies generally aligned better with the 53° track option than with tracks having other inclinations --- which seemed to suggest that, no matter which track inclination had been identified by a timings correlation exercise, the basic navigational alignment had first been determined by the 53° option. Furthermore, this seemed to apply, also, to the fixed, star-orientated, sets of tracks. In view of this, **for any given event, the listed R.A. value representing the orientation of each correlating track (of whatever inclination), is that corresponding to an inclination of 53°**, but this is a provisional measure and the issue requires further investigation. (One explanation might be that all approaches to Earth are made by delivery craft travelling on 53° inclined tracks ---and that the inclination is changed on arrival in proximity with the planet, to facilitate delivery of a probe to a pre-selected location. A similar procedure could be adopted, in reverse, on departure from the Earth.)

# APP. 6 Review of the Results. (App. Shts 1 to 7)

## App. 6.1 General Information and Comments.

App. Shts. 1 to 7 give the essential details for each of the 28 cases considered, there being four cases on each page.

On the L.H. side of a page are given the R.A. locations (in the Ecliptic Plane) being indicated by the orientation of the track lines associated with the given dates and timings at each location. Each orientation indicated in this way projects outwards on opposite sides of the Earth; and this results in two possible locations for any aligning solar system bodies to occupy. In R.A. terms, they are 12 hours apart. These two R.A. figures are presented and associated with the identified track orientation. The R.A. position of any solar system body which has been found to be in the vicinity of an expected R.A. position, and the name of that body, are also printed out. In this way, numerical appraisals can be made. The position of the Sun in the sky on each occasion enables the orientation of the Terminator to be determined.

On the R.H. side of each page there are four diagrams, each of which portrays the printed information presented to the left of it. The diagrams are plan views of the solar system, the Earth being placed at the centre of each. The R.A. divisions have been distorted to account for the tilt of the Ecliptic Plane relative to the Celestial Plane. Whenever alignments have been identified, the complying solar system bodies have been drawn at their correct R.A. positions and dashed radial lines from the centre of the Earth have been drawn in to represent the anticipated alignments. However, the relative distances from the Earth are only approximately represented and the diagrams are by no means to scale.

Before going on to comment about individual cases, some general comments can be made.

It was observed, in the main report, that certain comets seemed to figure frequently during those alignment studies, and the same comments apply to this special exercise. Notably, **Comet Gale** has been found to be associated with CE3/4 events reported during the years 1952, 1964, 1976/7 and 1988, the gaps between them being roughly in accordance with that comet's period of 11.24 years.

On each occasion, Gale was situated between 8.3 and 8.9 AUs from the Sun and close to 6.00 hr R.A. Given the stable orbit of Gale, it is perhaps not too surprising that it has been invariably associated with the fixed 11:00 hr R.A. orientation --- but it may be important to note that 6.00 hr R.A. also marks the intersection of the Ecliptic and Galactic Planes. (The latter is the plane of our galaxy, the Milky Way.) In contrast, in the main report, Gale was associated mostly with tracks with 21:30hr R.A. orientation, during several years between those listed above, which involved only CE3/4 activity. This, also, may be a significant finding.

**Comet Grigg-Skjellerup** featured 6 times, during the years 1976, 1978, 1979, 1981, 1995 and 1996. It was found to have been aligned in all four orientations: 21:30 hr R.A. (2), 'sr' (2), 'ss' (1) and 11:00 hr R.A. (1). Its distance from the Sun had always been between 3.1 and 4.9 AUs on those occasions, which means that it would not have been visible from Earth.

**Comet Machholz** was a feature of events recorded during the years 1966, 1976, 1981, 1982 and 1997, these spacings being roughly multiples of its period of 5.24 years.

Its correspondence with the sought-for alignments was generally not very good, but on one occasion it was in a conforming conjunction with Neptune on a sunrise track bearing (16. 7. 1981) and on two other occasions (when it was not signalled) it was close to the 18:00 hr R.A. position, which marks both the 11:00 hr track orientation and the Ecliptic/Galactic Planes' intersection in that part of the sky. It's distance from the Sun had varied between 2.3 and 4.3 AUs.

**Hartley 2** featured as a possible 21:30 hr R.A. track marker in 1975 and 1976, when it was 5.5 and 5.9 AUs from the Sun. The closely-bound comet **Encke** (period 3.28 years) featured only twice, once as a sunrise track marker and again as a good, but not sought-for,11:00 hr R.A. track marker**. Halley's Comet** (period 76.3 years) also featured twice: in 1980, when it provided a good marker for a signalled sunrise track and, less significantly, during September 1990.

## App. 6.2 Presentation of App. Shts 1 to 7

Results sheets 1 to 7 follow, and are presented in the manner described in App. 6.1. Thereafter, special elements of this exercise will discussed.

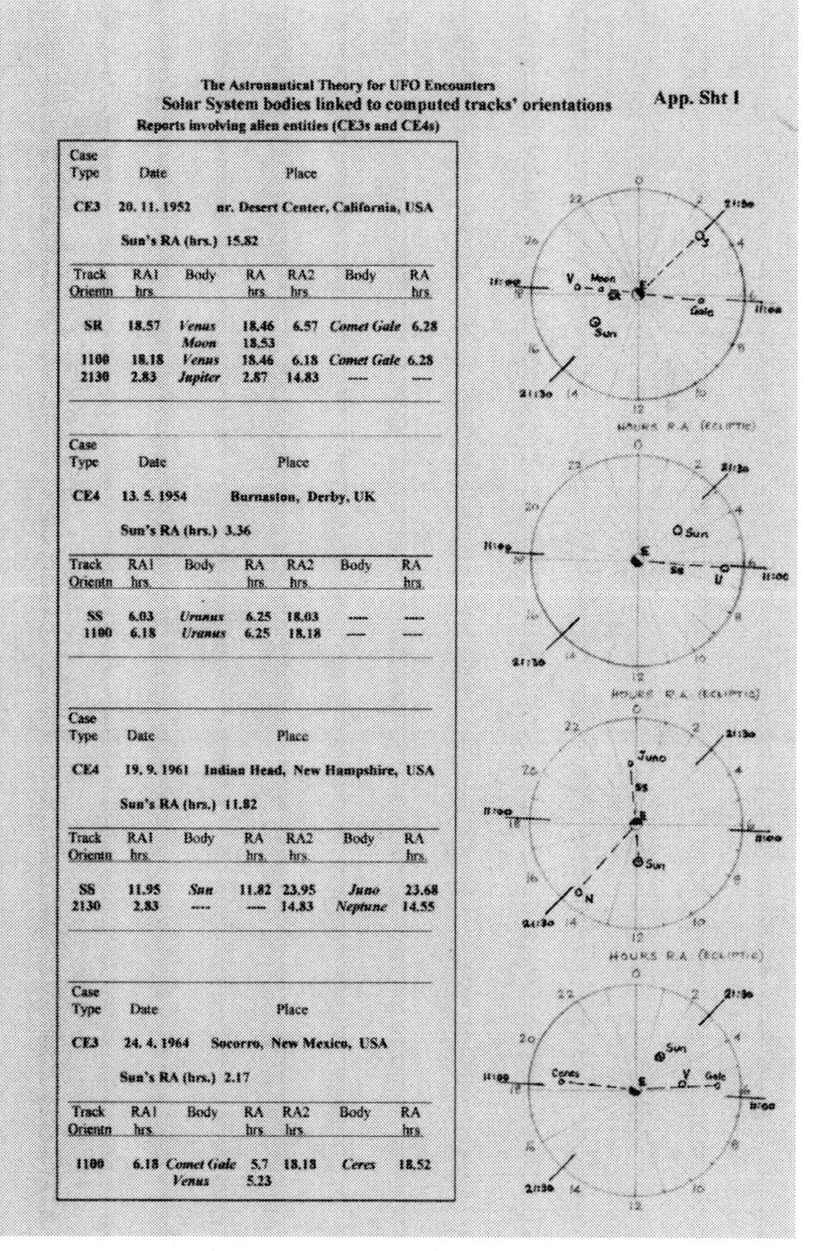

The Astronautical Theory for UFO Encounters

## Solar System bodies linked to computed tracks' orientations

**App. Sht 1**

**Reports involving alien entities (CE3s and CE4s)**

| Case Type | Date | Place |
|---|---|---|
| **CE3** | **20. 11. 1952** | **nr. Desert Center, California, USA** |

**Sun's RA (hrs.) 15.82**

| Track Orientn | RA1 hrs. | Body | RA hrs. | RA2 hrs. | Body | RA hrs. |
|---|---|---|---|---|---|---|
| **SR** | **18.57** | ***Venus*** | **18.46** | **6.57** | ***Comet Gale*** | **6.28** |
| | | ***Moon*** | **18.53** | | | |
| **1100** | **18.18** | ***Venus*** | **18.46** | **6.18** | ***Comet Gale*** | **6.28** |
| **2130** | **2.83** | ***Jupiter*** | **2.87** | **14.83** | — | — |

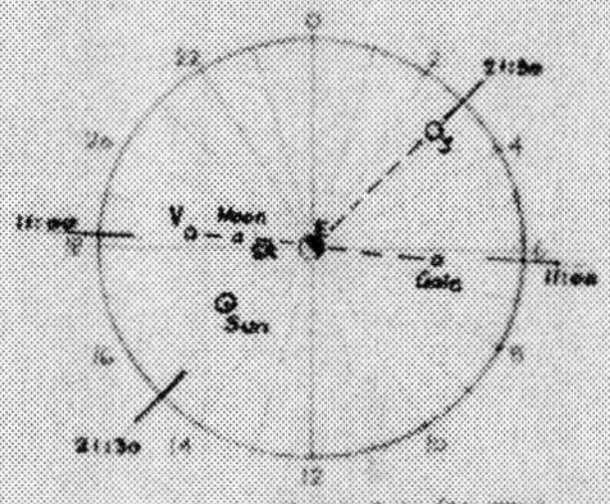


| Case Type | Date | Place |
|---|---|---|
| **CE4** | **13. 5. 1954** | **Burnaston, Derby, UK** |

**Sun's RA (hrs.) 3.36**

| Track Orientn | RA1 hrs. | Body | RA hrs. | RA2 hrs. | Body | RA hrs. |
|---|---|---|---|---|---|---|
| **SS** | **6.03** | ***Uranus*** | **6.25** | **18.03** | — | — |
| **1100** | **6.18** | ***Uranus*** | **6.25** | **18.18** | — | — |

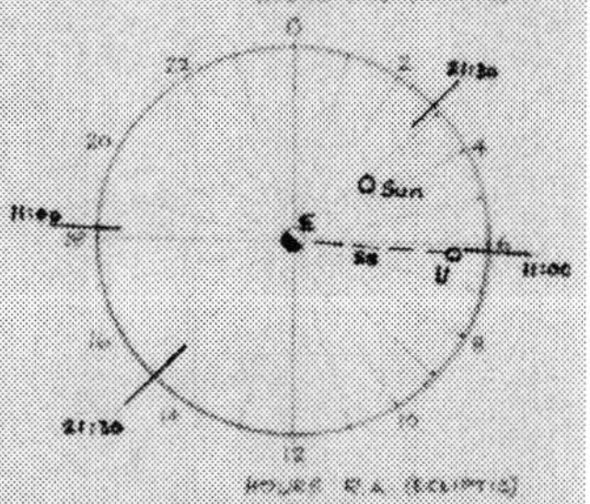


| Case Type | Date | Place |
|---|---|---|
| **CE4** | **19. 9. 1961** | **Indian Head, New Hampshire, USA** |

**Sun's RA (hrs.) 11.82**

| Track Orientn | RA1 hrs. | Body | RA hrs. | RA2 hrs. | Body | RA hrs. |
|---|---|---|---|---|---|---|
| **SS** | **11.95** | ***Sun*** | **11.82** | **23.95** | ***Juno*** | **23.68** |
| **2130** | **2.83** | — | — | **14.83** | ***Neptune*** | **14.55** |

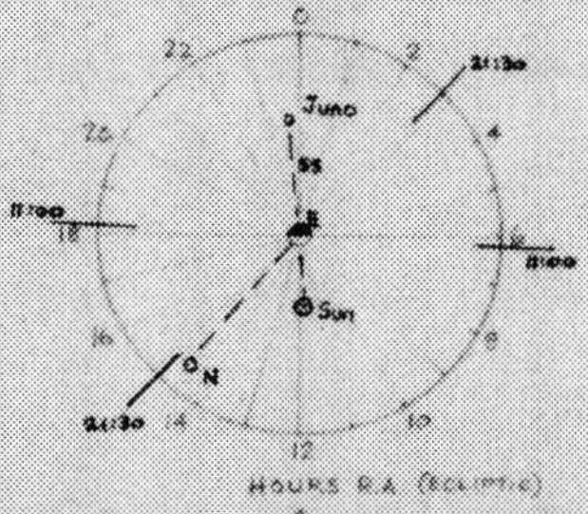


| Case Type | Date | Place |
|---|---|---|
| **CE3** | **24. 4. 1964** | **Socorro, New Mexico, USA** |

**Sun's RA (hrs.) 2.17**

| Track Orientn | RA1 hrs. | Body | RA hrs. | RA2 hrs. | Body | RA hrs. |
|---|---|---|---|---|---|---|
| **1100** | **6.18** | ***Comet Gale*** | **5.7** | **18.18** | ***Ceres*** | **18.52** |
| | | ***Venus*** | **5.23** | | | |

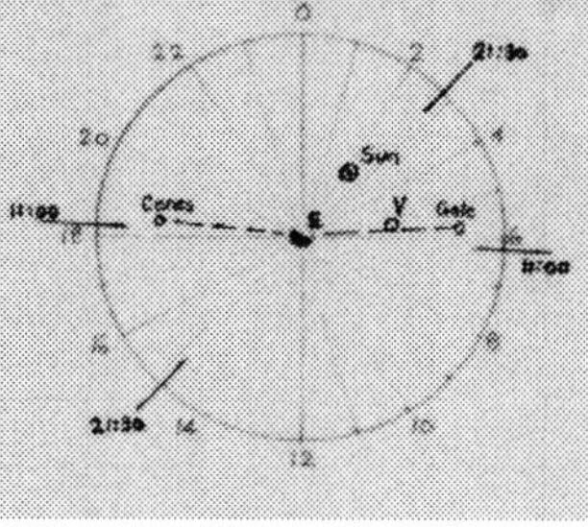


**App. Sht 1**

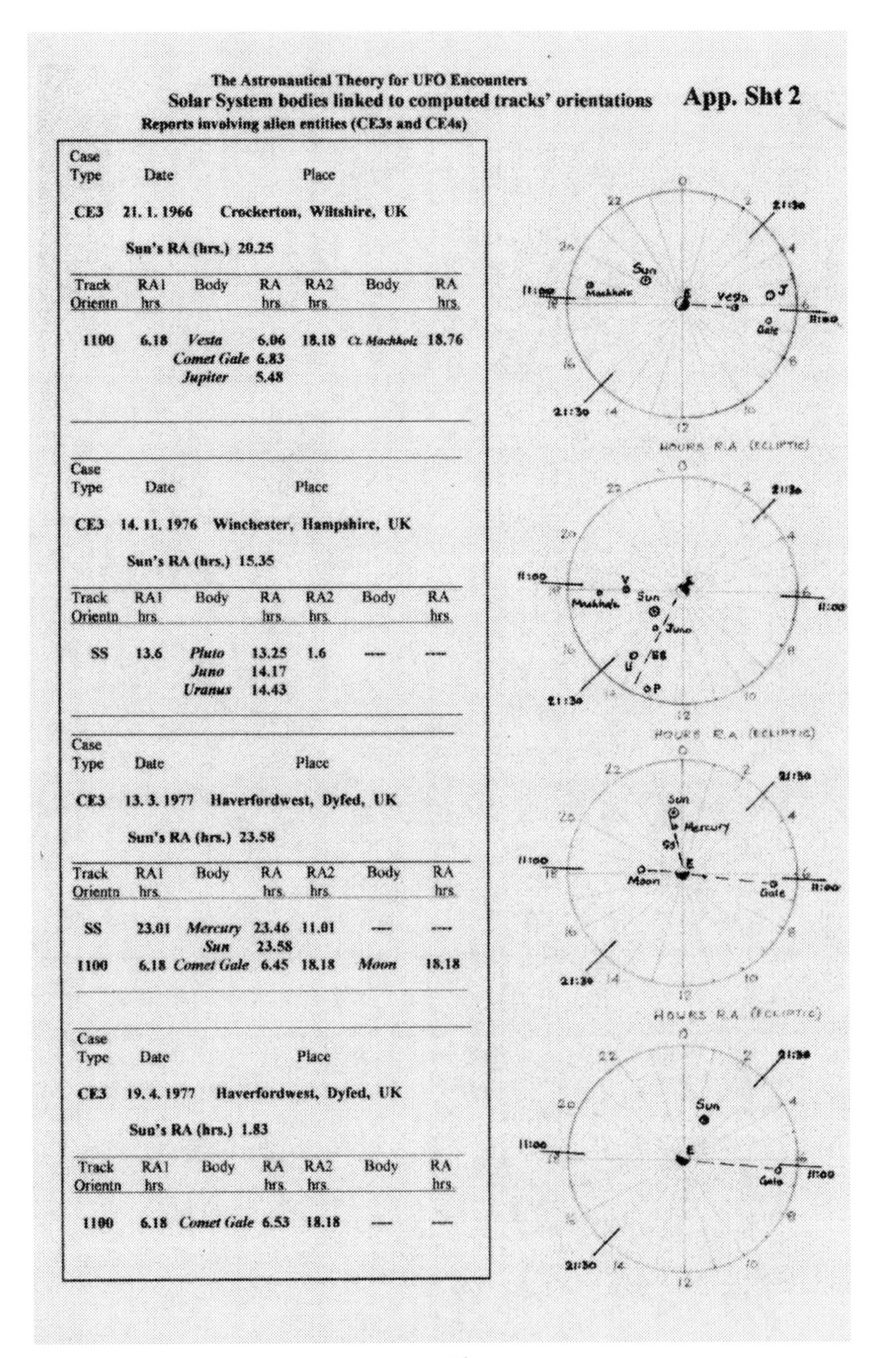

The Astronautical Theory for UFO Encounters

**Solar System bodies linked to computed tracks' orientations** **App. Sht 2**

**Reports involving alien entities (CE3s and CE4s)**

| Case Type | Date | Place |
|---|---|---|
| CE3 | 21. 1. 1966 | Crockerton, Wiltshire, UK |

Sun's RA (hrs.) 20.25

| Track Orientn | RA1 hrs. | Body | RA hrs. | RA2 hrs. | Body | RA hrs. |
|---|---|---|---|---|---|---|
| 1100 | 6.18 | *Vesta* | 6.06 | 18.18 | *Ct. Machholz* | 18.76 |
| | | *Comet Gale* | 6.83 | | | |
| | | *Jupiter* | 5.48 | | | |

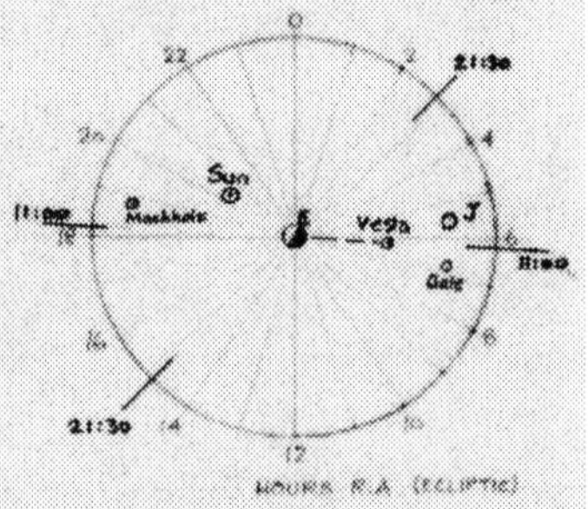


| Case Type | Date | Place |
|---|---|---|
| CE3 | 14. 11. 1976 | Winchester, Hampshire, UK |

Sun's RA (hrs.) 15.35

| Track Orientn | RA1 hrs. | Body | RA hrs. | RA2 hrs. | Body | RA hrs. |
|---|---|---|---|---|---|---|
| SS | 13.6 | *Pluto* | 13.25 | 1.6 | — | — |
| | | *Juno* | 14.17 | | | |
| | | *Uranus* | 14.43 | | | |

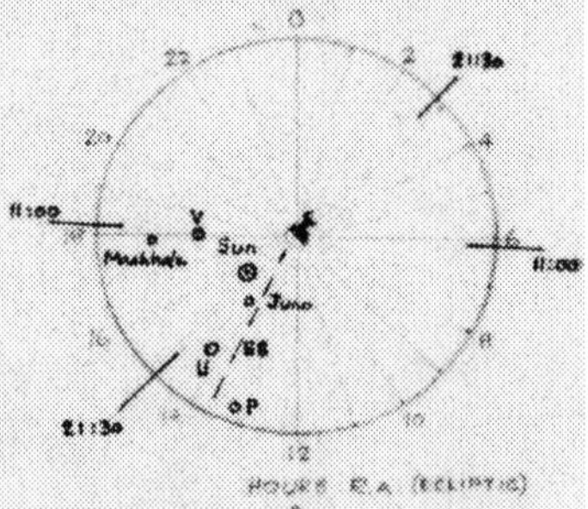


| Case Type | Date | Place |
|---|---|---|
| CE3 | 13. 3. 1977 | Haverfordwest, Dyfed, UK |

Sun's RA (hrs.) 23.58

| Track Orientn | RA1 hrs. | Body | RA hrs. | RA2 hrs. | Body | RA hrs. |
|---|---|---|---|---|---|---|
| SS | 23.01 | *Mercury* | 23.46 | 11.01 | — | — |
| | | *Sun* | 23.58 | | | |
| 1100 | 6.18 | *Comet Gale* | 6.45 | 18.18 | *Moon* | 18.18 |

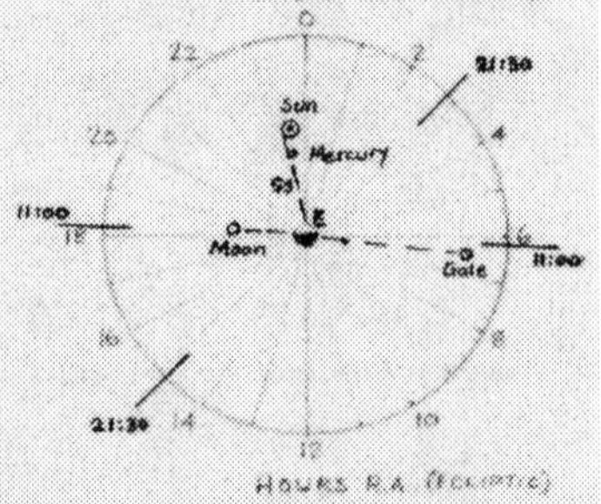


| Case Type | Date | Place |
|---|---|---|
| CE3 | 19. 4. 1977 | Haverfordwest, Dyfed, UK |

Sun's RA (hrs.) 1.83

| Track Orientn | RA1 hrs. | Body | RA hrs. | RA2 hrs. | Body | RA hrs. |
|---|---|---|---|---|---|---|
| 1100 | 6.18 | *Comet Gale* | 6.53 | 18.18 | — | — |

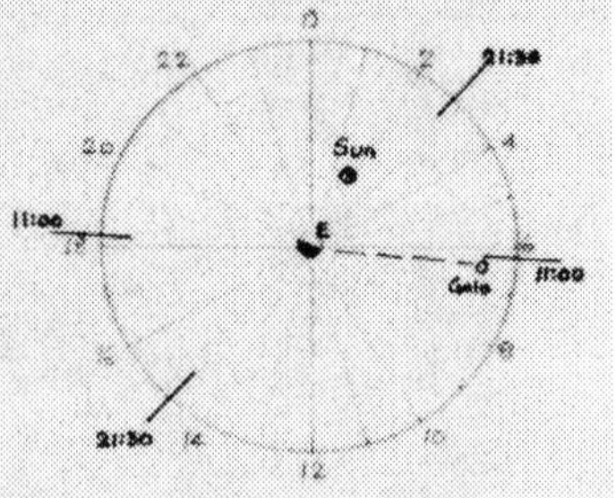


**App. Sht 2**

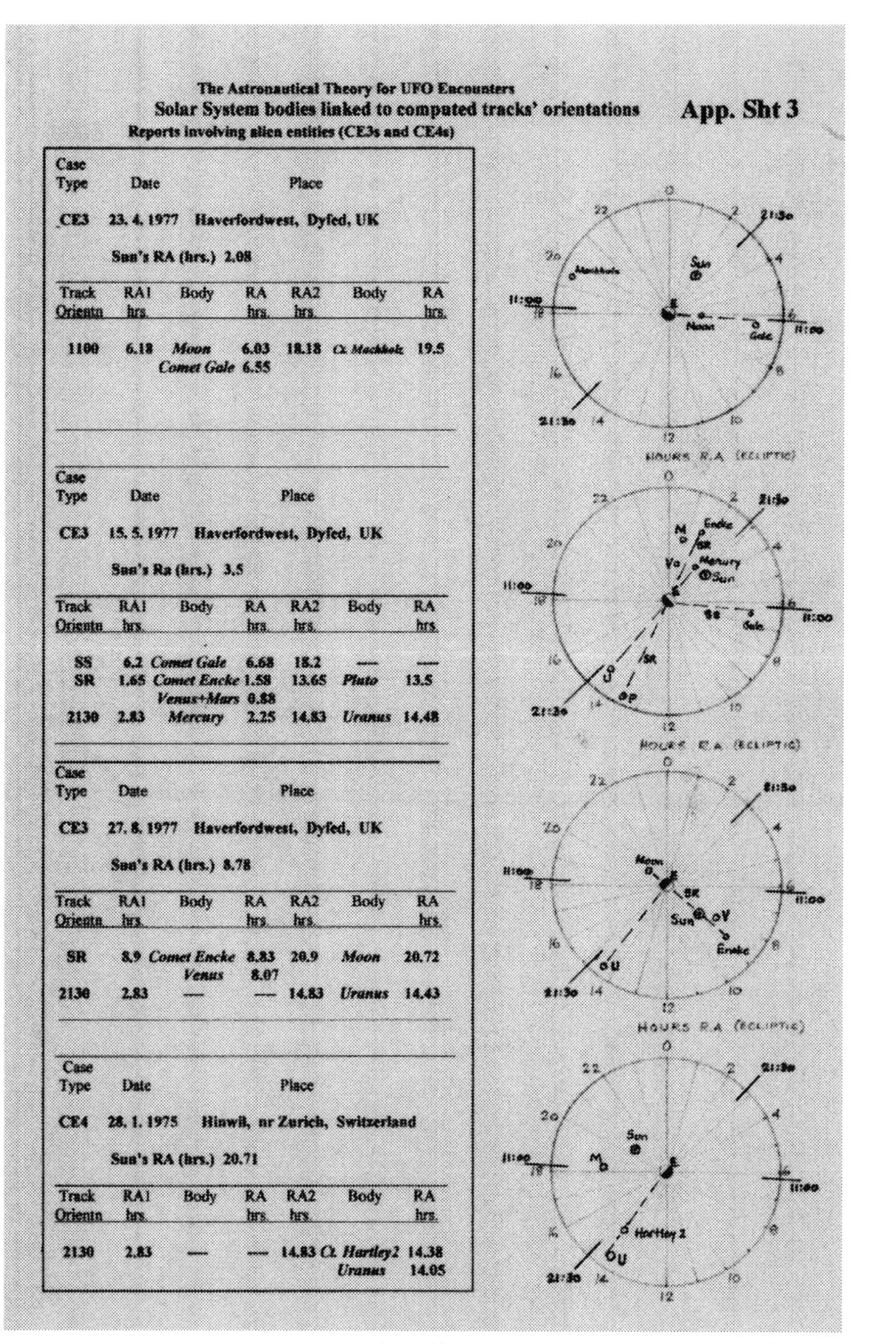

The Astronautical Theory for UFO Encounters

**Solar System bodies linked to computed tracks' orientations** **App. Sht 3**

**Reports involving alien entities (CE3s and CE4s)**

| Case Type | Date | Place |
|---|---|---|
| CE3 | 23. 4. 1977 | Haverfordwest, Dyfed, UK |

Sun's RA (hrs.) 2.08

| Track Orientn | RA1 hrs. | Body | RA hrs. | RA2 hrs. | Body | RA hrs. |
|---|---|---|---|---|---|---|
| 1100 | 6.18 | *Moon* | 6.03 | 18.18 | *Ct. Machholz* | 19.5 |
| | | *Comet Gale* | 6.55 | | | |

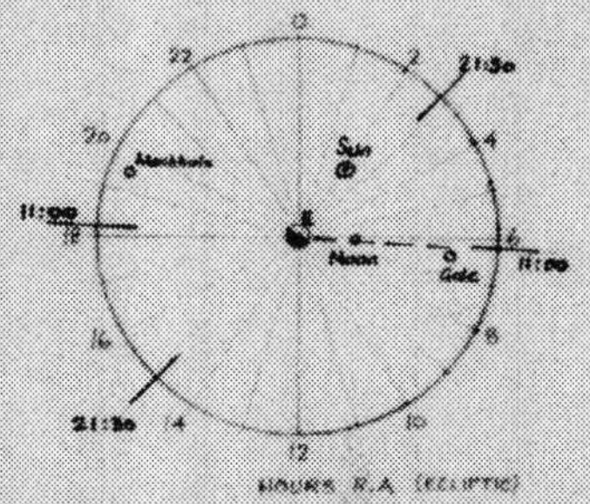


| Case Type | Date | Place |
|---|---|---|
| CE3 | 15. 5. 1977 | Haverfordwest, Dyfed, UK |

Sun's Ra (hrs.) 3.5

| Track Orientn | RA1 hrs. | Body | RA hrs. | RA2 hrs. | Body | RA hrs. |
|---|---|---|---|---|---|---|
| SS | 6.2 | *Comet Gale* | 6.68 | 18.2 | — | — |
| SR | 1.65 | *Comet Encke* | 1.58 | 13.65 | *Pluto* | 13.5 |
| | | *Venus+Mars* | 0.88 | | | |
| 2130 | 2.83 | *Mercury* | 2.25 | 14.83 | *Uranus* | 14.48 |

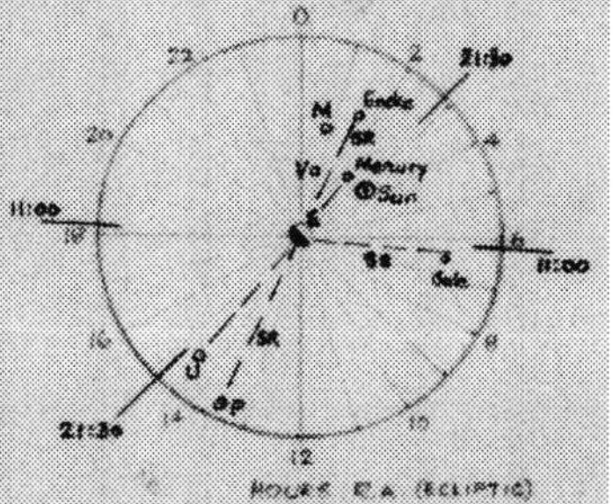


| Case Type | Date | Place |
|---|---|---|
| CE3 | 27. 8. 1977 | Haverfordwest, Dyfed, UK |

Sun's RA (hrs.) 8.78

| Track Orientn | RA1 hrs. | Body | RA hrs. | RA2 hrs. | Body | RA hrs. |
|---|---|---|---|---|---|---|
| SR | 8.9 | *Comet Encke* | 8.83 | 20.9 | *Moon* | 20.72 |
| | | *Venus* | 8.07 | | | |
| 2130 | 2.83 | — | — | 14.83 | *Uranus* | 14.43 |

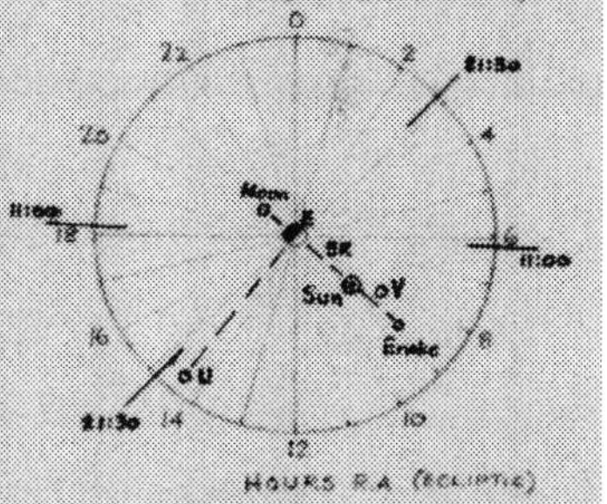


| Case Type | Date | Place |
|---|---|---|
| CE4 | 28. 1. 1975 | Hinwil, nr Zurich, Switzerland |

Sun's RA (hrs.) 20.71

| Track Orientn | RA1 hrs. | Body | RA hrs. | RA2 hrs. | Body | RA hrs. |
|---|---|---|---|---|---|---|
| 2130 | 2.83 | — | — | 14.83 | *Ct. Hartley2* | 14.38 |
| | | | | | *Uranus* | 14.05 |

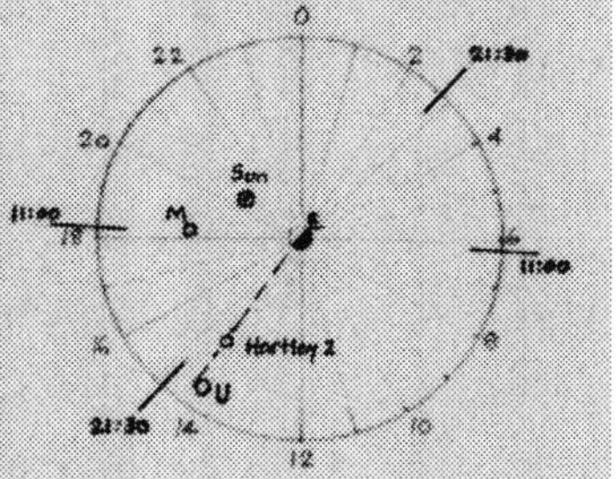


**App. Sht 3**

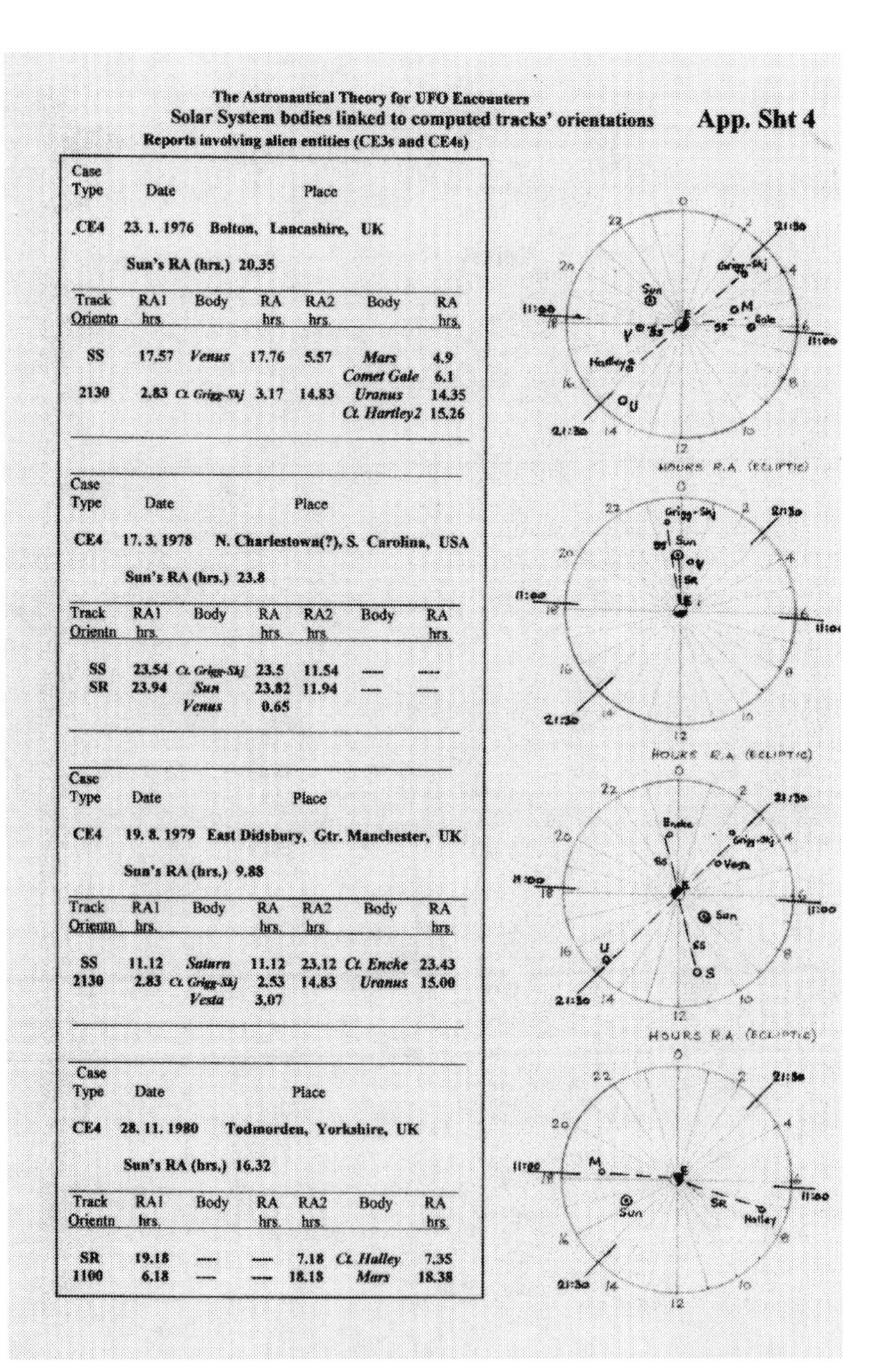

The Astronautical Theory for UFO Encounters

**Solar System bodies linked to computed tracks' orientations** **App. Sht 4**

**Reports involving alien entities (CE3s and CE4s)**

| Case Type | Date | Place |
|---|---|---|
| CE4 | 23. 1. 1976 | Bolton, Lancashire, UK |

Sun's RA (hrs.) 20.35

| Track Orientn | RA1 hrs. | Body | RA hrs. | RA2 hrs. | Body | RA hrs. |
|---|---|---|---|---|---|---|
| SS | 17.57 | *Venus* | 17.76 | 5.57 | *Mars* | 4.9 |
| | | | | | *Comet Gale* | 6.1 |
| 2130 | 2.83 | *Ct. Grigg-Skj* | 3.17 | 14.83 | *Uranus* | 14.35 |
| | | | | | *Ct. Hartley2* | 15.26 |

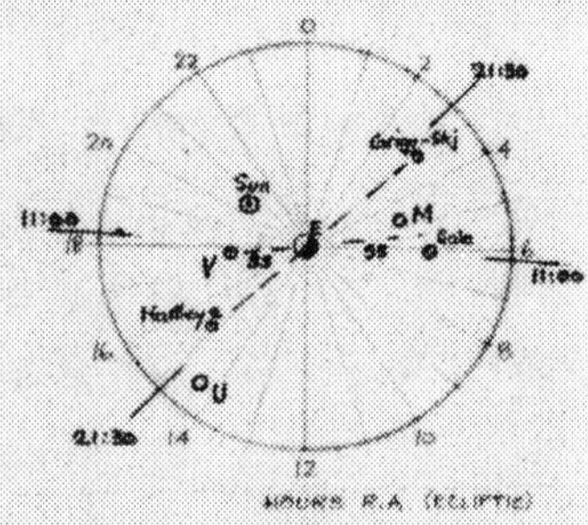


| Case Type | Date | Place |
|---|---|---|
| CE4 | 17. 3. 1978 | N. Charlestown(?), S. Carolina, USA |

Sun's RA (hrs.) 23.8

| Track Orientn | RA1 hrs. | Body | RA hrs. | RA2 hrs. | Body | RA hrs. |
|---|---|---|---|---|---|---|
| SS | 23.54 | *Ct. Grigg-Skj* | 23.5 | 11.54 | ---- | ---- |
| SR | 23.94 | *Sun* | 23.82 | 11.94 | ---- | ---- |
| | | *Venus* | 0.65 | | | |

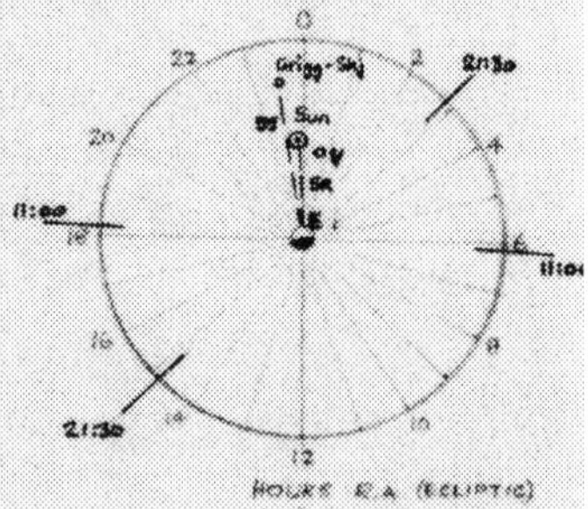


| Case Type | Date | Place |
|---|---|---|
| CE4 | 19. 8. 1979 | East Didsbury, Gtr. Manchester, UK |

Sun's RA (hrs.) 9.88

| Track Orientn | RA1 hrs. | Body | RA hrs. | RA2 hrs. | Body | RA hrs. |
|---|---|---|---|---|---|---|
| SS | 11.12 | *Saturn* | 11.12 | 23.12 | *Ct. Encke* | 23.43 |
| 2130 | 2.83 | *Ct. Grigg-Skj* | 2.53 | 14.83 | *Uranus* | 15.00 |
| | | *Vesta* | 3.07 | | | |

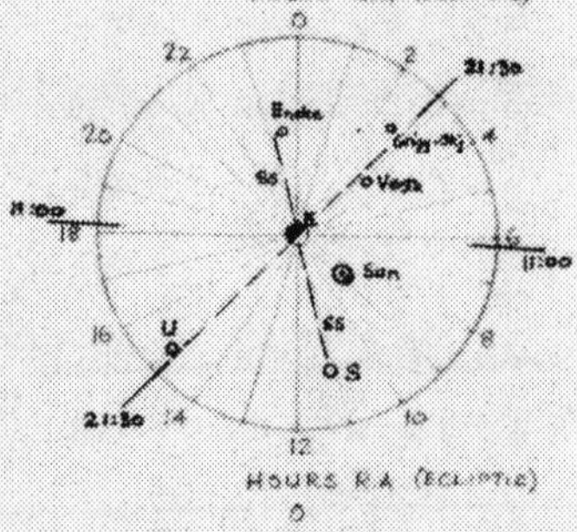


| Case Type | Date | Place |
|---|---|---|
| CE4 | 28. 11. 1980 | Todmorden, Yorkshire, UK |

Sun's RA (hrs.) 16.32

| Track Orientn | RA1 hrs. | Body | RA hrs. | RA2 hrs. | Body | RA hrs. |
|---|---|---|---|---|---|---|
| SR | 19.18 | ---- | ---- | 7.18 | *Ct. Halley* | 7.35 |
| 1100 | 6.18 | ---- | ---- | 18.18 | *Mars* | 18.38 |

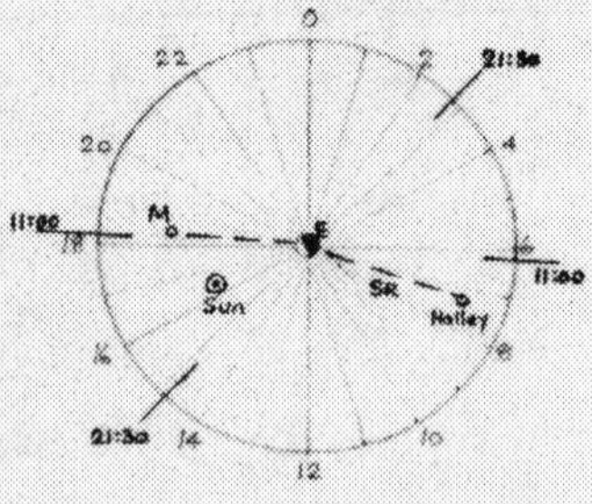


**App. Sht 4**

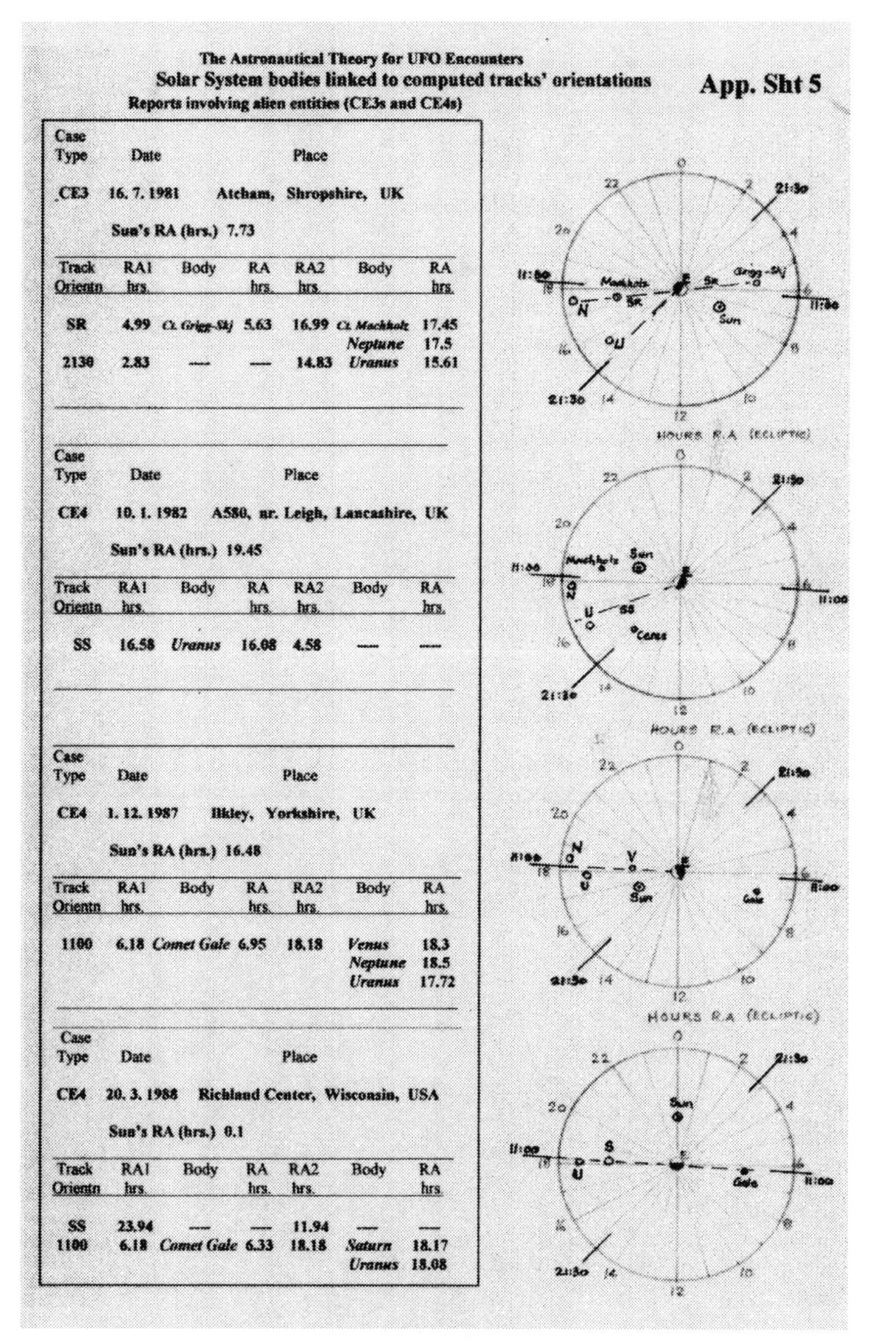

**The Astronautical Theory for UFO Encounters**

## Solar System bodies linked to computed tracks' orientations

**App. Sht 5**

**Reports involving alien entities (CE3s and CE4s)**

| Case Type | Date | Place |
|---|---|---|
| **CE3** | **16. 7. 1981** | **Atcham, Shropshire, UK** |

**Sun's RA (hrs.) 7.73**

| Track Orientn | RA1 hrs. | Body | RA hrs. | RA2 hrs. | Body | RA hrs. |
|---|---|---|---|---|---|---|
| **SR** | **4.99** | ***Ct. Grigg-Skj*** | **5.63** | **16.99** | ***Ct. Machholz*** | **17.45** |
| | | | | | ***Neptune*** | **17.5** |
| **2130** | **2.83** | — | — | **14.83** | ***Uranus*** | **15.61** |

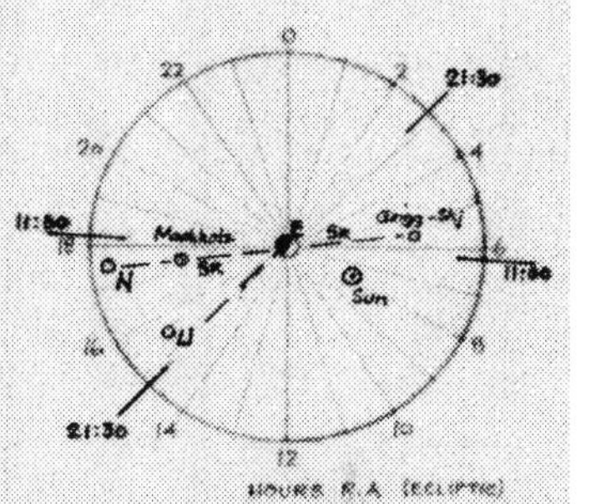


| Case Type | Date | Place |
|---|---|---|
| **CE4** | **10. 1. 1982** | **A580, nr. Leigh, Lancashire, UK** |

**Sun's RA (hrs.) 19.45**

| Track Orientn | RA1 hrs. | Body | RA hrs. | RA2 hrs. | Body | RA hrs. |
|---|---|---|---|---|---|---|
| **SS** | **16.58** | ***Uranus*** | **16.08** | **4.58** | — | — |

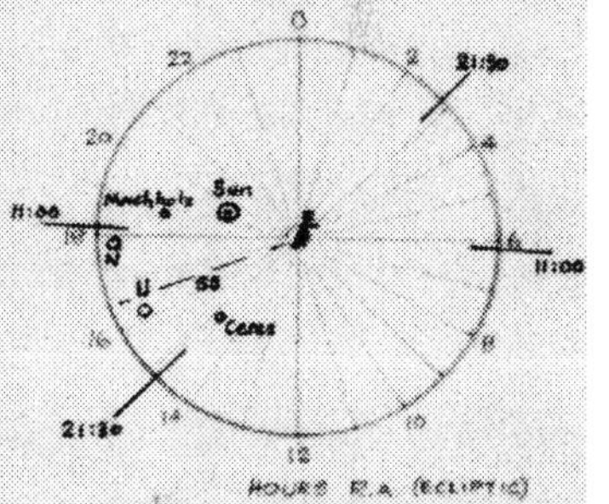


| Case Type | Date | Place |
|---|---|---|
| **CE4** | **1. 12. 1987** | **Ilkley, Yorkshire, UK** |

**Sun's RA (hrs.) 16.48**

| Track Orientn | RA1 hrs. | Body | RA hrs. | RA2 hrs. | Body | RA hrs. |
|---|---|---|---|---|---|---|
| **1100** | **6.18** | ***Comet Gale*** | **6.95** | **18.18** | ***Venus*** | **18.3** |
| | | | | | ***Neptune*** | **18.5** |
| | | | | | ***Uranus*** | **17.72** |

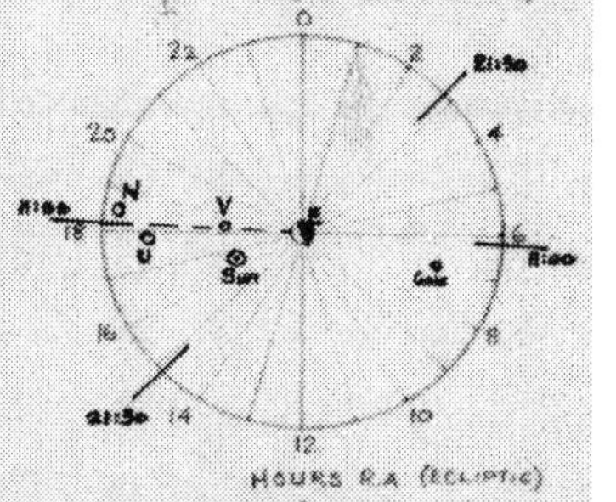


| Case Type | Date | Place |
|---|---|---|
| **CE4** | **20. 3. 1988** | **Richland Center, Wisconsin, USA** |

**Sun's RA (hrs.) 0.1**

| Track Orientn | RA1 hrs. | Body | RA hrs. | RA2 hrs. | Body | RA hrs. |
|---|---|---|---|---|---|---|
| **SS** | **23.94** | — | — | **11.94** | — | — |
| **1100** | **6.18** | ***Comet Gale*** | **6.33** | **18.18** | ***Saturn*** | **18.17** |
| | | | | | ***Uranus*** | **18.08** |

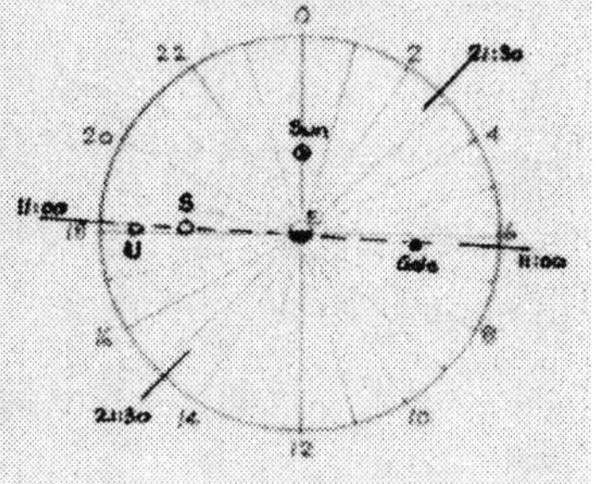


**App. Sht 5**

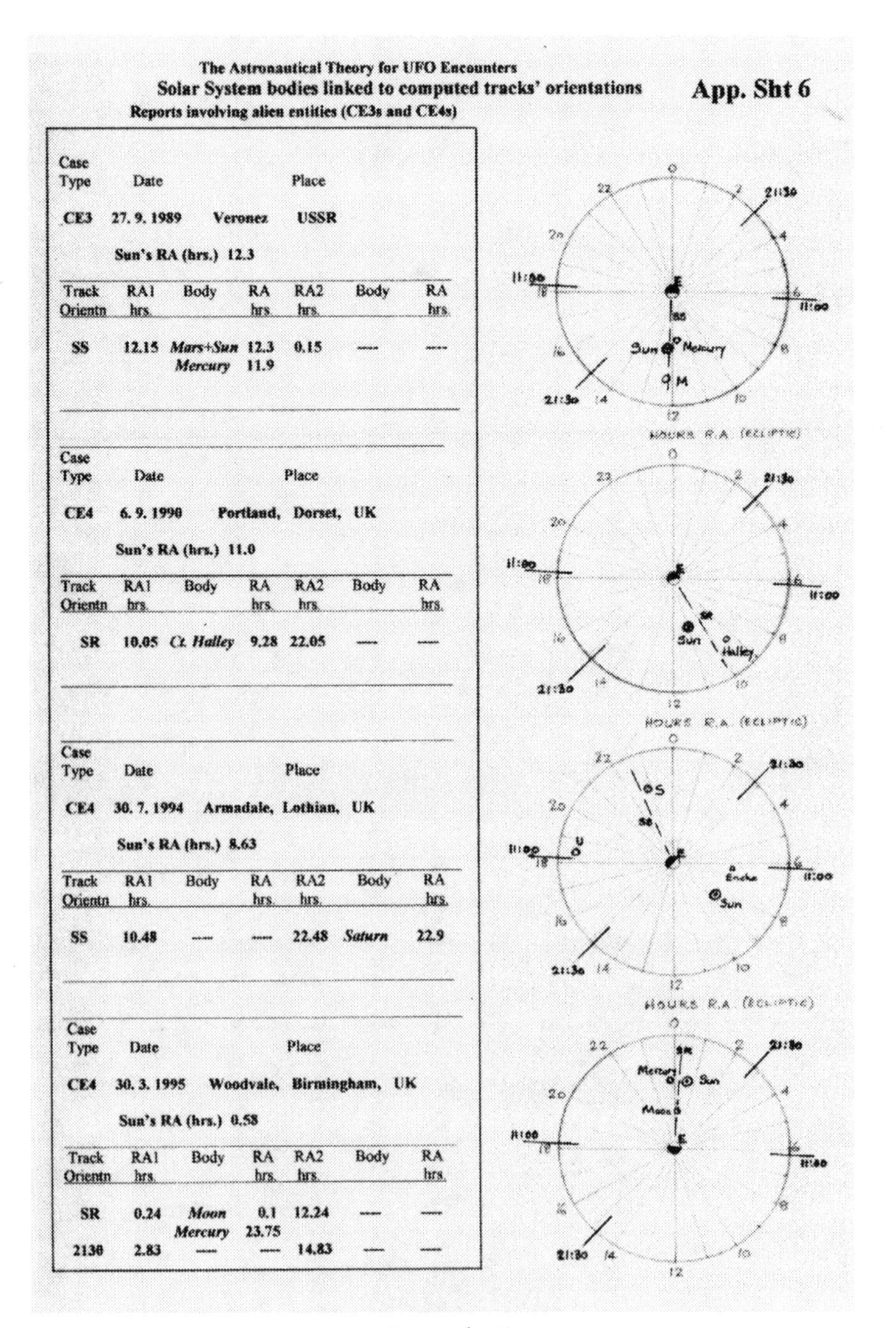

**The Astronautical Theory for UFO Encounters**

## Solar System bodies linked to computed tracks' orientations

**App. Sht 6**

**Reports involving alien entities (CE3s and CE4s)**

| Case Type | Date | Place | |
|---|---|---|---|
| **CE3** | **27. 9. 1989** | **Veronez** | **USSR** |

**Sun's RA (hrs.) 12.3**

| Track Orientn | RA1 hrs. | Body | RA hrs. | RA2 hrs. | Body | RA hrs. |
|---|---|---|---|---|---|---|
| **SS** | **12.15** | ***Mars+Sun*** | **12.3** | **0.15** | **----** | **----** |
| | | ***Mercury*** | **11.9** | | | |

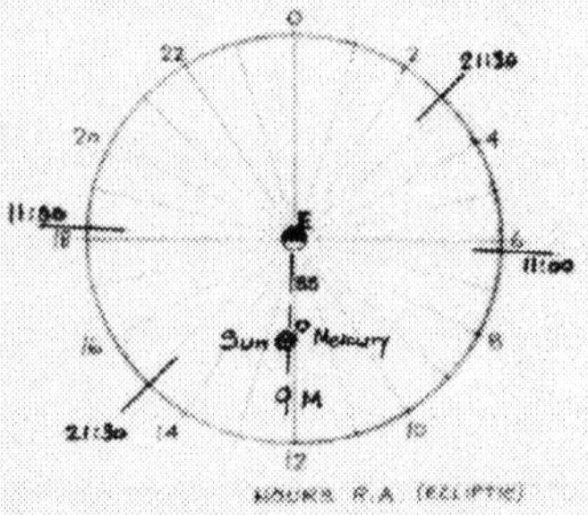


| Case Type | Date | Place | | |
|---|---|---|---|---|
| **CE4** | **6. 9. 1990** | **Portland,** | **Dorset,** | **UK** |

**Sun's RA (hrs.) 11.0**

| Track Orientn | RA1 hrs. | Body | RA hrs. | RA2 hrs. | Body | RA hrs. |
|---|---|---|---|---|---|---|
| **SR** | **10.05** | ***Ct. Halley*** | **9.28** | **22.05** | **----** | **----** |

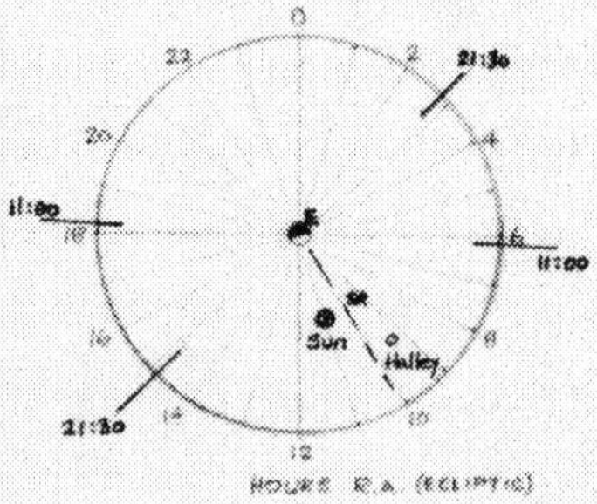


| Case Type | Date | Place | | |
|---|---|---|---|---|
| **CE4** | **30. 7. 1994** | **Armadale,** | **Lothian,** | **UK** |

**Sun's RA (hrs.) 8.63**

| Track Orientn | RA1 hrs. | Body | RA hrs. | RA2 hrs. | Body | RA hrs. |
|---|---|---|---|---|---|---|
| **SS** | **10.48** | **----** | **----** | **22.48** | ***Saturn*** | **22.9** |

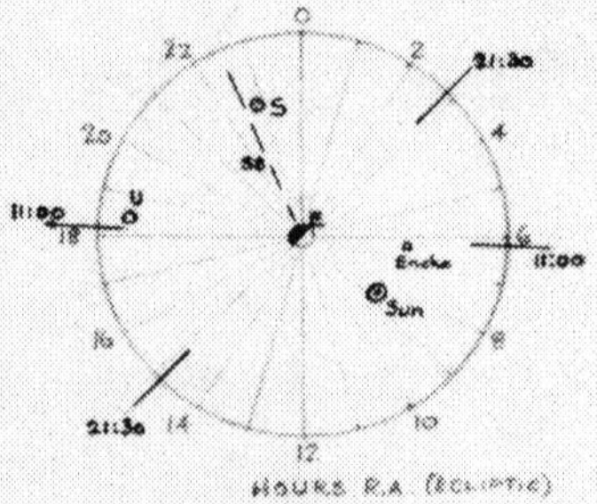


| Case Type | Date | Place | | |
|---|---|---|---|---|
| **CE4** | **30. 3. 1995** | **Woodvale,** | **Birmingham,** | **UK** |

**Sun's RA (hrs.) 0.58**

| Track Orientn | RA1 hrs. | Body | RA hrs. | RA2 hrs. | Body | RA hrs. |
|---|---|---|---|---|---|---|
| **SR** | **0.24** | ***Moon*** | **0.1** | **12.24** | **----** | **----** |
| | | ***Mercury*** | **23.75** | | | |
| **2130** | **2.83** | **----** | **----** | **14.83** | **----** | **----** |

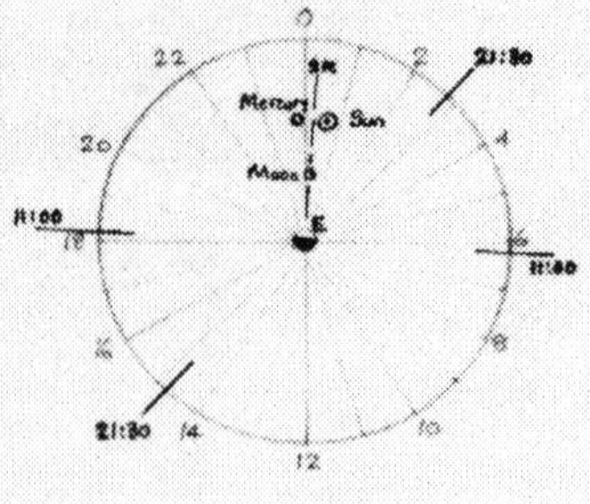


**App. Sht 6**

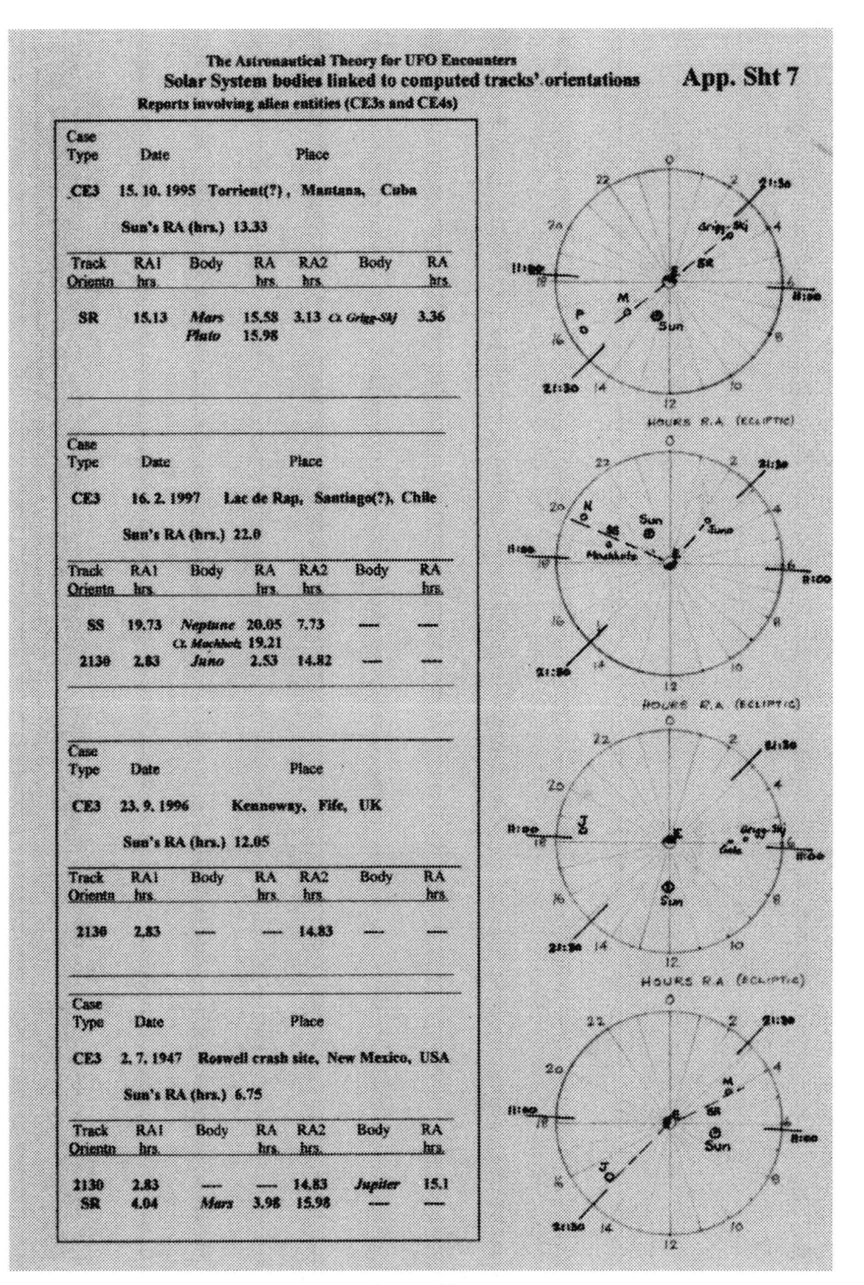

**The Astronautical Theory for UFO Encounters**

**Solar System bodies linked to computed tracks' orientations** **App. Sht 7**

**Reports involving alien entities (CE3s and CE4s)**

| Case Type | Date | Place |
|---|---|---|
| CE3 | 15. 10. 1995 | Torrient(?), Mantana, Cuba |

Sun's RA (hrs.) 13.33

| Track Orientn | RA1 hrs. | Body | RA hrs. | RA2 hrs. | Body | RA hrs. |
|---|---|---|---|---|---|---|
| SR | 15.13 | *Mars* | 15.58 | 3.13 | *α Grigg-Skj* | 3.36 |
| | | *Pluto* | 15.98 | | | |

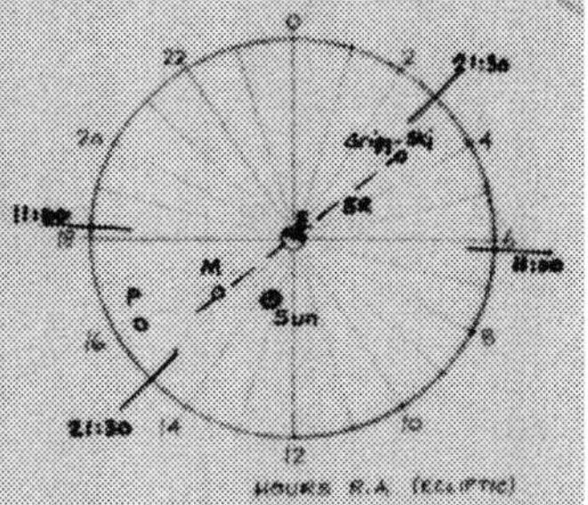


| Case Type | Date | Place |
|---|---|---|
| CE3 | 16. 2. 1997 | Lac de Rap, Santiago(?), Chile |

Sun's RA (hrs.) 22.0

| Track Orientn | RA1 hrs. | Body | RA hrs. | RA2 hrs. | Body | RA hrs. |
|---|---|---|---|---|---|---|
| SS | 19.73 | *Neptune* | 20.05 | 7.73 | — | — |
| | | *α Machhol*z | 19.21 | | | |
| 2130 | 2.83 | *Juno* | 2.53 | 14.82 | — | — |

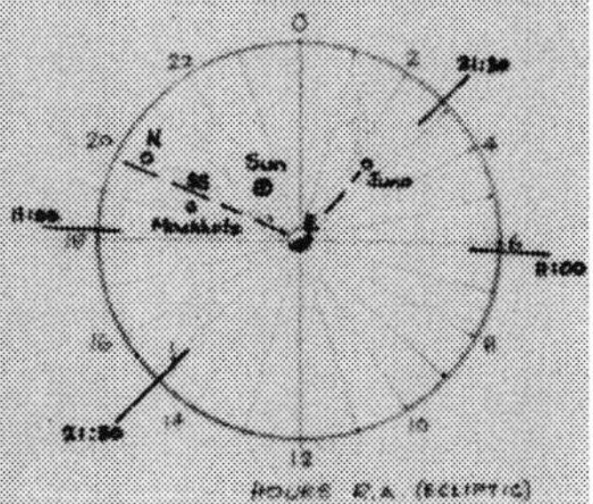


| Case Type | Date | Place |
|---|---|---|
| CE3 | 23. 9. 1996 | Kennoway, Fife, UK |

Sun's RA (hrs.) 12.05

| Track Orientn | RA1 hrs. | Body | RA hrs. | RA2 hrs. | Body | RA hrs. |
|---|---|---|---|---|---|---|
| 2130 | 2.83 | — | — | 14.83 | — | — |

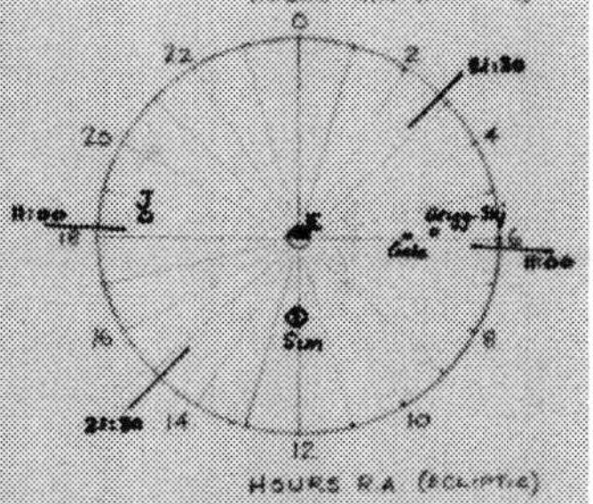


| Case Type | Date | Place |
|---|---|---|
| CE3 | 2. 7. 1947 | Roswell crash site, New Mexico, USA |

Sun's RA (hrs.) 6.75

| Track Orientn | RA1 hrs. | Body | RA hrs. | RA2 hrs. | Body | RA hrs. |
|---|---|---|---|---|---|---|
| 2130 | 2.83 | — | — | 14.83 | *Jupiter* | 15.1 |
| SR | 4.04 | *Mars* | 3.98 | 15.98 | — | — |

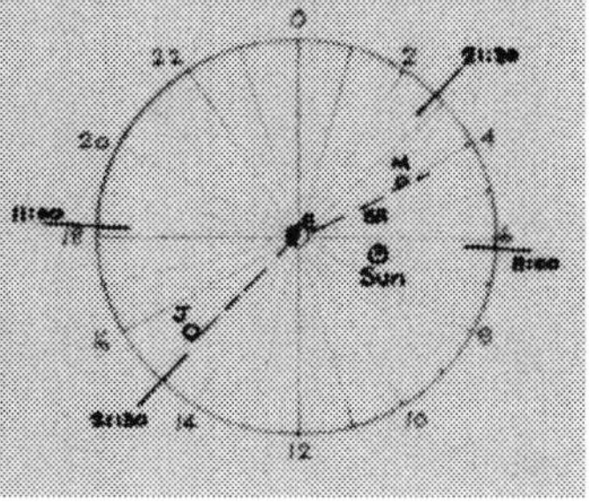

**App. Sht 7**

## App. 6.3 Consideration of Significant Results.

*Referring first to* **App. Sht 1:**

The first case considered is the Adamski report of 20th November, 1952. As can be seen from the numerical and diagrammatic presentations, both of the signalled sunrise and 21:30 hr R.A. track orientations were very closely marked on that day. In view of Adamski's claim that his 'visitor' had come from Venus, it seems to be very remarkable that the sunrise track orientation was indeed marked by **Venus, in close conjunction with the Moon.** It is also worthy of note that **Comet Gale** was in close opposition in the same alignment. (The expected 'stargate' association of the sunrise and 11:00 hr R.A. orientations also becomes very evident.) The 21:30 hr R.A. track orientation was very well marked by **Jupiter**.

The times associated with the two identified orientations suggest an arrival which referenced Venus and a departure referencing Jupiter. As will be seen, this is one of the best results obtained, overall, and seems to provide appreciable validation of the reported event.

The Burnaston (UK) event of 13th May, 1954 is considered next. It involved the reported abduction of a retired naval commander and occurred during an interesting 'stargate' date period, when Sunset and 11:00 hr R.A tracks coincided. Uranus provided an excellent marker for that common orientation, as shown.

Next case considered on App. Sht. 1 is the famous Barney and Betty Hill encounter of 19th September, 1961. The minor planet **Juno**, in opposition to the **Sun**, provided a

good marker for a sunset orientated approach track, whilst **Neptune** provided an acceptable marker for a 21:30 hr R.A. orientated departure track.

The departure time for the craft seen at Socorro on 24th April, 1964, had signalled that an 11:00 hr R.A. track had been adopted. A near conjunction of **Venus** and **Comet Gale** and the near opposition of minor planet **Ceres** could have provided the markers referenced for a departure from Earth on that occasion.

*Referring* **App. Sht 2**

The British high-profile case of 14th November, 1976, involved a car-stop and an alleged CE3 experience for the two occupants. This was another 'stargate' date, but the timing given did not indicate links with the characteristic sunrise and 11:00 hr R.A. tracks. Instead, a sunset arrival track had been indicated (with possible departure with a 21:30hr R.A. orientation.) It can be seen that **Pluto** and **Juno** were closest to such a required alignment for the sunset orientation and **Juno** in near conjunction with **Uranus** might have provided provisional 21:30 hr R.A. markers. Meanwhile, elsewhere in the sky, Venus and Comet Machholz provided better markers for the 11:00 hr R.A. /sunrise orientation, which was not signalled. This suggests that **perhaps the actual timings given for an encounter do not necessarily mark the true beginning and true end of surface exploration activity.**

(Further indications of that kind are observed on **App. Sht 3**, the Hinwil, Switzerland, event, when Mars was in excellent alignment with the un-signalled 11:00 hr R.A. orientation; on **App. Sht 6**, Armadale, Lothian, when Uranus was similarly aligned; and on **App. Sht 7**, Kennoway, Fife, when Jupiter, Gale and Grigg-Skjellerup were also all closely linked with that 11:00 hr R.A. line.)

*Referring* **App. Sht 4**

Another high profile British case producing excellent alignments with the signalled orientations is that of 19th August, 1979, at East Didsbury. This was a CE4 involving a woman and her two children, and both the arrival and the departure of the craft were witnessed. The timings given were in perfect agreement with those predicted. The arrival and departure seemed to be sunset and 21:30 hr R.A. orientated, respectively. The sunset orientation is seen to have been closely marked by **Saturn** with **Comet Encke** in close opposition. The other (star-linked) orientation was closely marked by **Uranus**, with **Vesta** and **Grigg-Skjellerup** placed in approximate opposition positions.

**Mars** and **Comet Halley** seem to have provided the markers for the Todmorden, Yorkshire, CE4 event, which involved a police patrolman in an immobilised patrol car.

*Referring***App. Sht 5**

The alignments obtained for the Atcham, Shropshire, CE4 of 16th July, 1981 are also worthy of note. This was another car-stop event and involved three young women. The estimated times for the happening correlated well with predictions, but as three orientations could have produced those timings, identification of the most likely two was only resolved after numerical processing and by selecting the closest predictions (in minutes) to the times given. This left 21:30 hr R.A. as the arrival orientation (with the sunset orientation discarded) and a sunrise-orientated signalled departure.

**Neptune** and **Machholz** were in conjunction and displaced by about ½ hr R.A. from the nominal 53° sunrise orientation, as shown, and they were in close opposition to **Grigg-Skjellerup. Uranus** was about 0.8 hr R.A. displaced from 21:30 hr R.A. orientation. This is, therefore, a case which challenges a few assumption --- for example, the presumptions made about 53° inclined tracks being always used to define the orientation adopted.

Another puzzle is presented by the Leigh, Lancashire, happening of 10th January, 1982. This involved the stopping of another car with its two women occupants. The retrospective times given for this event were very much in accordance with the predictions, both of which (arrival and departure) were produced by sunset-orientated tracks. However, as the diagram shows, **Uranus** was displaced about ½ hr R.A. from the expected line, **Ceres** was almost exactly aligned with the (un-signalled) 21:30 hr R.A. orientation and **Neptune** was close to the (un-signalled)11:00 hr R.A. orientation. Again, the question is raised as to whether this event was part of a much more extended exploration activity.

Good positions are evident in the diagram for the Richland Center, Wisconsin, event of 20th March, 1988, but only for one of the two orientations signalled. **Uranus** and **Saturn** were in close conjunction and **Comet Gale** was in close opposition along the 11:00 hr R.A. line.

*Referring* **App. Sht 6.**

The Veronez, USSR, event of 27th September, 1989, involved a landing in a public park with many witnesses. The given time linked it with a sunset track prediction. As can be seen from the numbers and diagram provided, the sunset orientation was very closely linked to the position of the **Sun** and **Mars**, in conjunction and, approximately, with **Mercury.**

The question arises as to whether that terminator-linked track was established from the direction of the Sun or towards it.

*Referring* **App. Sht 7.**

Finally, the Roswell crash of 2nd July, 1947, will be considered. Assuming that the given information is trustworthy, the object seems to have arrived on a 21:30 hr R.A. orientated track, perhaps an hour before the crash, and was in the process of departing, following a sunrise-linked track, when catastrophe befell it. The numbers and the diagram show that

**Jupiter** provided an excellent marker for its arrival and that **Mars** had probably been providing its departure reference point in space.

## APP. 7 Closing Summary (Appendix)

From the evidence presented here, it would seem that procedures for exploration activities undertaken by **living ETIs** (rather than the usual automatons) may have been expanded since the 1950s. It could be that as human detection technology improved, more distant and less conspicuous markers had to be adopted, and these have included the minor planets and comets --- though there is the indication that **Comet Gale**, at least, was being referenced back in 1952. The presence of that comet in the 6:00 hr R.A. area of the sky, during some of the major events just considered, creates the suspicion that it may be referenced at regular 11 year intervals by **occupied** craft arriving into the solar system from elsewhere in our galaxy. From the findings of the main report, it seems also that the same comet may be used as a marker for activities following the 21:30 hr R.A. tracks in the years between. All this seems to emphasise that Gale is a comet which merits constant and close surveillance. However, the major planets still feature in the activities when they are conveniently placed and, at such times, should also be monitored.

## APP. 8 The prediction of future activity periods.

It is now suggested that further study of the results of this exercise, taken together with those from the main report, could provide effective guidelines for predicting the periods of future activities by the ETIs. To facilitate this it will be important to create methods for determining when solar system bodies are positioned in close alignment with the four identified track orientations as each year progresses.

**App. Fig. 5** has been created for that task. It provides vital information on the changing 53° terminator-linked track intersections with the Ecliptic Plane and the position of the Sun throughout each year. Also shown are the R.A.s associated with 53° star-linked tracks. All that remains to be done, using this technique, is to superimpose plots of the changing positions of selected solar system bodies. Solar system conjunctions, and dates when the defined tracks and solar system bodies are closely associated, become clearly evident. Recognition of oppositions and their interactions with the track orientations, however, requires very detailed and careful scrutiny of the graph produced for a given year.

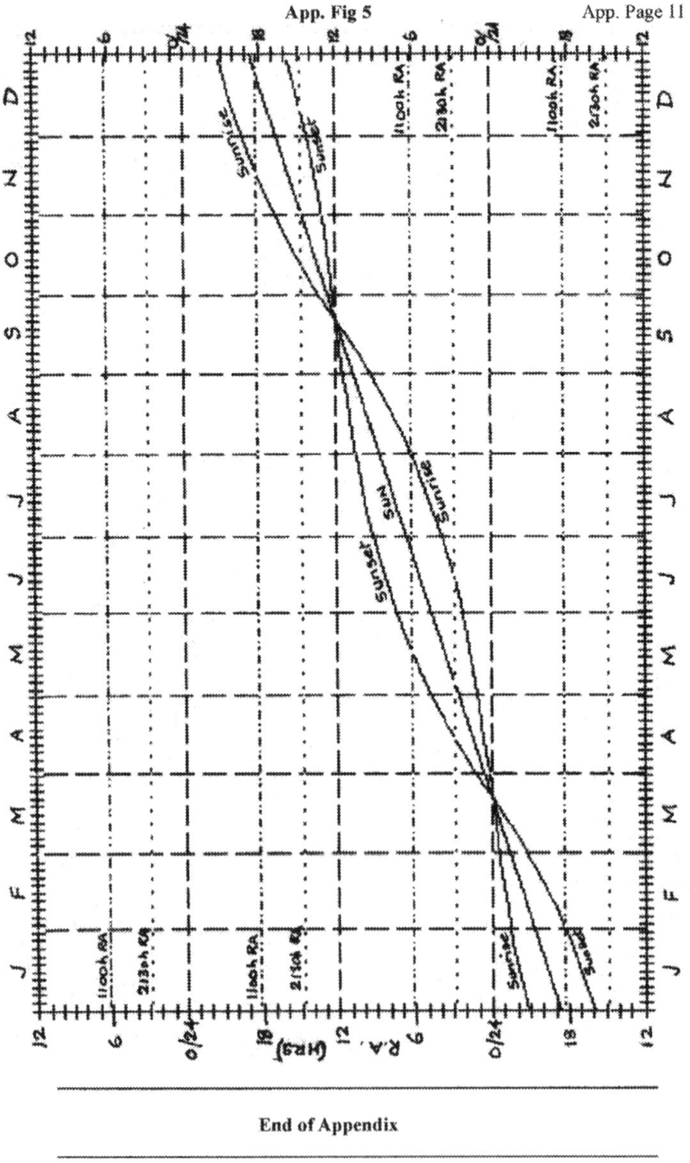


**App. FIG 5.jpg**

**End of Appendix**

# INDEX

## D



## E



## F



## G



## H

## N



## O



## P



## Q



## R

**V**



**W**



**Y**



**Z**